# Exchanging Data between SAS® and Microsoft Excel

## Tips and Techniques to Transfer and Manage Data More Efficiently

William E. Benjamin, Jr.

support.sas.com/bookstore

The correct bibliographic citation for this manual is as follows: Benjamin, William E., Jr. 2015. *Exchanging Data Between SAS® and Microsoft Excel: Tips and Techniques to Transfer and Manage Data More Efficiently.* Cary, NC: SAS Institute Inc.

**Exchanging Data Between SAS® and Microsoft Excel: Tips and Techniques to Transfer and Manage Data More Efficiently**

Copyright © 2015, SAS Institute Inc., Cary, NC, USA

ISBN 978-1-60764-985-4 (Hardcopy)
ISBN 978-1-62959-690-7 (EPUB)
ISBN 978-1-62959-691-4 (MOBI)
ISBN 978-1-62959-689-1 (PDF)

All rights reserved. Produced in the United States of America.

**For a hard-copy book:** No part of this publication may be reproduced, stored in a retrieval system, or transmitted, in any form or by any means, electronic, mechanical, photocopying, or otherwise, without the prior written permission of the publisher, SAS Institute Inc.

**For a web download or e-book:** Your use of this publication shall be governed by the terms established by the vendor at the time you acquire this publication.

The scanning, uploading, and distribution of this book via the Internet or any other means without the permission of the publisher is illegal and punishable by law. Please purchase only authorized electronic editions and do not participate in or encourage electronic piracy of copyrighted materials. Your support of others' rights is appreciated.

**U.S. Government License Rights; Restricted Rights:** The Software and its documentation is commercial computer software developed at private expense and is provided with RESTRICTED RIGHTS to the United States Government. Use, duplication or disclosure of the Software by the United States Government is subject to the license terms of this Agreement pursuant to, as applicable, FAR 12.212, DFAR 227.7202-1(a), DFAR 227.7202-3(a) and DFAR 227.7202-4 and, to the extent required under U.S. federal law, the minimum restricted rights as set out in FAR 52.227-19 (DEC 2007). If FAR 52.227-19 is applicable, this provision serves as notice under clause (c) thereof and no other notice is required to be affixed to the Software or documentation. The Government's rights in Software and documentation shall be only those set forth in this Agreement.

SAS Institute Inc., SAS Campus Drive, Cary, North Carolina 27513-2414.

April 2015

SAS® and all other SAS Institute Inc. product or service names are registered trademarks or trademarks of SAS Institute Inc. in the USA and other countries. ® indicates USA registration.

Other brand and product names are trademarks of their respective companies.

# Contents

Preface ..................................................................................................................... xi
About This Book ..................................................................................................... xiii
About The Author .................................................................................................. xvii

**Chapter 1: Easy Data Movement between SAS and Microsoft Excel ................. 1**
1.1 Introduction ........................................................................................................ 1
1.2 Examination of Excel Files ................................................................................. 2
    1.2.1 Purpose ...................................................................................................... 2
    1.2.2 Excel Data Types ....................................................................................... 2
    1.2.3 General Excel Workbook Limitations ........................................................ 2
    1.2.4 Excel Workbook Formatting Groups ......................................................... 3
    1.2.5 Excel Data Ranges .................................................................................... 4
1.3 Examples of Copy-and-Paste Techniques ........................................................ 4
    1.3.1 Highlight, "Cut" or "Copy," and Then "Paste" ........................................ 5
    1.3.2 Convert Text Data to Excel Column Data Fields ..................................... 5
    1.3.3 Copy Data to the SAS Enhanced Editor Window for Use in a SAS Program ......................... 7
    1.3.4 Save Multiple Lines of Text in a Single Excel Cell .................................. 8
    1.3.5 Converting Excel Tables to Text ............................................................... 9
1.4 Accessing Excel Data from the SAS Explorer Window and Toolbar ............... 9
    1.4.1 SAS Explorer Window and Toolbar Processing Method Descriptions .... 10
    1.4.2 Picking the Export Wizard from the SAS Explorer Window "Export" Menu ....................... 11
    1.4.3 Using the "Copy Contents to Clipboard" Option of the SAS Explorer Window ................... 11
    1.4.4 Selecting the "Save as Html" Option of the SAS Explorer Window ..... 12
    1.4.5 Using the "View in Excel" Option to Copy Data to Excel Files via HTML ........................... 13
    1.4.6 SAS Toolbar File Option, the Gateway to the SAS Export / Import Wizards .................... 15
    1.4.7 Choosing the "Export Data" SAS Toolbar (Export Wizard) File Option .............................. 16
    1.4.8 Electing the "Import Data" SAS Toolbar (Import Wizard) File Option.............................. 16
    1.4.9 Using the Export / Import Wizards in a 32/64-Bit Mixed Environment ............................... 17
1.5 Chapter Summary ............................................................................................. 18

**Chapter 2: Use PROC EXPORT to Write SAS Data to External Files and Excel Workbooks .................................................................................................. 19**
2.1 Introduction ....................................................................................................... 19
2.2 Purpose ............................................................................................................. 20
2.3 Syntax of the SAS EXPORT Procedure ........................................................... 20
2.4 Data Access Methods for Excel Files Supported by PROC EXPORT ............. 21
2.5 Overview of the Examples ................................................................................ 22

2.6 List of Examples ........................................................................................................................ 23
    Example 2.1 SAS Code to Export Data to an Excel 4 or Excel 5 Format File .............................. 23
    Example 2.2 PROC EXPORT Using the DBMS=DLM Option ...................................................... 24
    Example 2.3 PROC EXPORT Using the DBMS=EXCEL Option ................................................... 25
    Example 2.4 PROC EXPORT Using the DBMS=EXCELCS Option ............................................... 27
    Example 2.5 SAS Code to Export Data to an Excel File with No Column Headers ..................... 28
    Example 2.6 SAS Code to Export Data to a Network Windows Computer ................................ 28
2.7 Conclusion ................................................................................................................................ 29

## Chapter 3: Use PROC IMPORT to Read External Data Files and Excel Workbooks into SAS .................................................................................................................................. 31

3.1 Introduction .............................................................................................................................. 31
3.2 Purpose .................................................................................................................................... 32
3.3 Syntax of the SAS IMPORT Procedure ..................................................................................... 32
3.4 Data Access Methods for Excel Files Supported by PROC IMPORT ....................................... 33
3.5 Overview of the Examples ....................................................................................................... 34
3.6 List of Examples ....................................................................................................................... 34
    Example 3.1 PROC IMPORT Using the DBMS=EXCEL4 or EXCEL5 Option ............................... 35
    Example 3.2 PROC IMPORT Using the DBMS=DLM Option ...................................................... 35
    Example 3.3 PROC IMPORT Using the DBMS=EXCEL Option ................................................... 37
    Example 3.4 PROC IMPORT Using the DBMS=EXCELCS Option ............................................... 40
    Example 3.5 PROC IMPORT Using the DBMS=XLS or XLSX to Select Columns ....................... 42
    Example 3.6 PROC IMPORT Using the DBMS=XLS or XLSX to Select Rows ............................ 43
    Example 3.7 PROC IMPORT Using the DBMS=XLS or XLSX to Select Excel Ranges ............... 44
3.7 Conclusion ................................................................................................................................ 45

## Chapter 4: Using the SAS LIBNAME to Process Excel Files .................................... 47

4.1 Introduction .............................................................................................................................. 47
4.2 Purpose .................................................................................................................................... 48
4.3 Excel-Specific Features of the SAS LIBNAME Statement ....................................................... 48
4.4 Syntax of the SAS LIBNAME Statement .................................................................................. 49
4.5 LIBNAME Statement ENGINE CONNECTION OPTION Descriptions ...................................... 50
    4.5.1 HEADER Option to Read Variable Names ....................................................................... 50
    4.5.2 MIXED Option to Select Data Types ................................................................................ 50
    4.5.3 PATH Option to Define Physical File Locations .............................................................. 51
    4.5.4 VERSION Option to Identify Excel File Version ............................................................... 52
    4.5.5 PROMPT Option to Interactively Assign a Libref ........................................................... 52
    4.5.6 Other Common SAS PC File LIBNAME Options ............................................................. 53
4.6 Excel-Specific Dataset Options ............................................................................................... 53
4.7 UNIX, LINUX, and 64-Bit Windows Connection Options ........................................................ 54
4.8 Overview of the Examples ....................................................................................................... 55
    4.8 List of Examples ................................................................................................................... 55
4.9 Examples .................................................................................................................................. 56
    Example 4.1 Using the Engine Connection HEADER Option ..................................................... 56
    Example 4.2 Using the Engine Connection MIXED Option ........................................................ 57
    Example 4.3 Using the Engine Connection PATH Option .......................................................... 58

    Example 4.4 Using the Engine Connection VERSION Option .................................................................. 58
    Example 4.5 Using Named Literals with the LIBNAME Statement ...................................................... 59
    Example 4.6 Using PROC CONTENTS to Examine an Excel Workbook .............................................. 60
    Example 4.7 Using Dataset Options to Process Date and Time Values ............................................ 62
    Example 4.8 Using Dataset Options to Process Variable Type Conversions .................................... 63
    Example 4.9 Processing on 64-Bit Operating Systems ....................................................................... 64
4.10 Conclusion .................................................................................................................................................. 65

# Chapter 5: SAS Enterprise Guide Methods and Examples ................................................ 67

5.1 Introduction ................................................................................................................................................. 67
5.2 Purpose ........................................................................................................................................................ 68
5.3 Typical Methods to Access Excel from SAS Enterprise Guide ............................................................... 68
5.4 Overview of the Examples ......................................................................................................................... 68
5.5 List of Examples .......................................................................................................................................... 68
5.6 Examples ...................................................................................................................................................... 69
    Example 5.1 Using the Export Method with Enterprise Guide ............................................................ 69
    Example 5.2 Using the "Send To" Method .............................................................................................. 71
    Example 5.3 Using the "Send To" Method to Output a Graph or Report .......................................... 71
    Example 5.4 Using the "Export" Method to Output a Graph or Report ............................................ 75
    Example 5.5 Using "Open" or "Import" Toolbar Options to Read Excel Workbooks ...................... 77
    Example 5.6 Using the "Import Data" Toolbar Option to Read a Range of Cells ............................ 80
5.7 Conclusion ................................................................................................................................................... 84

# Chapter 6: Using JMP to Share Data with Excel ................................................................. 85

6.1 Introduction ................................................................................................................................................. 85
6.2 Purpose ........................................................................................................................................................ 85
6.3 Methods of Sharing Data between JMP and Excel ................................................................................. 86
6.4 List of Examples .......................................................................................................................................... 87
6.5 Examples ...................................................................................................................................................... 87
    Example 6.1 Within Excel, Set the JMP Preferences for Loading Excel Data .................................. 87
    Example 6.2 Reading Data from Excel to JMP ...................................................................................... 88
    Example 6.3 Writing Data from JMP to Excel ........................................................................................ 89
6.6 Conclusion ................................................................................................................................................... 90

# Chapter 7: SAS Add-In for Microsoft Office (Excel) .......................................................... 91

7.1 Introduction ................................................................................................................................................. 91
7.2 Purpose ........................................................................................................................................................ 91
7.3 Methods of Sharing Data Using SAS Add-In for Microsoft Office ......................................................... 92
7.4 List of Examples .......................................................................................................................................... 94
7.5 Examples ...................................................................................................................................................... 94
    Example 7.1 Open a SAS Dataset Using SAS Add-In for Microsoft Office ....................................... 94
    Example 7.2 Open a SAS Report Dataset (*.srx) Using SAS Add-In for Microsoft Office ........... 99
7.6 Conclusion ................................................................................................................................................. 105

## Chapter 8: Creating Output Files with ODS for Use by Excel ............ 107

8.1 Introduction ............ 108
8.2 Purpose ............ 108
8.3 An Introduction to SAS Tagset Templates That Create Files for Excel ............ 109
    8.3.1 How to Locate a Tagset Template ............ 110
8.4 Difference Between an ODS Tagset and an ODS Destination ............ 111
8.5 Syntax of the ODS CSV and CSVALL Output Processes ............ 111
8.6 CSV and CSVALL Tagset Options ............ 111
8.7 Overview of CSV and CSVALL Examples ............ 113
8.8 CSV and CSVALL Examples to Write *.csv Files ............ 113
    Example 8.8.1 Simple CSV and CSVALL File Default Output Differences ............ 113
    Example 8.8.2 CSV and CSVALL Title and Footnote Output Differences ............ 115
    Example 8.8.3 Write Currency Values as Unformatted Numbers ............ 118
    Example 8.8.4 Change Delimiters When Outputting Data with CSV Tagset ............ 120
    Example 8.8.5 Save Leading Zeroes in Character Fields Sent to Excel ............ 123
8.9 Syntax of ODS MSOFFICE2K Output Processes to Write HTML Files ............ 124
8.10 MSOFFICE2K Tagset Template Options ............ 125
8.11 Overview of MSOFFICE2K Examples ............ 126
8.12 MSOFFICE2K Examples to Write HTML Files ............ 126
    Example 8.12.1 Generating an HTML Output File with No Options ............ 126
    Example 8.12.2 Generating an HTML File Using the Summary_Vars Option ............ 127
8.13 Syntax of the ODS EXCELXP Tagset Template Output Processes ............ 128
8.14 ODS EXCELXP Tagset Options ............ 130
8.15 Overview of EXCELXP Examples ............ 132
8.16 EXCELXP Examples to Write XML Files ............ 133
    Example 8.16.1 Generating an XML Output File with No Options ............ 133
    Example 8.16.2 Adjusting Column Width Using Tagset Template Options ............ 134
    Example 8.16.3 Tagset Option to Hide Columns While Writing the File ............ 135
    Example 8.16.4 Apply an Excel "AUTOFILTER" to Selected Output Columns ............ 136
    Example 8.16.5 Using Multiple Options to Produce a "Ready-to-Print" Spreadsheet ............ 137
    Example 8.16.6 Creating a Table of Contents in an Excel Workbook ............ 138
    Example 8.16.7 Methods of Naming Excel Worksheets ............ 140
    Example 8.16.8 Splitting One Report onto Multiple Excel Worksheets ............ 141
    Example 8.16.9 Methods of Placing Labels in Excel Worksheet Names ............ 142
    Example 8.16.10 Use SHEET_INTERVAL= BYGROUP to Create Worksheets ............ 143
    Example 8.16.11 Use SHEET_INTERVAL= PROC to Create Worksheets ............ 144
    Example 8.16.12 Build Separate Worksheets with Titles on Each Sheet ............ 146
8.17 The New ODS Destination EXCEL for Writing Workbooks ............ 147
8.18 Conclusion ............ 148

## Chapter 9: Accessing Excel with OLE DB or ODBC Application Program Interfaces (API Methods) ............ 149

9.1 Introduction ............ 149
9.2 Purpose ............ 149
9.3 Concept of the OLE DB or ODBC API Processes ............ 149

9.4 Guidelines for Setting Up OLE DB or ODBC Connections .................................................. 150
9.5 List of Examples ........................................................................................................................ 150
9.6 Examples .................................................................................................................................... 151
    Example 9.1 Assign a Libref to an Excel Worksheet with the OLE-DB Dialog Box .................. 151
    Example 9.2 Using LIBNAME Prompt Mode to Build an OLE-DB Connection ...................... 152
    Example 9.3 Using an OLE-DB init_string to Open an Excel Workbook .............................. 154
    Example 9.4 Using PROC CONTENTS to Verify Excel to OLE DB Connection ..................... 154
9.7 Conclusion ................................................................................................................................. 156

## Chapter 10: Using PROC SQL to Access Excel Files .................................................. 157

10.1 Introduction ............................................................................................................................. 157
10.2 Purpose .................................................................................................................................... 158
10.3 Basic Syntax of the SQL Procedure ..................................................................................... 158
10.4 A Simple Explanation of SQL "PASS-THROUGH" Processing ......................................... 160
10.5 Overview of the Examples .................................................................................................... 160
    10.5.1 List of Examples ........................................................................................................... 160
10.6 Examples .................................................................................................................................. 160
    Example 10.1 LIBNAME Assignments to Access Excel Using PROC SQL ............................. 160
    Example 10.2 Create an Excel File, Read It with SQL, and Then Compare the Files ............. 161
    Example 10.3 Use PROC SQL to Read a Subset of Records from an Excel Workbook .......... 162
    Example 10.4 Use PROC SQL Pass-Through Facilities to Process an Excel File .................... 162
    Example 10.5 Read a Pre-defined Range of Cells from an Excel Workbook ........................... 163
    Example 10.6 Calculate a New Variable within the SQL Code and Sort the Output ............... 165
    Example 10.7 Examine the Contents and Structure of an Excel Workbook with a "PCFILES::" Special Query ........................................................................................................................ 165
10.7 Conclusion ............................................................................................................................... 166

## Chapter 11: Using DDE to Read and Write to Excel Workbooks ......................... 167

11.1 Introduction ............................................................................................................................. 167
11.2 Purpose .................................................................................................................................... 167
11.3 Basic Concept of the DDE Client-Server Environment .................................................... 168
    11.3.1 How the DDE Client-Server Relationship Works ...................................................... 168
    11.3.2 General DDE Syntax and Options ............................................................................. 168
11.4 List of User-Written SAS Macros That Can Enhance DDE Processing ......................... 171
    11.4.1 SAS Macro to Start Excel ............................................................................................ 171
    11.4.2 SAS Macro to SAS to Issue Commands to Excel ...................................................... 172
    11.4.3 SAS Macro to Define a Range of Excel Cells for Processing .................................. 172
    11.4.4 SAS Macro to Save the Contents of an Excel Workbook ........................................ 174
    11.4.5 SAS Macro to Close Excel Workbook ....................................................................... 174
    11.4.6 SAS Macro to Write All or Selected Variables to an Excel Output Workbook ...... 175
11.5 List of Examples ...................................................................................................................... 177
11.6 Examples .................................................................................................................................. 177
    Example 11.6.1 The Hello World Project ................................................................................ 177
    Example 11.6.2 The Hello World Project When the Excel Workbook Is Closed ..................... 179
    Example 11.6.3 The Hello World Project Using NOTAB and LRECL= Options ..................... 180

Example 11.6.4 Writing "Hello World" to an Excel File Using DDE Macros ............................ 182
Example 11.6.5 Writing a SAS Dataset to an Excel File Using the SAS_2_EXCEL DDE Macro 184
11.7 Conclusion .................................................................................................................... 187

## Chapter 12: Building a System of Excel Macros Executable by SAS .................. 189
12.1 Introduction ................................................................................................................... 189
12.2 Purpose ......................................................................................................................... 190
12.3 General Design of a Tool to Control Excel Macros from SAS ....................................... 190
    12.3.1 Prepare a SAS File and Execute Excel to Process the Output ............................. 191
    12.3.2 Prepare Excel to Open the File Output by SAS .................................................... 192
    12.3.3 Prepare Excel Macros to Reformat the Excel Workbooks .................................... 194
12.4 Automate the Tool So That SAS Creates a Formatted Excel Output Workbook ........... 197
    12.4.1 Eliminate the Manual Steps from the Processing ................................................. 197
    12.4.2 Create a SAS Output File with More Data and Control Information ..................... 202
    12.4.3 Create an Excel Macro to Process the Output SAS File ....................................... 203
    12.4.4 Build an Excel Graph Using an Excel Macro ......................................................... 207
12.5 Conclusion .................................................................................................................... 209

## Chapter 13: Building a System of Microsoft Windows Scripts to Control Excel Macros .................................................................................................................. 211
13.1 Introduction ................................................................................................................... 211
13.2 Purpose ......................................................................................................................... 212
13.3 Guidelines for Building and Using a VBS/VBA Macro Library ...................................... 214
    13.3.1 Create Naming Conventions for Storing and Executing VBS/VBA Macros .......... 214
    13.3.2 Set Up Workstation Options .................................................................................. 215
    13.3.3 Where to Store VBS/VBA Scripts and Macros ...................................................... 217
    13.3.4 SAS Code to Execute a Visual Basic Script .......................................................... 219
    13.3.5 Build a Parameter-Driven VBS Script to Control the Execution of Excel .............. 220
    13.3.6 Build a Control Macro for Each Excel Report ....................................................... 223
13.4 Conclusion .................................................................................................................... 229

## Chapter 14: Create an Excel Workbook That Runs SAS Programs .................... 231
14.1 Introduction ................................................................................................................... 231
14.2 Purpose ......................................................................................................................... 232
14.3 Guidelines for Building an Excel User Form Interface .................................................. 233
    14.3.1 Common Excel and Excel User Form Terms ........................................................ 233
    14.3.2 Introduction to the Integrated Development Environment (IDE) .......................... 235
    14.3.3 Using the Integrated Development Environment (IDE) Toolbox Menu ................ 236
    14.3.4 Building a Sample Integrated Development Environment (IDE) Menu ................ 237
    14.3.5 Linking the Integrated Development Environment (IDE) Menu and the Data ..... 239
    14.3.6 Storing Control Information in the Excel Workbook Worksheets ......................... 240
    14.3.7 Set Up Control Variables to Access Data Stored in the Workbook ..................... 241
    14.3.8 Learn How to Make the Excel UserForm Execute ................................................ 245
14.4 Excel VBA Routines to Make the Workbook UserForm Active ..................................... 248
    14.4.1 Initialize the User Form ......................................................................................... 248
    14.4.2 Write the User Parameters to a File in a Working Directory ................................ 253
    14.4.3 Copy Source Program from a Production Directory to the Working Directory .... 253

14.4.4 Verify the Output Batch File Points to the Correct SAS Run Time Module ...................... 254
14.4.5 A Routine to Save the Changes and Exit the Program ....................................................... 255
14.4.6 Directory Structure Associated with the Processing .......................................................... 255
14.4.7 Common Issues That Might Occur ...................................................................................... 257
14.4.8 Prepare a VBA Macro to Process Your Output Report ....................................................... 258
14.5 Conclusion ...................................................................................................................................... 259
**Index** .................................................................................................................................................**261**

x

# Preface

Over time, I have learned that, to keep life simple, you give the boss data in the format he or she wants. In other words, if the boss wants an Excel workbook, you deliver an Excel workbook with all the data and formatting requested. If the data comes from Excel, you need to find a way to read the data from the Excel files. For this book, I wrestled with the question of how to describe merging the SAS and Excel distinct skill sets in a way that would be understandable to both groups. This book is primarily written for people who need to transfer data between SAS and Excel. That usually means that you are well-versed in one application, but not the other. My choice of words was very careful when I used the word "BETWEEN."

I have included detailed explanations for some tasks that seem relatively simple so that users in their first week on the job can find a simple transfer method they can understand. I then build and show more complex examples that help more experienced users. There is a difference between needing to transfer data quickly and being able to repeatedly produce an identically formatted report month after month. I once saw someone fired because it took them three days to write a CSV-formatted file without asking anyone for help. This experience fortified my desire to find a way to help someone else avoid the same fate. I feel that detailed explanations are critical to the development of the skills described within these pages.

The topics presented here address *data transfer issues* rather than *data analysis tools*. I focus on how to move the data, and I let other authors tell you how to analyze the data. Each product, SAS and Excel, has a rich set of features requiring many skills. This document focuses upon data preparation and data transfer techniques of both SAS software and Excel in order to describe some of the many ways to transfer data between the applications. The examples shown here highlight the data exchange methods, but spend little time manipulating the data.

The transfer of data between the files that SAS can read and files that Microsoft Excel can read can be accomplished in many ways. These methods range from the very simple to the very complex and require different skill sets to accomplish the transfers. Once again, I note that the word "between" was chosen carefully to avoid limiting the tools explained or the examples used. This book defines some of the methods and enumerates various skills needed to complete these data transfers in both directions. The transformation of data into information has been going on since the first shepherd sold his sheep at the local market, and the buyer marked notches on a stick to record the amount due. Mankind has been looking for better ways since then. Within this book, many methods are shown to provide users of both SAS and Excel with the methods that fit their individual needs, giving even the most advanced SAS or Excel users at least one new tool for transferring information between SAS and Excel.

## Audience

I have written this book to address the needs of analysts, programmers, and others who use SAS and Excel at any user level. Included in this book are simple commands that aid novice users to get started with simple tasks. I used a step–by-step approach to increase the complexity of the commands presented and the tasks explained. The book will stretch and teach even the most advanced users of both SAS and Excel. While some basic knowledge of both SAS and Excel is assumed on the part of the reader, I have attempted to present this material in such a way that the book can be used either to gain knowledge by reading it sequentially or to find resource assistance from the table of contents or the index that to pinpoint a needed explanation.

## Approach

In my 40 plus years of programming computers, I have noticed that people who write system code (like compilers, operating systems, and software languages) for other people tend to focus on the current release and look forward to work for the next release. But this is done privately, because the next release is always intended to improve the current release. On the other hand, applications programmers tend to look at the current release and work to see how they can make the code from the past release step up to the current release. For them (the vast majority of all programmers), the next release is behind the curtain and will perpetually be delayed.

This book takes a tutorial approach to presenting the information. I also assume that old methods do not die quickly. The concept of being "backward-compatible" is something that both SAS and Microsoft practice. Therefore, older formats that are not actively supported are documented because they are still in use, and current software is still compatible with them. I feel that older methods and simple techniques still have a place in the programming life cycle. Young, new, or inexperienced programmers will always need to have the simple processes shown and explained so their experience level can also grow. Therefore, in this book a problem or method is presented with an explanation of procedures to solve the problem or method. Detailed examples are presented and explained. Because of the number of SAS versions available and the way that they interact individually with Excel, this book presents several different versions of SAS and Excel within the examples.

This book is for programmers who are always upgrading their code to catch up with the current release.

I would be remiss if I failed to thank my wife for the patience she has shown over the last many years as this project has slowly progressed to completion. Without her understanding and encouragement it may not have been possible.

# About This Book

## Purpose

I wrote this book to help SAS users of all skill levels find out how to move data between SAS and Microsoft Excel. My years of programming experience have helped me decode the mysteries of vendor-supplied system documentation. I wanted to gather that information together and present it in an easy-to-understand tutorial format with the prime emphasis on examples. I have also scattered in my observations on the world of programming in general and pieced together an array of examples that include both simple and complex task descriptions.

## Is This Book for You?

Whatever your skill level, I hope you will find examples that will teach you something. In every class I teach or paper I present, I always ask if anyone learned anything. I want you to be able to find a place on your desk for this book, use it as you progress through the skills presented, and gain expertise to easily move your data.

## Prerequisites

This book is designed for you to use without need for prerequisites. If you can open the SAS program and copy data using your mouse, then you can get started. I do not attempt to teach you how to write SAS programs or build an Excel spreadsheet, but I present methods to move data between the two data storage tools.

## Scope of This Book

This book attempts to show you how to move data "BETWEEN" SAS and Excel. I have attempted to use as many differing techniques as I could within the limited space available. As I worked my way through the chapters, I created examples that progressively increased in power and complexity.

But, what I do not do is show you very much about how to use the data after it is moved or copied into either Excel or SAS. Within this book I have covered many ways that show you how to shuffle your data between SAS and Excel. I hope I have also opened ways to manipulate the worksheets after they have been written. I have tried to keep the data simple and only change the methods. In fact, nearly every example uses the same SAS dataset, as noted below.

## About the Examples

### Software Used to Develop the Book's Content

Because SAS users are likely to be working with different SAS versions, I have included examples that use several versions of SAS software. Most of the examples use SAS 9.4. Some JMP examples and SAS Enterprise Guide examples are also shown. Examples of Excel screens also vary across several versions of Excel, from Excel 2003 to Excel 2013. The examples in the book cover the transition from the xls workbooks to the xlsx workbooks and the way SAS has adapted to those Excel changes.

## Example Code and Data

The primary dataset used for examples in this book is the SASHELP.SHOES SAS dataset; it is used as an exported file to Excel and then as input from Excel. The SASHELP.SHOES dataset is shipped with every version of SAS and is therefore convenient for all users.

You can access the example code and data for this book by accessing my author page at http://support.sas.com/publishing/authors. Select the name of the author, look for the cover thumbnail of this book, and select Example Code and Data to display the SAS programs that are included in this book.

For an alphabetical listing of all books for which example code and data is available, see http://support.sas.com/bookcode. Select a title to display the book's example code.

If you are unable to access the code through the website, email saspress@sas.com.

## Additional Help

Although this book illustrates many analyses regularly performed in businesses across industries, questions specific to your aims and issues may arise. To fully support you, SAS Institute and SAS Press offer you the following help resources:

- For questions about topics covered in this book, contact the author through SAS Press:
    - Send questions by email to saspress@sas.com; include the book title in your correspondence.
    - Submit feedback on the author's page at http://support.sas.com/author_feedback.
- For questions about topics in or beyond the scope of this book, post queries to the relevant SAS Support Communities at https://communities.sas.com/welcome.
- SAS Institute maintains a comprehensive website with up-to-date information. One page that is particularly useful to both the novice and seasoned SAS user is the SAS Knowledge Base. Search for relevant notes in the "Samples and SAS Notes" section of the Knowledge Base at http://support.sas.com/resources.
- Registered SAS users or their organizations can access SAS Customer Support at http://support.sas.com. Here you can pose specific questions to SAS Customer Support. Under *Support*, click *Submit a Problem*. You will need to provide an email address to which replies can be sent, identify your organization, and provide a customer site number or license information. This information can be found in your SAS logs.

## Keep in Touch

We look forward to hearing from you. We invite questions, comments, and concerns. If you want to contact us about a specific book, please include the book title in your correspondence.

### Contact the Author through SAS Press

- By email: saspress@sas.com
- Via the web: http://support.sas.com/author_feedback

### Purchase SAS Books

For a complete list of books available through SAS, visit sas.com/store/books.

- Phone: 1-800-727-0025
- Email: sasbook@sas.com

## Subscribe to the SAS Training and Book Report

Receive up-to-date information about SAS training, certification, and publications via email by subscribing to the SAS Training & Book Report monthly eNewsletter. Read the archives and subscribe today at http://support.sas.com/community/newsletters/training!

## Publish with SAS

SAS is recruiting authors! Are you interested in writing a book? Visit http://support.sas.com/saspress for more information.

# About The Author

William E. Benjamin, Jr., owns Owl Computer Consultancy, LLC, and works as a consultant, trainer, and author. William has been a SAS user for over 30 years and a consultant since 2007. He received an MBA from Western International University and a BS in computer science from Arizona State University. He has written and presented papers for SAS Global Forum, as well as many regional and local SAS users groups.

Learn more about this author by visiting his author page at http://support.sas.com/publishing/authors/benjamin.html. There you can download free book excerpts, access example code and data, read the latest reviews, get updates, and more.

# Chapter 1: Easy Data Movement between SAS and Microsoft Excel

1.1 Introduction ..................................................................................................... 1
1.2 Examination of Excel Files ............................................................................. 2
    1.2.1 Purpose ................................................................................................ 2
    1.2.2 Excel Data Types ................................................................................. 2
    1.2.3 General Excel Workbook Limitations .................................................. 2
    1.2.4 Excel Workbook Formatting Groups ................................................... 3
    1.2.5 Excel Data Ranges .............................................................................. 4
1.3 Examples of Copy-and-Paste Techniques ..................................................... 4
    1.3.1 Highlight, "Cut" or "Copy," and Then "Paste" ...................................... 5
    1.3.2 Convert Text Data to Excel Column Data Fields ................................. 5
    1.3.3 Copy Data to the SAS Enhanced Editor Window for Use in a SAS Program ......... 7
    1.3.4 Save Multiple Lines of Text in a Single Excel Cell .............................. 8
    1.3.5 Converting Excel Tables to Text .......................................................... 9
1.4 Accessing Excel Data from the SAS Explorer Window and Toolbar ............. 9
    1.4.1 SAS Explorer Window and Toolbar Processing Method Descriptions ............. 10
    1.4.2 Picking the Export Wizard from the SAS Explorer Window "Export" Menu ......... 11
    1.4.3 Using the "Copy Contents to Clipboard" Option of the SAS Explorer Window ... 11
    1.4.4 Selecting the "Save as Html" Option of the SAS Explorer Window ................. 12
    1.4.5 Using the "View in Excel" Option to Copy Data to Excel Files via HTML ......... 13
    1.4.6 SAS Toolbar File Option, the Gateway to the SAS Export / Import Wizards .... 15
    1.4.7 Choosing the "Export Data" SAS Toolbar (Export Wizard) File Option ........... 16
    1.4.8 Electing the "Import Data" SAS Toolbar (Import Wizard) File Option ............. 16
    1.4.9 Using the Export / Import Wizards in a 32/64-Bit Mixed Environment ............. 17
1.5 Chapter Summary ......................................................................................... 18

## 1.1 Introduction

It may seem a bit odd to start a book about SAS programming with examples of Excel files and their limitations. But, because this information is at the front of the book, it will be easy to locate when you want to figure out why some of your data seems to be missing or how much data this Excel format can hold.

Increased computer memory and speed have spurred the growth of computer capabilities. For example, the software known as either the Joint Engine Technology (JET) database engine or the Access Connectivity Engine (ACE) are built into the Microsoft Windows operating systems. These Microsoft database access engines are used to access data for several Microsoft products including Microsoft Excel. These Microsoft database engines provide an interface to Excel (and other Microsoft products) that can be used by SAS and other database interface tools to access data in Excel workbooks.

When you are looking at a computer monitor or a printed page, both a view of a SAS file and the display of an Excel worksheet are very similar. Each image has an array of rows and columns with data values. SAS calls them observations and variables while Excel calls them rows and columns. While most SAS users change all the values of individual elements of a SAS dataset with a SAS program or procedure, the

Excel spreadsheet user often modifies one element (or cell) at a time. The major difference between SAS datasets and Excel worksheets is that even though a SAS user can open a SAS dataset in "edit" mode and change the values of individual elements, SAS will enforce data type restrictions. A character value cannot be placed into a variable defined as a numeric variable. A number can be placed into a character variable, but it will be converted to a character value. Excel has no such restriction. This puts a burden of examining the output data on the user. This examination is often skipped and data losses can occur.

## 1.2 Examination of Excel Files

### 1.2.1 Purpose

Let's take a minute to look at why we need to know anything about Excel files. Excel files have changed over time. In the beginning, they had one worksheet and very few features or capabilities and proprietary data formats. Today, the features, options, and capabilities of Excel are impressive. We will start here with the simple things and, by the end of the book, show you not only how SAS can send data to Excel, but also how Excel can interact with SAS.

### 1.2.2 Excel Data Types

This section provides general descriptions of Excel Workbook sheet elements with row and column limitations, along with general descriptions of the file formats, data types, and range definitions.

#### Formulas

Excel cells can contain formulas. While these are not actually data types, the action taken by a formula may represent and be displayed as either a number or a character value. The SAS EXCELXP tagset can process character strings and output them to Excel as a formula within an Excel workbook.

#### Character

Character data fields can contain any text values and are usually alphanumeric data strings that may contain characters, words, numbers, Excel formulas, date and/or time values including special characters. Up to 32,767 (32KB-1) characters may be stored in one cell.

#### Numeric

Numeric values are usually numbers like the digits 0-9, but the data in these cells may also include formulas. The numeric character type also includes several special values that indicate error conditions like #NULL!, #N/A, #VALUE!. Numeric dates are represented as numbers the integer portion is the number of days and the fractional portion is the number of minutes and seconds within 24 hours. The counter resets at midnight to 0.0, noon is 0.5, and one second before midnight is represented as 0.999988.

#### Other Excel Data Display Formats

Excel, like SAS, can display numeric data in several different formats. These include displays such as a currency, percentage, accounting, numbers with integer and fractional parts, several date formats, and several numeric formats. Other display formats exist but the data values underlying the display are either character or numeric.

#### Other Excel File Elements

In addition to the data elements of an Excel file, an Excel workbook can also contain charts, graphs, and other images. The SAS Enterprise Guide features described in this book can generate SAS graphs that can be sent to Excel and displayed as Excel worksheet elements.

### 1.2.3 General Excel Workbook Limitations

Table 1.2.1 describes Excel workbook limitations and characteristics for versions of Microsoft Excel formats. As noted in the table, the size and capacity of Excel workbook files has grown over the years. However, care must be taken when processing large Excel files because the Microsoft JET and ACE engines that read and write to Excel workbooks have limitations that may not allow for processing of all data in an Excel workbook when using a SAS LIBNAME statement.

**Table 1.2.1: Microsoft Excel Workbook File Limitations.**

| Version | Limited to One Sheet | Rows | Columns | Extension |
|---|---|---|---|---|
| EXCEL4 | YES | 16,384 | 256 * | .xls |
| EXCEL 5 | NO | 16,384 | 256 * | .xls |
| EXCEL 95 | NO | 16,384 | 256 * | .xls |
| EXCEL 97 | NO | 65,536 * | 256 * | .xls |
| EXCEL 2000 | NO | 65,536 * | 256 * | .xls |
| EXCEL 2002 | NO | 65,536 * | 256 * | .xls |
| EXCEL 2003 | NO | 65,536 * | 256 * | .xls |
| EXCEL 2007 | NO | 1,048,576 * | 16,384 * | .xlsx, xlsb, xlsm |
| EXCEL 2010 | NO | 1,048,576 * | 16,384 * | .xlsx, xlsb, xlsm |
| EXCEL 2013 | NO | 1,048,576 * | 16,384 * | .xlsx, xlsb, xlsm |

\* The limits listed here are the limits for rows and columns of Excel files. However, the Microsoft JET and ACE engines used by the LIBNAME statement to access Excel files for reading and writing data are limited to 65,535 rows and 255 columns. Files larger than 255 columns and 65,535 rows should be processed without using the LIBNAME statement or any method that uses the Microsoft JET or ACE engines to refer to the Excel files.

### 1.2.4 Excel Workbook Formatting Groups

**File Format Groups**

- Excel 4 is unique with only one sheet and is the oldest of supported Excel formats. This format may not be supported in future versions of Excel.
- Excel 5 and Excel 95 share the same format.
- Excel 97, Excel 2000, Excel 2002, and Excel 2003 share the same format.
- Excel 2007, Excel 2010, and Excel 2013 share the same format and are the newest versions of Excel files.
- Excel 4 – Excel 2003 are binary formats, but as noted above they are not the same format.

**NOTE:** Newer Excel software can read and write any older Excel format.

Excel 2007, Excel 2010, and Excel 2013 file formats:

- .xlsx format does not allow Visual Basic for Applications (VBA) macros to be stored within the spreadsheet.
- .xlsm format allows (VBA) macros to be stored within the spreadsheet.
- .xlsb format stores the data in a binary format.
- .xlsx, .xlsm, and .xlsb formatted files are .ZIP files of XML documents that comprise the Excel workbook. .xlsb formatted files also contain binary formatted elements. While these files may also contain style sheets, images and other components, this book generally addresses the data contents unless otherwise specified.

Most Excel programs will open delimited or .xml files with a double click, but these are not Excel formatted files—they are text files that Excel knows how to format.

### Excel-Readable Files

There are several general file types that Excel can open and directly read the data into an Excel workbook, such as *.csv, *.tsv, *.txt, *.htm, *.html, and *.xml files. These are text files and can be edited with any text editor, as long as you know the file format. SAS processes that read and write to Excel workbook files generally can write out data files in several formats that Excel can read and convert to an Excel spreadsheet, such as *.csv, *.tsv, *.txt, *.htm, *.html, and *.xml files. These will be discussed in turn through this book, but here is a simple explanation of their general formats.

### Simple File Formats

Files with a .csv, .tsv, or .txt file extension are strings of text usually separated by a comma, tab, or blank, respectively. In these files, numbers are entered as text values. However, including commas in numbers greater than 999 may cause Excel to treat the number as two data items and put the values into two cells. These files may also contain quoted strings to avoid separating the character values into multiple cells. The actual results you see may differ based upon the computer configuration you are using.

### Complex File Formats

Files with a .htm, .html, or .xml file extension are strings of text that have had complex formatting rules applied to the contents of the file. These formatting rules provide instructions to the program opening the file that tell that program how to store, display, or format the data being read from the input file. These instructions may also include applying colors to the text and background or inserting sheet formatting to the worksheet.

## 1.2.5 Excel Data Ranges

Excel workbook files can contain groups of cells called named "ranges" that are not complete worksheets. The name is assigned to a range of cells by selecting the cells and using the name box usually found on the left side of the formula bar just below the control bar or ribbon on the Excel screen. SAS can read these named ranges by using the name of the Excel worksheet range. By default the LIBNAME statement defines a worksheet with a dollar sign "$" at the end of the name. Therefore, full Excel worksheets need to be defined to SAS as name literals (in the form 'sheetname$'n ) when processed by SAS. A name literal is a special SAS syntax structure. The named ranges need to be defined in Excel before SAS can read data from them. Excel named ranges or worksheet names without blanks or invalid characters for SAS names do not need to be referenced as a name literal.

## 1.3 Examples of Copy-and-Paste Techniques

One of the first methods of moving data around that people learn to use when starting to use a computer is to "CUT" or "COPY" something and then "PASTE" it where you want it to appear. This works well for many small applications. Moving data from Word documents to text documents, to Excel spreadsheets, or even to the SAS Enhanced Editor window is pretty simple. The difference between cut and copy is that the cut action removes the data from the original location while copy does not.

But how does that information fit into a book about moving data between SAS and Excel? Well, let's take a look at how this technique can help shuffle data around. Some of these examples will be used in later chapters to build more extensive tools. I use the Microsoft Windows environment. Other operating systems may have modified version of these procedures or not allow them at all. Each of the following methods is similar because they generally use the operating system to capture data with a cut or copy command and store it until a location can be identified to accept the data from the "paste" command. Different methods are shown here because the output of the "paste" command can be altered by the method used in the examples. The results of the "paste" command can therefore present differing consequences.

## 1.3.1 Highlight, "Cut" or "Copy," and Then "Paste"

When you are using Microsoft Windows, this method has only a few simple steps.

- Hold down the left mouse button and highlight the area you want to cut or copy.
- Release the left mouse button and leave the area highlighted.
- When using a Microsoft Windows (and some other systems) right-click on the highlighted area.
- Choose the menu option to either "cut" or "copy" the highlighted information.
- Move the cursor to the desired location and right-click.
- Choose the menu option to "paste".

Microsoft Windows operating systems allow key stroke combinations to substitute for the menu choices:

- key combination "CNTL" and "x" substitute for "cut"
- key combination "CNTL" and "c" substitute for "copy"
- key combination "CNTL" and "v" substitute for "paste"

**NOTE:** The mouse devices on some computers have their left and right mouse button functions inverted.

## 1.3.2 Convert Text Data to Excel Column Data Fields

Here, we will select a small part of the SASHELP.SHOES file as an example. I recommend this method for moving a few lines of data at a time, and I use this technique frequently. In SAS 9.3 and beyond, the default is to output HTML and display it in the Results window. I usually go to the SAS **Tools▶Options▶Preferences** window and, on the Results tab, select both "Create Listing" and "Create HTML." That way I can go to the Output window and highlight the data printed by the PROC PRINT statement below. The SAS WHERE statement limits the output observations from the region "Asia" for products "Boot", "Sandal", and "Slipper".

```
PROC PRINT DATA= sashelp.shoes
  (WHERE=(region = "Asia" and
          product IN ("Boot", "Sandal", "Slipper")));
RUN;
```

**Figure 1.3.1: SAS Output Window Showing the Output from the PROC PRINT Statement Above.**

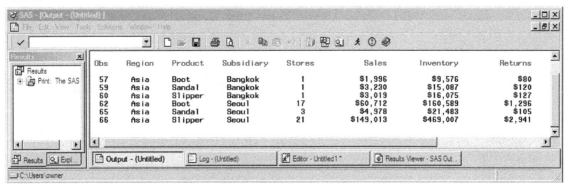

The data in the SAS Output window can then be copied from the SAS Output window and pasted into column A of an Excel spreadsheet. See Figure 1.3.2 below. Microsoft Excel applications have a wizard to convert text values to Excel columns. The output in the SAS Output window will be separated by spaces.

**Figure 1.3.2: Excel Worksheet with the Copied SAS Data Pasted into Column A of the Worksheet.**

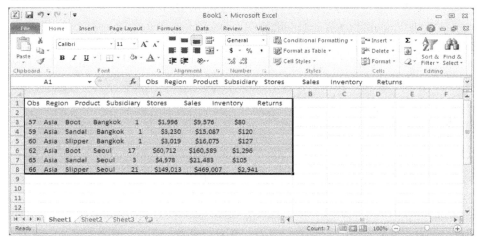

The Excel wizard to convert text to columns allows the data to be selected as either delimited or fixed-length fields. When you choose the delimited option, a screen similar to Figure 1.3.3 appears. Since we know that the output is delimited by spaces, the settings below allow us to convert the data from the SAS Output window to Excel columns as shown in Figure 1.3.4.

**Figure 1.3.3: Excel Convert Text to Columns Wizard Window 2.**

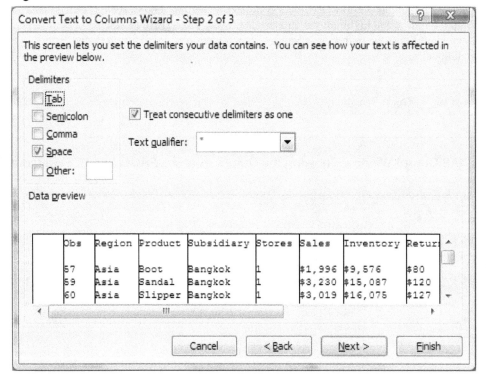

When the "Excel Text to Column Wizard" completes, the leading spaces and blank line from the SAS Output window may find their way into the Excel spreadsheet. If these are not needed, they can be removed quickly.

**Figure 1.3.4: Excel Worksheet after Converting the Text in Column A to Columns of Data.**

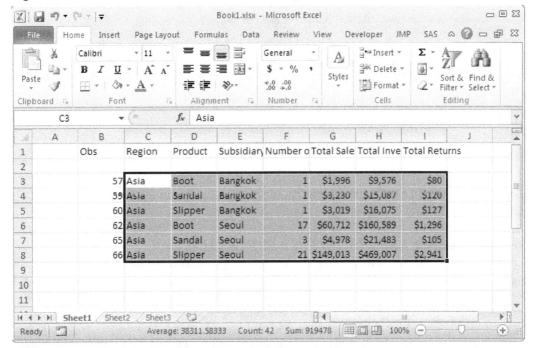

### 1.3.3 Copy Data to the SAS Enhanced Editor Window for Use in a SAS Program

This example highlights and copies the data in Figure 1.3.4 cells C3 to I8 into the SAS code segment below. When data is copied from Excel, tab characters are used to separate the data columns. This code shows the use of the INFILE command with the DLM option to identify the SPACE character as the delimiter within the SAS code. The irregular spacing shown on the lines following the "CARDS;" statement is caused by the character fields not being the same length and the characters shifting columns of data. The TRUNCOVER option causes the data to read correctly.

```
DATA Copy_n_paste;

INFILE DATALINES DLM = ' ' TRUNCOVER;
INPUT   Region  $ Product  $  Subsidiary  $ Stores    Sales $ Inventory  $
Returns $;
CARDS;
Asia     Boot   Bangkok            1         $1,996       $9,576         $80
Asia     Sandal Bangkok    1         $3,230       $15,087        $120
Asia     Slipper       Bangkok     1         $3,019       $16,075        $127
Asia     Boot   Seoul     17        $60,712      $160,589     $1,296
Asia     Sandal Seoul     3         $4,978       $21,483      $105
Asia     Slipper       Seoul     21        $149,013     $469,007     $2,941
;
RUN;
```

Running the sample SAS code that appears above Figure 1.3.1 will produce the SAS dataset with six observations and seven variables shown below in Figure 1.3.5.

### Figure 1.3.5: SAS Dataset Produced by Running Example 3.

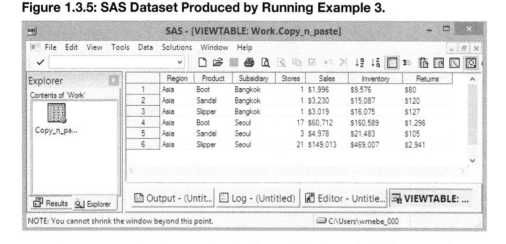

## 1.3.4 Save Multiple Lines of Text in a Single Excel Cell

For this example we will place multiple lines of SAS code into a single Excel cell. Later in the book we will find that this will be useful because we can recall the information and use it to send the SAS code back to a running SAS program. We will start by highlighting and copying the sample code that appears in Section 1.3.3 after the first paragraph. We place it into Excel by selecting cell A1, then selecting the formula entry field, and pasting the text into that field. The result is shown below after the formula entry field and cell A1 are expanded to show the full contents. Figure 1.4.1a shows the Excel worksheet after this is complete.

### Figure 1.4.1a: Several Lines of SAS Code Pasted into a Single Excel Cell of a Worksheet.

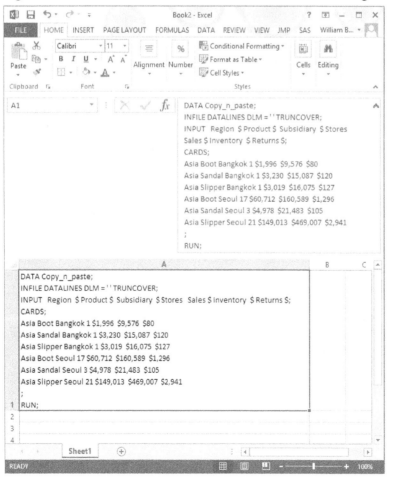

### 1.3.5 Converting Excel Tables to Text

This example starts by deleting the blank line in row 2 of the Excel spreadsheet from Figure 1.3.4 and copying the data from Figure 1.3.4 (cols C to I) into a Microsoft Word document and making the table borders visible. The second step is to place that same data into the SAS Enhanced Editor window and then copy the data from there into Microsoft Word. The results are shown in Figure 1.4.2a.

**Figure 1.4.2a: A Microsoft Word Document with Two Copy and Paste Outputs.**

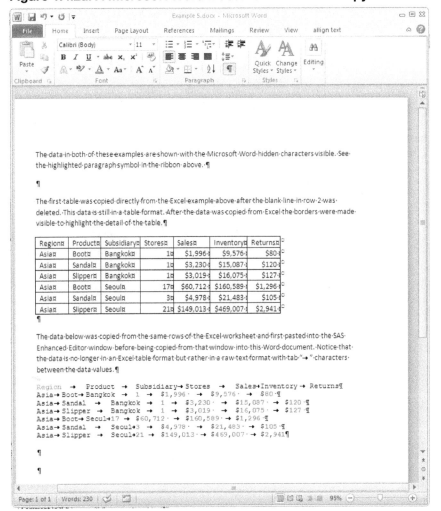

## 1.4 Accessing Excel Data from the SAS Explorer Window and Toolbar

I could have called this section "**Accessing Excel Data Using the SAS Export/Import Wizards,**" but not all of the methods I describe here use the wizards. When using SAS without SAS/ACCESS Interface to PC Files software, the conversion methods are very limited. The next six examples show how to access the SAS Export wizard, Import wizard, and other conversion methods directly from the SAS Explorer window and the SAS toolbar. I will show you how to do this without writing SAS code to do the conversion. Only two of these routines write SAS code for you. These methods allow you to save the code wherever you can write a data file. These two methods are the Export/Import wizards within SAS. They can be found on the "File" menu of the toolbar. As the names imply, the Export wizard sends data to Excel, and the Import wizard reads data from Excel. The examples shown here will use the SASHELP.SHOES data set. They send data to and read data from Excel without writing any SAS code. Within this book I treat the Export/Import wizards differently than PROC EXPORT and PROC IMPORT. The reason I treat them differently is because the wizards do the work for you and provide only a limited number of option selections, while the procedures require more knowledge about the software and the options that are available for use to read or write the SAS and Excel files. Not all versions of SAS will write to Excel files formatted for use by Excel 2007, Excel 2010, Excel 2013, or later versions of Excel.

### 1.4.1 SAS Explorer Window and Toolbar Processing Method Descriptions

The data transfer methods described in the remainder of this chapter start from the SAS Explorer window. These methods can be used to prepare data to be viewed with the Excel application. A right-click on the SAS file icon in the SAS Explorer window displays a menu from which one of the following options can be selected. The first option starts the SAS Export wizard, and the other three options create HTML code that can be opened by Excel or other applications.

- Select the "Export" option to run the SAS Export wizard.
- Select the "Copy Contents to Clipboard" option to copy HTML text and paste to an Excel file.
- Select the "Save as Html" option to copy data an HTML file that Excel can read.
- Select the "View in Excel" option to copy data to Excel files via HTML.

From the SAS toolbar there are two "File" options that provide access to Excel files. These are the IMPORT and EXPORT options. As their names imply, the first option reads data from Excel into a SAS dataset, while the second option writes data to an Excel workbook. These two options work with Excel formatted files.

- Menu option "Export Data" runs the Export wizard to copy data to an Excel file.
- Menu option "Import Data" runs the Import wizard to copy data from an Excel file.

I had been using SAS for nearly 25 years before someone showed me that I could right-click on a SAS dataset icon in the SAS Explorer window and have a menu pop up with output options for SAS files. Figure 1.4.1b below is that menu. The process is to open the SAS Explorer window, and then right-click on a SAS dataset. A menu will appear with options that can be selected. The next four examples discuss the options that can either read or write data for an Excel workbook.

**Figure 1.4.1b: SAS Explorer Window "Right-Click" Options.**

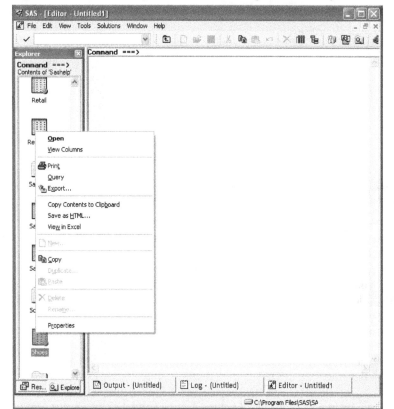

The pop-up menu shown in Figure 1.4.1b lists items that are explained in the following sections. Each of these menu items represents a method of transferring data between SAS and Excel. In some cases other options also exist, but here we are looking for methods that work to get data into Excel. As a shortcut, once the menu is displayed, type the underlined letter to invoke the option.

## 1.4.2 Picking the Export Wizard from the SAS Explorer Window "E*x*port" Menu

### Example 1.4.2

Selecting the "Export" option displays the following Export wizard menu. This is the first of several menus that will guide you through the SAS Export process. SAS versions have similar tools, but each version of SAS may have slightly different features available based upon the underlying hardware, operating system, and SAS software version you are using. But all of the versions will guide you through the menus and offer to save the code into a file for reuse later.

**Figure 1.4.2b: First Menu of the SAS Export Wizard Tool for SAS 9.2 TS Level 1M0.**

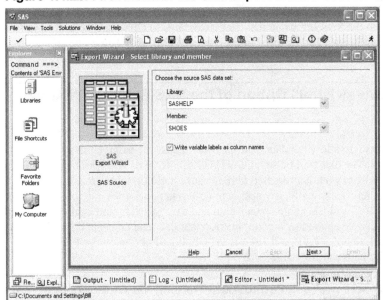

The screen image shown in Figure 1.4.2b is from SAS 9.2 and does not present the options for 64-bit hardware or software. The Export wizard will generally walk you through the process selecting the SAS LIBNAME and data file, and then ask you questions about the output format you wish to use and the file name and location where you want to store the output file. While the Export wizard menus are relatively self-explanatory, you do need to know the output format you want and the location to which you can write the output file. More information about the SAS Export wizard can be found in the SAS documentation for each version of SAS.

## 1.4.3 Using the "Copy Contents to Clip*b*oard" Option of the SAS Explorer Window

### Example 1.4.3

This option generates HTML code that contains the data from the SAS dataset and places that data on the Windows clipboard. This HTML data can be pasted into any file or application that will accept a paste command. When this is saved as an HTML file or a text (.txt) file, the resulting file can be opened with

Excel or a browser. The *.txt file might need to be renamed to *.htm or *.html to be opened with Excel.

```
1      filename _temp_ clipbrd;
2      ods noresults;
3      ods listing close;
4      ods html file=_temp_ rs=none style=minimal;
NOTE: Writing HTML Body file: _TEMP_
5      proc print data=Sashelp.'Shoes'N noobs;
6      run;

NOTE: There were 395 observations read from the data set SASHELP.SHOES.
NOTE: PROCEDURE PRINT used (Total process time):
       real time           0.51 seconds
       cpu time            0.29 seconds

7      ods html close;
8      ods results;
9      ods listing;
10     filename _temp_;
NOTE: Fileref _TEMP_ has been deassigned.
```

This information was printed onto the SAS log when the "Copy Contents to Clipboard" option was selected. This code was generated by SAS and resulted in the output HTML data being placed onto the Windows clipboard using the SAS FILENAME statement to define the output location.

## 1.4.4 Selecting the "Save as Html" Option of the SAS Explorer Window

### Example 1.4.4

This option offers a "Save as" menu to allow the data from the SAS dataset to be saved in a file on any device attached to your computer. The data will be saved in HTML file format. The options on the left side of the "Save As" menu may vary from what is presented here because these options are unique to the software installed on your computer. Specifically, my option to access network files may not be the same as yours. When the output file directory and file name are selected and the "Save" button is clicked, the output file is actually created. This file can be opened by any application that can read an HTML file. Notice that the only real difference between Examples 1.4.3 and 1.4.4 is the Clipboard option on the FILENAME statement instead of an actual file path and file name.

**Figure 1.4.3: "Save As" SAS Screen Presented to Allow Naming the Output Location of an HTML File.**

The following information was printed onto the SAS log when the "Save as HTML" option was selected. This code was generated by SAS and resulted in the output HTML data being placed into a file on my computer using the SAS FILENAME statement to define the output location.

```
1    filename _temp_  "C:\My_HTLM_Output\Shoes.html";
2    ods noresults;
3    ods listing close;
4    ods chtml file=_temp_ rs=none;
NOTE: Writing CHTML Body file: _TEMP_
5    proc print data=Sashelp.'Shoes'N noobs;
6    run;

NOTE: There were 395 observations read from the data set SASHELP.SHOES.
NOTE: PROCEDURE PRINT used (Total process time):
      real time           0.15 seconds
      cpu time            0.09 seconds
```

```
7    ods chtml close;
8    ods results;
9    ods listing;
10   filename _temp_;
NOTE: Fileref _TEMP_ has been deassigned.
```

## 1.4.5 Using the "View in Excel" Option to Copy Data to Excel Files via HTML

### Example 1.4.5

The "View in Excel" option also generates an HTML file. This file is saved in your SAS Work directory as a text file of HTML commands with the extension *.xls. The file is not visible in the SAS Explorer window and Excel is invoked to open the file. The file will typically have a name similar to "#LNxxxxxx.xls". This naming convention dates back to when all file names were limited to 8 characters in length and had a 3-character extension. This file-naming structure is known as the "8.3" format for names that were in use before SAS and Microsoft Windows could support long file names. The 3-byte extension (xls) allows Excel to open the file without hesitation before Excel 2007. The newer versions check the contents of the file and, if the file name ends in .xls but contains HTML or XML formatted commands for Excel, then a message is displayed asking you to verify that you want to proceed (see Figure 1.4.4).

**Figure 1.4.4: Warning Message Shown when Excel 2007 or Later Opens the Generated HTML File.**

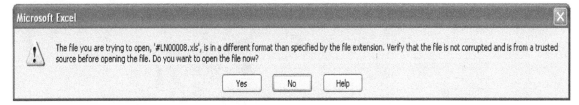

The SAS log output below is similar to the listing output in Example 1.4.4, but has the addition of the Display Manager command at the bottom of the log. The "winexecfile" command caused the named file to be opened with the application that is associated with the file extension. In this case, it would be Excel.

```
1    filename _temp_ "C:\Users\owner\AppData\Local\Temp\SAS Temporary
Files\_TD3228\#LN00008.xls";
2    ods noresults;
3    ods listing close;
4    ods html file=_temp_ rs=none style=minimal;
NOTE: Writing HTML Body file: _TEMP_
5    proc print data=Sashelp.'Shoes'N label noobs;
6    run;

NOTE: There were 395 observations read from the data set SASHELP.SHOES.
NOTE: PROCEDURE PRINT used (Total process time):
      real time           0.23 seconds
      cpu time            0.12 seconds

7    ods html close;
8    ods results;
9    ods listing;
10    filename _temp_;
NOTE: Fileref _TEMP_ has been deassigned.
11    dm "winexecfile ""C:\Users\owner\AppData\Local\Temp\SAS Temporary
Files\_TD3228\#LN00008.xls"" ";
```

On computers with Excel 2007 or newer versions installed, the pop-up screen appears to allow you to verify that you still want to open the file even though the file is not formatted as an Excel file. When you click on "Yes" to view the data using Microsoft Excel, the output HTML file is opened by Excel. The SAS option "style=minimal" is applied to the HTML formatted output. It has simple formatting with a header "The SAS System" in row one and the SAS variable labels in row three. When no labels exist, the variable names are placed into row three. This output method is handy for getting SAS data directly into Excel workbooks. But remember that big SAS files may exceed the limits of Excel worksheets, as noted in Table 1.2.1, Microsoft Excel Workbook File Limitations.

**Figure 1.4.5: View of the HTML File Opened by Excel.**

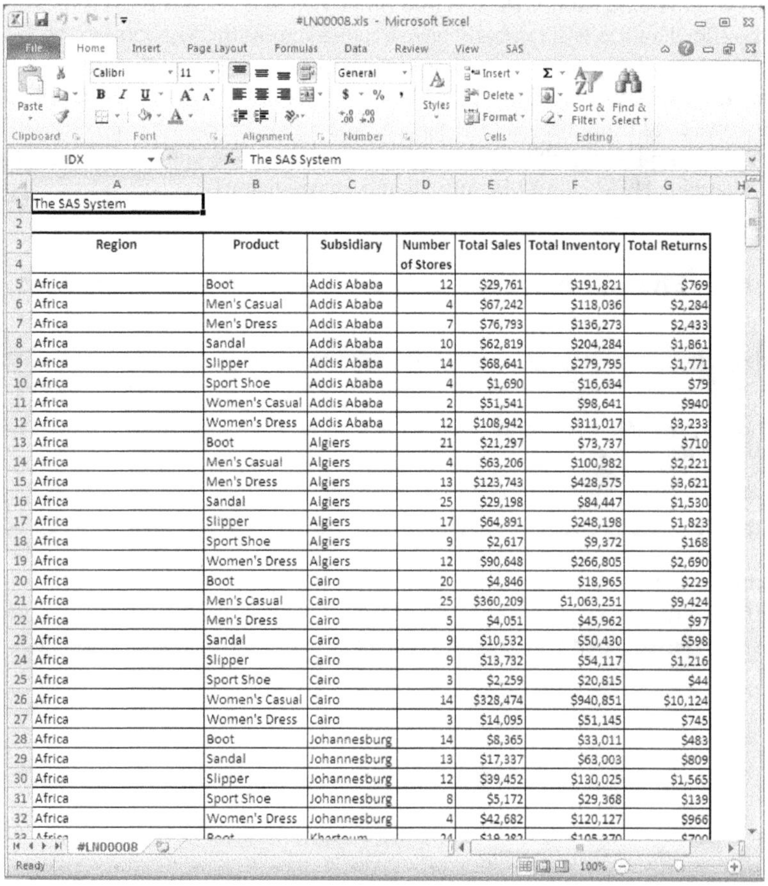


The SAS System

| Region | Product | Subsidiary | Number of Stores | Total Sales | Total Inventory | Total Returns |
|---|---|---|---|---|---|---|
| Africa | Boot | Addis Ababa | 12 | $29,761 | $191,821 | $769 |
| Africa | Men's Casual | Addis Ababa | 4 | $67,242 | $118,036 | $2,284 |
| Africa | Men's Dress | Addis Ababa | 7 | $76,793 | $136,273 | $2,433 |
| Africa | Sandal | Addis Ababa | 10 | $62,819 | $204,284 | $1,861 |
| Africa | Slipper | Addis Ababa | 14 | $68,641 | $279,795 | $1,771 |
| Africa | Sport Shoe | Addis Ababa | 4 | $1,690 | $16,634 | $79 |
| Africa | Women's Casual | Addis Ababa | 2 | $51,541 | $98,641 | $940 |
| Africa | Women's Dress | Addis Ababa | 12 | $108,942 | $311,017 | $3,233 |
| Africa | Boot | Algiers | 21 | $21,297 | $73,737 | $710 |
| Africa | Men's Casual | Algiers | 4 | $63,206 | $100,982 | $2,221 |
| Africa | Men's Dress | Algiers | 13 | $123,743 | $428,575 | $3,621 |
| Africa | Sandal | Algiers | 25 | $29,198 | $84,447 | $1,530 |
| Africa | Slipper | Algiers | 17 | $64,891 | $248,198 | $1,823 |
| Africa | Sport Shoe | Algiers | 9 | $2,617 | $9,372 | $168 |
| Africa | Women's Dress | Algiers | 12 | $90,648 | $266,805 | $2,690 |
| Africa | Boot | Cairo | 20 | $4,846 | $18,965 | $229 |
| Africa | Men's Casual | Cairo | 25 | $360,209 | $1,063,251 | $9,424 |
| Africa | Men's Dress | Cairo | 5 | $4,051 | $45,962 | $97 |
| Africa | Sandal | Cairo | 9 | $10,532 | $50,430 | $598 |
| Africa | Slipper | Cairo | 9 | $13,732 | $54,117 | $1,216 |
| Africa | Sport Shoe | Cairo | 3 | $2,259 | $20,815 | $44 |
| Africa | Women's Casual | Cairo | 14 | $328,474 | $940,851 | $10,124 |
| Africa | Women's Dress | Cairo | 3 | $14,095 | $51,145 | $745 |
| Africa | Boot | Johannesburg | 14 | $8,365 | $33,011 | $483 |
| Africa | Sandal | Johannesburg | 13 | $17,337 | $63,003 | $809 |
| Africa | Slipper | Johannesburg | 12 | $39,452 | $130,025 | $1,565 |
| Africa | Sport Shoe | Johannesburg | 8 | $5,172 | $29,368 | $139 |
| Africa | Women's Dress | Johannesburg | 4 | $42,682 | $120,127 | $966 |

## 1.4.6 SAS Toolbar File Option, the Gateway to the SAS Export / Import Wizards

The SAS view shown in Figure 1.4.6 is an image from Windows XP and SAS 9.3. It shows the toolbar "File" option selected and the resulting pop-up menu. This figure will be referenced for both this example and the example for the next section.

**Figure 1.4.6: View of the SAS Toolbar File Pop-Up Menu.**

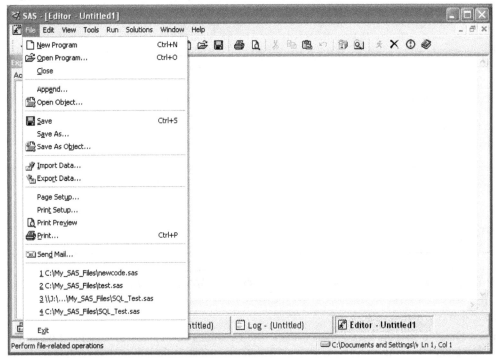

The actual menu that is shown using the "File" option in the SAS toolbar varies. What is shown depends upon the screen that is the currently active screen.

- When no SAS dataset is highlighted in the SAS Explorer window, the selection of the "File" option on the SAS toolbar will show a pop-up menu (Figure 1.4.6). That menu, among other things, has two options that read and write external files.
- When a SAS dataset is highlighted in the SAS Explorer window, the selection of the "File" option on the SAS toolbar will show a pop-up menu similar to Figure 1.4.6 except that the following options are not available.
    - New
    - Page Setup …
    - Print setup …
    - Print Preview
    - Print …

These two options "Export Data" and "Import Data" start the SAS Export wizard and the SAS Import wizard, respectively. The variations within the wizards are slight and depend upon SAS version and operating system installed on your computer.

Note that Example 1.4.7 and Example 1.4.8 below write data to and read data from the same external Excel file. If you want to use these exact examples you must use the SAS Export wizard to write the Excel file before you can use the SAS Import wizard to read it. The SAS Export wizard asks for the SAS dataset first, and then asks for the external file name and location. The SAS Import wizard asks for an external file to import and then a SAS dataset to create.

## 1.4.7 Choosing the "Export Data" SAS Toolbar (Export Wizard) File Option

### Example 1.4.7

Select "Export Data" to start the SAS Export wizard. The variations between SAS versions of the Export wizard are slight and depend upon the SAS version and operating system installed on your computer. This example writes data to an external Excel file. The SAS Export wizard asks for the SAS dataset first, and then asks for the external file name and location.

The SAS Export wizard defaults to pointing to your SAS datasets in your SAS WORK area, but allows you to select an active libref assigned to your SAS session. Any SAS dataset referenced by an active LIBNAME statement can be selected. If the SAS dataset you wish to use is not currently referenced by an assigned libref, then you need to cancel the SAS Export wizard and return to submit a LIBNAME statement. The wizard guides you through the process one screen at a time. Drop-down menu selections or browse buttons on each screen allow you to search the directories and files available. This tool uses methods similar to other Microsoft Windows search screens and should not present any navigation problems. The SAS Export wizard will allow you to save the generated code anywhere on your computer that you can write a file. A sample of the code generated appears below.

```
PROC EXPORT DATA= SASHELP.SHOES
            OUTFILE= "F:\My_Excel_File\test_file_1.xls"
            DBMS=EXCEL LABEL REPLACE;
     SHEET="Shoes_Sheet";
     NEWFILE=YES;
RUN;
```

## 1.4.8 Electing the "Import Data" SAS Toolbar (Import Wizard) File Option

### Example 1.4.8

Select "Import Data" to start the SAS Import wizard. The variations between SAS versions of the Import wizard are slight and depend upon SAS version and operating system installed on your computer. This example reads data from an external Excel file. The SAS Import wizard asks for the external file name and location first, and then asks for the SAS dataset name. The SAS Import wizard defaults to reading an Excel workbook, but many other file type options are also available to read, including user-defined file formats (see Figure 1.4.7). But, we will limit our work here to Excel file formats. The SAS Import wizard will allow you to select both the Excel workbook and worksheet name to import. Then you can select a location to save the generated code anywhere on your computer that you can write a file. The WORK libref is used as the default output libref to save your data that is read from the external file, and must be assigned before the Import wizard runs.

Chapter 1: Easy Data Movement between SAS and Excel  17

**Figure 1.4.7: SAS V9.3 Import Wizard Running on a 32-bit Windows XP Operating System.**

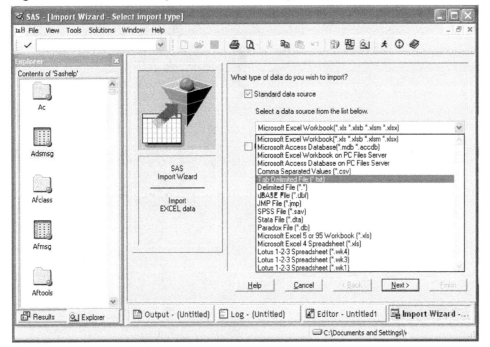

A sample of the code generated appears below. The code uses a DATAFILE statement to locate the input file. You use the wizard windows to select this file. It must exist when the Import wizard runs.

```
PROC IMPORT OUT= WORK.Sample_import
            DATAFILE= "F:\My_Excel_File\test_file_1.xls"
            DBMS=EXCEL LABEL REPLACE;
     RANGE="Shoes_Sheet_1";
     GETNAMES=YES;
     MIXED=NO;
     SCANTEXT=YES;
     USEDATE=YES;
     SCANTIME=YES;
RUN;
```

### 1.4.9 Using the Export / Import Wizards in a 32/64-Bit Mixed Environment

Special requirements exist when SAS, Excel, and your operating system do not all conform to the same "bit" size for executing software. Many different combinations exist and not all of them are compatible. For SAS 9.2 TS Level 2M3 and above running on a 64-bit Windows operating system with SAS/ACCESS Interface to PC Files installed, there are a few minor differences. There is an option to use SAS PC Files Server to read or write the Excel workbooks. The "Connect to MS Excel" window shown in Figure 1.4.8 will be displayed when either wizard needs to define a connection that uses SAS PC Files Server. This is required because SAS PC Files Server translates 64-bit data requests into instructions that the 32-bit access tools can understand. All other screens and output SAS code are generally the same. In both SAS 9.2 and SAS 9.3, SAS PC Files Server performs this way; however, the two versions of SAS PC Files Server have the same name and therefore cannot reside on the same computer. They also use different default port numbers. SAS 9.2 uses the port number 8621, and SAS 9.3 uses the port number 9621. SAS 9.4 can also operate in these mixed environments.

**Figure 1.4.8: The "Connect to MS Excel" Options Screen May Appear in a 32/64 Bit Mixed Environment.**

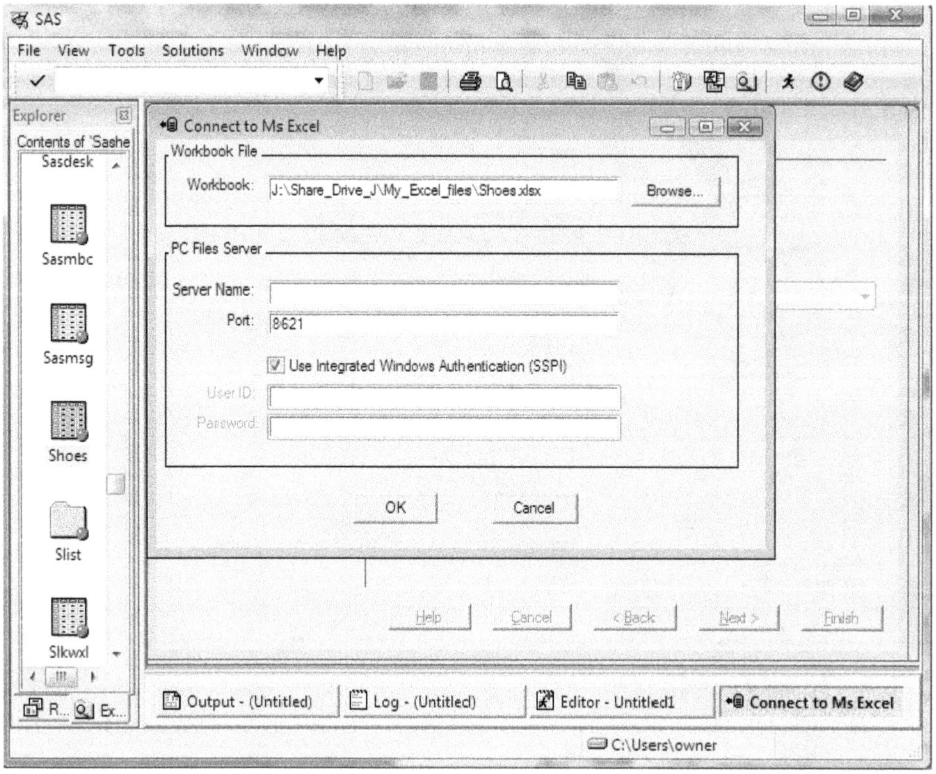

To get an accurate assessment of the software installed on your computer, contact your IT department or SAS Technical Support.

## 1.5 Chapter Summary

The examples in this chapter demonstrated how to move data between SAS datasets and Excel worksheets using the following methods.

- Cut or copy then paste
- Convert text data to Excel column data
- Copy Excel data into a SAS Enhanced Editor window
- Save multiple lines of text into a single Excel cell
- Convert Excel tables to text
- The Export wizard from the SAS Explorer window
- Copy contents to the Windows clipboard
- Save SAS data as HTML
- View a SAS file in Excel
- SAS Export wizard from the File toolbar tab
- SAS Import wizard from the File toolbar tab

We also touched upon some cautions about using SAS to read or write data to Excel when running your programs in a 32/64-bit mixed environment. These are introductory methods of moving data between the two systems. In the rest of the book, I will use and expand on these and other methods that these two powerful software tools use to transfer data and communicate with one another.

# Chapter 2: Use PROC EXPORT to Write SAS Data to External Files and Excel Workbooks

| | |
|---|---|
| 2.1 Introduction | 19 |
| 2.2 Purpose | 20 |
| 2.3 Syntax of the SAS EXPORT Procedure | 20 |
| 2.4 Data Access Methods for Excel Files Supported by PROC EXPORT | 21 |
| 2.5 Overview of the Examples | 22 |
| 2.6 List of Examples | 23 |
|     Example 2.1 SAS Code to Export Data to an Excel 4 or Excel 5 Format File | 23 |
|     Example 2.2 PROC EXPORT Using the DBMS=DLM Option | 24 |
|     Example 2.3 PROC EXPORT Using the DBMS=EXCEL Option | 25 |
|     Example 2.4 PROC EXPORT Using the DBMS=EXCELCS Option | 27 |
|     Example 2.5 SAS Code to Export Data to an Excel File with No Column Headers | 28 |
|     Example 2.6 SAS Code to Export Data to a Network Windows Computer | 28 |
| 2.7 Conclusion | 29 |

## 2.1 Introduction

This chapter will explain the syntax, usage, and the results that can be generated when using the SAS EXPORT procedure. Examples will range from simple to complex. This chapter builds upon the Chapter 1 explanation and examples of the SAS Export Wizard.

As we saw in the last chapter, the SAS Export Wizard optionally can create a SAS code file of the PROC EXPORT instructions used to write the output file. The SAS Export Wizard can execute SAS code to output a SAS dataset in various formats. The code generated by the wizard can be saved, modified, and reused. A SAS user can also write code to execute PROC EXPORT directly. The syntax of PROC EXPORT will be explained and the options listed below in a table. PROC EXPORT is a general purpose routine and is able to output data from SAS files in several different formats.

Because PROC EXPORT can write many of the file formats that Microsoft Excel can read directly, several examples will be shown that write to files that can be loaded into Excel worksheets. But, because the focus of this work is moving data from SAS to Excel and back, only options relative to creating Excel files will be explored. Some of the code examined here was shown in Chapter 1, but in this chapter we will examine the PROC EXPORT syntax in detail to show the flexibility of the SAS EXPORT procedure. One important aspect of PROC EXPORT is its ability to interface with an external Data Base Management System (DBMS). PROC EXPORT has a syntax argument called DBMS that makes this option available and permits access to many different output data formats. We will focus on the ability to interface with different versions of Excel.

Depending upon what operating system and version of SAS you are using, you may be able to write to some or all of the following formats. The details for reading these other formats are explained in the SAS documentation.

- Microsoft Access database files
- Microsoft Excel workbook files
- Lotus 1-2-3 spreadsheet files
- Paradox files
- SPSS files
- Stata files
- dBase files
- JMP files
- delimited files

## 2.2 Purpose

I will discuss the syntax of the PROC EXPORT and point you to the SAS online documentation for your version of SAS in this chapter. There will be several examples to show you how to write the code to use PROC EXPORT and the results that the examples produce. Because not everyone has the latest version of either SAS or Excel installed on his or her computer, I will not restrict my examples to those newest versions. This chapter will show you how to write SAS code to use PROC EXPORT. Because of the size and complexity of some of the reference tables, I suggest that you refer to *SAS/ACCESS Interface to PC Files: Reference* for the version of SAS that you have installed. In addition, the volume of the options does not allow for an example describing each option. While every effort has been made to reproduce the concepts accurately, the SAS documentation is the final authority.

## 2.3 Syntax of the SAS EXPORT Procedure

**PROC EXPORT** DATA=<*libref.*>*SAS data set* <(*SAS data set options*)>

OUTFILE="*filename*" | OUTTABLE="*tablename*"

<DBMS=*identifier*> <REPLACE> <LABEL>;

<*file-format-specific-statements*>;

**NOTE:** Some features relating to Microsoft Excel 2007, 2010, and 2013 for operating systems Microsoft Vista 64-bit, Microsoft Windows 7 and 8, LINUX, and UNIX may not be available in SAS versions prior to the third maintenance release of SAS 9.2. Other operating systems may not be compatible until later versions of SAS are released. SAS is not supported on some versions of the Microsoft Windows operating system.

Table 2.1 provides a high-level definition of the parts of the syntax for PROC EXPORT as listed above. See the SAS documentation for your version of SAS for more details about PROC EXPORT in the SAS software version you are using.

**Table 2.1: General Description of PROC EXPORT Syntax Options.**

| Argument / (Alias) | Required | Definition of the Function of the Argument |
|---|---|---|
| DATA= | Yes | Provide the input SAS dataset name. |
| SAS Data Set Options | No | Options like KEEP=, DROP=, RENAME=, WHERE=, and others may be provided. |
| OUTFILE/(FILE) | Yes | Provide the output file name. OUTTABLE (or TABLE) is for Microsoft Access databases only. |

| Argument / (Alias) | Required | Definition of the Function of the Argument |
|---|---|---|
| DBMS | No | See the examples below for specific options relating to the individual DBMS identifier values. Options are based upon the file types being processed and direct the actions of the SAS PROC EXPORT features. |
| LABEL /(DBLABEL) | No | When either "LABEL" or "DBLABEL" is present, then SAS will use the SAS variable's label value as the column title in row one of the output Excel file. |
| REPLACE | No | When "REPLACE" is present, then SAS will overwrite an existing output file. A new file will be created if the requested file name does not exist. |

## 2.4 Data Access Methods for Excel Files Supported by PROC EXPORT

The data access methods listed in Figure 2.1 are used to create output files Excel has the ability to read. When you select a DBMS mode, it determines which utility will be used to process the SAS dataset to create an output file. The output file may be a text file or an Excel spreadsheet. See the documents listed above for more details about the SAS software version you are using. Some of these data access methods (the DBMS=modes) require SAS/ACCESS Interface to PC Files software to function. You must have SAS/ACCESS Interface to PC Files licensed before you can export files directly to Microsoft Excel workbooks. Some features relating to Microsoft Excel 2007, Excel 2010, and Excel 2013 when using Microsoft Windows, LINUX, and UNIX operating systems may not be available in SAS versions prior to the third maintenance release of SAS 9.2. Because the number of SAS, Excel, and operating system versions is large, I once again suggest that you refer to *SAS/ACCESS Interface to PC Files: Reference* for the version of SAS that you have installed.

Here would also be a good place to introduce the SAS PC Files Server. This is a software tool used by SAS to convert data when the computing environment you are using includes hardware and software of different "bit" configurations. For a human like you and me, this would be equivalent to learning a new language with more letters and more complex words. For a computer, this change is from processing data units with 32-bits (1s and 0s) to using 64-bits. These are the elements that are changing: SAS software can be either 32- or 64-bit software, Excel programs can be 32- or 64-bit software, the computer operating system can be 32- or 64- bit software, or the computer can be 32- or 64-bit hardware. Here I will say contact your IT Department. The SAS PC Files Server helps clear up some of the confusion.

**Figure 2.1: DBMS Output Formats Generated.**

| DBMS Identifier | SAS/ACCESS Interface to PC Files Required | General Description of the DBMS Output File |
|---|---|---|
| CSV | N | Text file with a comma delimiter |
| TAB | N | Text file with a tab delimiter |
| DLM | N | Text file with a user defined delimiter |
| EXCEL | Y | Excel workbook referenced by a LIBNAME |
| EXCELCS | Y | Excel workbook referenced by using the SAS PC Files Server |
| EXCEL4 | Y | Excel workbook |
| EXCEL5 | Y | Excel workbook |

| DBMS Identifier | SAS/ACCESS Interface to PC Files Required | General Description of the DBMS Output File |
|---|---|---|
| XLS | Y | Excel workbook using file formats prior to Excel 2007 except Excel 4 and Excel 5 |
| XLSX | Y | Excel workbook using file formats 2007, 2010, and 2013 |

**NOTE:** The DBMS identifiers listed in Figure 2.2 are relative to the file formats that Microsoft Excel can read or write. The SAS documentation lists other DBMS identifiers that PROC EXPORT can write. See the SAS documentation for your version of SAS for other options to write file formats. Different versions of SAS may not be able to write to all of the versions of Excel listed in Figure 2.2.

**Figure 2.2: DBMS Output Methods of Accessing Excel Files.**

| DBMS Identifier | Method of Accessing Excel Files | Excel Version | Comments |
|---|---|---|---|
| EXCEL | LIBNAME statement | 5, 95, 97, 2000, 2002, 2003, 2007, 2010, 2013 | This DBMS option will use the LIBNAME statement. Access is limited to the first 255 columns. Some older versions may be limited to 65,535 rows. |
| EXCELCS | SAS PC Files Server | 5, 95, 97, 2000, 2002, 2003, 2007, 2010, 2013 | This DBMS option will use the SAS PC Files Server. Access is limited to the first 255 columns. Some older versions may be limited to 65,535 rows. |
| EXCEL4 or EXCEL5 | DBLOAD procedure | 4, 5, 95 | This is supported only on Microsoft Windows operating systems and is for SAS 6 compatibility. |
| XLS | XLS format | 97, 2000, 2002, 2003 | Chinese, Japanese, or Korean DBCS character sets may not be supported in all versions of SAS. |
| XLSX | XLSX format | 2007, 2010, and later formats | This does not support the Chinese, Japanese, or Korean DBCS character sets, which may not be supported in all versions of SAS or *.xlsb Excel files. |

## 2.5 Overview of the Examples

The examples in this chapter will cover several but not all of the DBMS options used with PROC EXPORT. I like to group the output processing for PROC EXPORT into general categories within the DBMS options. Furthermore, I feel I must place a caveat onto these groupings because both SAS and Microsoft Excel are mature products that have changed over time. While these categories are generally accurate, your SAS version, Excel version, and computer hardware may not support every DBMS option, and each DBMS option might operate slightly differently depending upon what software you have installed. So, make sure you verify what is available to you by looking in the SAS manual that relates to your environment.

- An example retained for backward compatibility with files in the Excel 4 and Excel 5 formats.
- Text file output options like CSV, TAB and DLM do not require SAS/ACCESS Interface to PC Files because the methods write out text files.
- Options that use PROC DBLOAD to output the data.
- Options that write directly to a formatted Excel file.

- LIBNAME options that do not use the SAS PC Files Server.
- LIBNAME options that do use the SAS PC Files Server.

The options that generate text files will show one example and explain the differences that make the other options work.

## 2.6 List of Examples

The following is a table listing the examples in this chapter and some comments about each example.

**Figure 2.3: List of Examples for PROC EXPORT.**

| Example Number | General Description |
| --- | --- |
| 2.1 | **SAS Code to Export Data to an Excel 4 or Excel 5 Format File.** This example is included for backward compatibility. The Excel formats Excel 4 and Excel 5 were the first two Excel formats and it is rare to find a computer using this Microsoft Excel software today. The example shows how to write to these old Excel formats. |
| 2.2 | **PROC EXPORT Using the DBMS=DLM Option.** This example shows how to use a delimiter to separate output values and eliminate the header row of variable names at the top of the file. This example is equivalent to DBMS=CSV and DBMS=TAB because you can provide your own delimiter. |
| 2.3 | **PROC EXPORT Using the DBMS=EXCEL Option.** The three parts of this example all write Excel workbooks that do not need the PC Files Server to be processed. The main point of these code routines is to show how to create multiple worksheets within one workbook, and to create a sheet with a mixed case name. Also note that only some of the input SAS dataset variables are output. |
| 2.4 | **PROC EXPORT Using the DBMS=EXCELCS Option.** This example is similar to Example 2.3, but here the code is executed on a 64-bit operating system using a 64-bit copy of SAS 9.3 and a 32-bit copy of Microsoft Excel. Since this computer operating system and SAS use a 64-bit configuration, but Excel uses a 32-bit configuration, PROC EXPORT requires the use of the PC Files Server. The "CS" part of DBMS=EXCELCS annotates this feature is in use. |
| 2.5 | **SAS Code to Export Data to an Excel File with No Column Headers.** This example writes an Excel worksheet with no column headers (variable names) in the output Excel worksheet. It also demonstrates that the SAS PROC EXPORT routine will not write an Excel sheet name with spaces. |
| 2.6 | **SAS Code to Export Data to a Network Windows Computer.** Writing SAS data to another computer that you have shared privileges. |

## Example 2.1 SAS Code to Export Data to an Excel 4 or Excel 5 Format File

The SAS EXPORT procedure maintains the backward compatibility features required to process Excel workbooks in the Excel 4 and Excel 5 formats. This example shows how to write Excel files in those formats. For Excel 4 workbooks, the sheet name is the same as the file name (without the .xls) and there is only one sheet in the workbook. For Excel5 formatted workbooks the sheet name is "Sheet1".

### Figure 2.4: SAS Code to Export to Version 4 and 5 Excel Files.

```
* Write an Excel4 formatted file;
PROC EXPORT DATA=sashelp.shoes
   OUTFILE='C:\My_Files\shoes_to_Excel_4_file.xls'
   DBMS=EXCEL4
   REPLACE;
RUN;

* Write an Excel5 formatted file;
PROC EXPORT DATA=sashelp.shoes
   OUTFILE='C:\My_Files\shoes_to_Excel_5_file.xls'
   DBMS=EXCEL5
   REPLACE;
RUN;
```

## Example 2.2 PROC EXPORT Using the DBMS=DLM Option

This example uses PROC EXPORT to write a delimited file readable by Microsoft Excel. This example does not produce an Excel file; however, it produces a comma-delimited *.csv formatted file that can be read by Microsoft Excel. Using the SAS External File Interface (EFI), the following code writes out a delimited file from the SASHELP.shoes data set. Any character could have been used here. Using a comma makes the output file the same as if the DBMS=CSV option were used without the DELILITER="," statement. Using DBMS =TAB would produce a similar file with Tab characters instead of commas.

### Figure 2.5: SAS Code to Export to CSV File.

```
PROC EXPORT DATA=sashelp.shoes
   OUTFILE='C:\My_Files\Shoes.csv'
   DBMS=DLM
   REPLACE;
   DELIMITER=',';
   PUTNAMES=NO;
RUN;
```

### Figure 2.6: The Output Log Produced by the PROC EXPORT Code.

```
1    PROC EXPORT DATA=sashelp.shoes
2        OUTFILE='C:\My_Files\Shoes.csv'
3        DBMS=DLM
4        REPLACE;
5        DELIMITER=',';
6        PUTNAMES=NO;
7    RUN;

8    /*****************************************************************
9    *    PRODUCT:   SAS
10   *    VERSION:   9.3
11   *    CREATOR:   External File Interface
12   *    DATE:      11JAN14
13   *    DESC:      Generated SAS Datastep Code
14   *    TEMPLATE SOURCE:  (None Specified.)
15   *****************************************************************/
16       data _null_;
17       %let _EFIERR_ = 0; /* set the ERROR detection macro variable */
18       %let _EFIREC_ = 0;     /* clear export record count macro variable */
19       file 'C:\My_Files\Shoes.csv' delimiter=',' DSD DROPOVER lrecl=32767;
20       set  SASHELP.SHOES    end=EFIEOD;
21          format Region $25. ;
22          format Product $14. ;
23          format Subsidiary $12. ;
24          format Stores best12. ;
25          format Sales dollar12. ;
26          format Inventory dollar12. ;
```

```
27              format Returns dollar12. ;
28           do;
29              EFIOUT + 1;
30              put Region $ @;
31              put Product $ @;
32              put Subsidiary $ @;
33              put Stores @;
34              put Sales @;
35              put Inventory @;
36              put Returns ;
37              ;
38           end;
39           if _ERROR_ then call symputx('_EFIERR_',1);
40           if _EFIEOD_ then call symputx('_EFIREC_',EFIOUT);
41        run;
```

The next listing is the first few lines of the output-delimited file. There are no headers listed in the first row because the "PUTNAMES=NO" command suppressed the header row. Notice also that fields with commas embedded in the data are enclosed in quotation marks.

**Figure 2.7: Output in a CSV Format with PUTNAMES=NO Set Active.**

```
Africa,Boot,Addis Ababa,12,"$29,761","$191,821",$769
Africa,Men's Casual,Addis Ababa,4,"$67,242","$118,036","$2,284"
Africa,Men's Dress,Addis Ababa,7,"$76,793","$136,273","$2,433"
Africa,Sandal,Addis Ababa,10,"$62,819","$204,284","$1,861"
Africa,Slipper,Addis Ababa,14,"$68,641","$279,795","$1,771"
Africa,Sport Shoe,Addis Ababa,4,"$1,690","$16,634",$79
Africa,Women's Casual,Addis Ababa,2,"$51,541","$98,641",$940
Africa,Women's Dress,Addis Ababa,12,"$108,942","$311,017","$3,233"
Africa,Boot,Algiers,21,"$21,297","$73,737",$710
Africa,Men's Casual,Algiers,4,"$63,206","$100,982","$2,221"
Africa,Men's Dress,Algiers,13,"$123,743","$428,575","$3,621"
Africa,Sandal,Algiers,25,"$29,198","$84,447","$1,530"
Africa,Slipper,Algiers,17,"$64,891","$248,198","$1,823"
Africa,Sport Shoe,Algiers,9,"$2,617","$9,372",$168
Africa,Women's Dress,Algiers,12,"$90,648","$266,805","$2,690"
Africa,Boot,Cairo,20,"$4,846","$18,965",$229
```

## Example 2.3 PROC EXPORT Using the DBMS=EXCEL Option

This example runs PROC EXPORT three times. We do not need to define the SASHELP library, since it is defined and initiated automatically by SAS at the beginning of each session. Each execution of a code module places one output sheet into the Excel file. The first two times, we create sheets in the same workbook showing examples of the options SHEET= and DBDSOPTS=. The "SHEET=" option defines the Excel worksheet name. The second code module shows how to assign a mixed-case sheet name. Option "DBDSOPTS=" allows several different options, but only the "KEEP=" option is shown.

The third time that PROC EXPORT is run, the output file is deleted and only one sheet is left in the file. This is done with the "NEWFILE=YES" option. Quotes are not required around the sheet name as long as it is a valid SAS name, containing neither spaces nor special characters. Unquoted sheet names will be all capital letters when the code finishes running. This configuration did require SAS/ACCESS Interface to PC Files, but did not require the PC Files Server. The data in all three of these worksheets is the same.

The three parts of this example all write Excel workbooks that do not need the PC Files Server to be processed. The main point of these code routines is to show how to create multiple worksheets within one workbook, and to create a sheet with a mixed-case name. Also note that only some of the input SAS dataset variables are output.

### Step One – Write an Excel worksheet with the sheet name in Excel converted to capital letters.

Write the file 'c:\My_files\shoes.xls' with only four variables from the SHOES dataset in the SASHELP library. If this Excel workbook exists and the worksheet "MY_PAGE" also exists, the worksheet will be overwritten. If the workbook does not exist, it will be created.

#### Figure 2.8: SAS Code Export Data to an XLS File.

```
* Write the MY_PAGE sheet, the sheet name is all capital letters;
* since the sheet name entered on the SHEET= statement is not quoted;
* only variables region, sales, returns, and inventory will be output;

PROC EXPORT DATA=sashelp.shoes
   OUTFILE='C:\My_Files\shoes.xls'
   DBMS=EXCEL
   REPLACE;
   SHEET=My_Page;
   DBDSOPTS='keep=region sales returns inventory';
RUN;
```

### Step Two – Write an Excel worksheet with a mixed-case sheet name and spaces converted to underscores by SAS for Excel.

There are two differences between step one and step two which involve the output Excel worksheet name. The first is that the sheet name contains mixed case letters and underscores when viewed in the Excel file. The second difference is there will be two sheets in the output Excel workbook.

#### Figure 2.9: SAS Code Export Data to an XLS File with Mixed-Case Sheet Names.

```
* Write a sheet called My_New_Page, the sheet name will be mixed case   ;
* and the sheet name has underscores instead of spaces in the           ;
* name because SAS generated sheet names do not have spaces however     ;
* when a sheet name is created or renamed by Excel you can place a      ;
* space into the sheet name. A sheet name generated by SAS can contain  ;
* a single quote if the sheet name is enclosed in double quotes         ;

PROC EXPORT DATA=sashelp.shoes
   OUTFILE='C:\My_Files\shoes.xls'
   DBMS=EXCEL
   REPLACE;
   SHEET='My New Page';
   DBDSOPTS='keep=region sales returns inventory';
RUN;
```

### Step Three – Overwrite an existing Excel Workbook and output one sheet with a name in uppercase letters.

Step two and step three have different results when you output an Excel workbook. The first is that the sheet name contains uppercase letters. The second difference is that the old Excel workbook will be replaced and the new workbook will have one worksheet.

**Figure 2.10: SAS Code Export Data to an XLS File with Uppercase Sheet Names.**

```
* Write sheet called MY_PAGE, the sheet name will be all capital letters;
* Old output file will be deleted and the new file will have one sheet;

PROC EXPORT DATA=sashelp.shoes
   OUTFILE='C:\My_Files\shoes.xls'
   DBMS=EXCEL
   REPLACE;
   SHEET=My_Page;
   NEWFILE=YES;
   DBDSOPTS='keep=region sales returns inventory';
RUN;
```

## Example 2.4 PROC EXPORT Using the DBMS=EXCELCS Option

This code was executed on a Windows 64-bit operating system using 64-bit SAS 9.3, but the Excel program was a 32-bit based copy of Excel. Since this computer uses 64-bit hardware, PROC EXPORT requires the use of SAS/ACCESS to PC Files Server software. The option NEWFILE= is not valid in that configuration. Windows 64-bit operating systems and remote connections both require the use of the PC Files Server to execute. This code uses the DBMS=EXCELCS and the SERVICE=SASPCFILE statements.

Also note that SAS 9.2 and SAS 9.3 use different SAS PC Files Server executable modules, but maintain the same names and are therefore not compatible with each other. They also use different defaults ports when the processing occurs. See the documentation for your version of SAS software for complete details. In addition, be aware that some of these features may not be available in SAS versions prior to the third maintenance release of SAS 9.2. The PROC EXPORT DBMS=EXCELCS statement will write to *.xls, *.xlsb, and *.xlsx files but not to *.xlsm file formats.

**Figure 2.11: SAS Code Export Data to an xls file with Uppercase Sheet Names.**

```
* Write a sheet called MY_PAGE, the sheet name will be all capital;
* letters since the sheet name entered on the SHEET= statement is not;
* quoted;

* The PROC EXPORT DBMS=EXCELCS statement will write to *.xls, *.xlsb, ;
* *.xlsx but not to *.xlsm file formats.;

PROC EXPORT DATA=sashelp.shoes
OUTFILE='C:\My_Files\shoes.xlsb'
DBMS=EXCELCS
REPLACE;
SHEET=My_Page;
SERVICE=SASPCFILE;
RUN;
```

**Figure 2.12a: The Result of the PROC EXPORT Execution.**

```
1     PROC EXPORT DATA=sashelp.shoes
2        OUTFILE='C:\My_Files\shoes.xlsb'
3        DBMS=EXCELCS
4        REPLACE;
5        SHEET=My_Page;
6        SERVICE=SASPCFILE;
7     RUN;

NOTE: "MY_PAGE" range/sheet was successfully created.
NOTE: PROCEDURE EXPORT used (Total process time):
      real time           12.43 seconds
      cpu time             0.64 seconds
```

As shown in the examples in this chapter, the **file-format-specific-statements** are listed after the PROC EXPORT statement. In other words, the statements after the first semicolon are independent statements that modify PROC EXPORT. While no spaces are shown between the option names and the "=" sign, spaces are usually permitted in this syntax.

## Example 2.5 SAS Code to Export Data to an Excel File with No Column Headers

This example outputs an Excel file using PROC EXPORT with the DBMS=XLS option. For this output, either the "REPLACE" option or the NEWFILE= file format statements are valid. The PUTNAMES=NO statement puts only the data without row or column headers. While the sheet name, because it is enclosed in quotes, remains case sensitive, the spaces are replaced with underscores. SAS PROC EXPORT will not write a sheet name with spaces.

**Figure 2.12b: SAS Code That Outputs the Result without a Header Row.**

```
PROC EXPORT DATA=sashelp.shoes
   OUTFILE='C:\My_Files\shoes_to_xls.xls'
   REPLACE
   DBMS=XLS;
   NEWFILE=YES;
   PUTNAMES=NO;
   SHEET='My Sheet Name';
RUN;
```

**Figure 2.13: Screen Image from the Results of PROC EXPORT.**

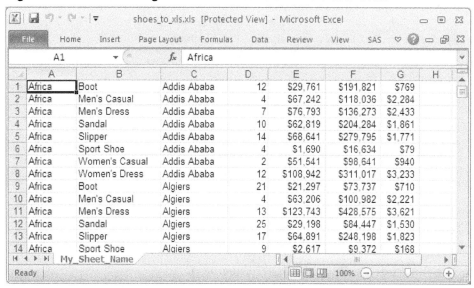

## Example 2.6 SAS Code to Export Data to a Network Windows Computer

When writing SAS data to another computer, you need access to the other computer and shared ability to write to the disk volume within the described path name.

**Figure 2.14: SAS Code to Export Data to Another Computer.**

```
PROC EXPORT DATA=sashelp.shoes
   OUTFILE='\\owl-2000-xp\E_Drive\My_Files\shoes_to_xls.xls'
   REPLACE
   DBMS=XLS;
   NEWFILE=YES;
   SHEET='My Sheet Name';
RUN;
```

## 2.7 Conclusion

The examples in this chapter covered a wide range of conditions, from simple text files that Excel can read to complex examples that span multiple computers. I have not covered all options and all conditions when you would need to write data using PROC EXPORT. Below is a simple list of the examples by title for a quick review.

- SAS Code to Export Data to an Excel4 or Excel5 Format File
- PROC EXPORT using the DBMS=DLM option
- PROC EXPORT using the DBMS=EXCEL option
- PROC EXPORT Using the DBMS=EXCELCS Option
- SAS Code to Export Data to an Excel File with No Column Headers
- SAS Code to Export Data to a Network Windows Computer

# Chapter 3: Use PROC IMPORT to Read External Data Files and Excel Workbooks into SAS

| | |
|---|---|
| 3.1 Introduction | 31 |
| 3.2 Purpose | 32 |
| 3.3 Syntax of the SAS IMPORT Procedure | 32 |
| 3.4 Data Access Methods for Files Supported by PROC IMPORT | 33 |
| 3.5 Overview of the Examples | 34 |
| 3.6 List of Examples | 34 |
|     Example 3.1 PROC IMPORT Using the DBMS=EXCEL4 or EXCEL5 Option | 35 |
|     Example 3.2 PROC IMPORT Using the DBMS=DLM Option | 35 |
|     Example 3.3 PROC IMPORT Using the DBMS=EXCEL Option | 37 |
|     Example 3.4 PROC IMPORT Using the DBMS=EXCELCS Option | 40 |
|     Example 3.5 PROC IMPORT Using the DBMS=XLS or XLSX to Select Columns | 42 |
|     Example 3.6 PROC IMPORT Using the DBMS=XLS or XLSX to Select Rows | 43 |
|     Example 3.7 PROC IMPORT Using the DBMS=XLS or XLSX to Select Excel Ranges | 44 |
| 3.7 Conclusion | 45 |

## 3.1 Introduction

This chapter builds upon the Chapter 1 explanation and examples of the SAS Import Wizard, and will explain the syntax, usage, and the results that can be generated when using the SAS IMPORT procedure, specifically PROC IMPORT.

PROC IMPORT is a general purpose routine and is able to read data from text files and Excel workbook files which can exist in several different formats. The ability to read files of many formats makes PROC IMPORT extremely useful. The primary focus of this chapter will be upon reading Excel files. However, some examples will show how to read text files with delimiters because Excel can write files with those formats. The syntax of PROC IMPORT will be explained and the options listed below in Table 3.3.1. One important aspect of PROC IMPORT is its ability to interface with an external Data Base Management System (DBMS). PROC IMPORT has a syntax argument called DBMS that makes this option available and permits access to many different input data formats. Options exist to enable the transfer of data between SAS and many other file formats, but because the focus of this work is moving data from SAS to Excel and back, only options relative to Excel will be explored.

Depending upon which operating system and version of SAS you are using, you may be able to read some or all of the following formats. The details for reading these other formats are explained in the SAS documentation.

- Microsoft Access database files
- Microsoft Excel workbook files
- Lotus 1-2-3 spreadsheet files
- Paradox files
- SPSS files
- Stata files
- dBase files

- JMP files
- delimited files

## 3.2 Purpose

I will discuss the syntax of the SAS IMPORT procedure and point you to the SAS online documentation for your version of SAS in this chapter. There will be several examples to show you how to write the code to use PROC IMPORT and the results that the examples produce. Because not everyone has the latest version of either SAS or Excel installed on his or her computer, I will not restrict my examples to those newest versions. This chapter will show you how to write SAS code to use PROC IMPORT. Because of the size and complexity of some of the reference tables I suggest that you refer to *SAS/ACCESS Interface to PC Files: Reference* for the version of SAS that you have installed.

## 3.3 Syntax of the SAS IMPORT Procedure

PROC IMPORT

DATAFILE= <*'filename'*> | DATATABLE= <*'tablename'*> (Not used for Microsoft Excel files)
<DBMS>= <*data-source-identifier*>
<OUT>= <*libref.SAS data-set-name*> <*SAS data-set-option(s)*>
<REPLACE>;
<*file-format-specific-statements*>;

**NOTE:** Some features relating to Microsoft Excel 2007, 2010, and 2013 for operating systems Microsoft Vista 64 bit, Microsoft Windows 7 and 8, LINUX, and UNIX, may not be available in SAS versions prior to the third maintenance release of SAS 9.2. Other operating systems may not be compatible until later versions of SAS are released. SAS is not supported on some versions of the Microsoft Windows operating system.

Table 3.3.1 provides a high-level definition of the parts of the syntax for PROC IMPORT as listed above. See SAS/ACCESS to PC Files: Reference for more details about PROC EXPORT in the SAS software version you are using.

**Table 3.3.1: General Description of PROC IMPORT Syntax Options.**

| Argument / (Alias) | Required | Definition of the Function of the Argument |
|---|---|---|
| OUTFILE/(FILE) | Yes | Provide the output file name. DATATABLE is not used for Excel files. |
| SAS Data Set Options | No | Options like KEEP=, DROP=, RENAME=, WHERE=, and others may be provided. |
| OUT= | Yes | Provide the output SAS dataset name. |
| DBMS | No | See Tables below for specific options relating to the individual DBMS <identifier> values. Options are based upon the file types being processed and direct the actions of the SAS PROC IMPORT features. |
| REPLACE | No | When "REPLACE" is present then SAS will overwrite an existing output file. A new file will be created if the requested file name does not exist. |

## 3.4 Data Access Methods for Excel Files Supported by PROC IMPORT

The data access methods listed in Figure 3.4.1 are used to read data files Excel has the ability to create. Selecting a DBMS mode determines which utility will be used to process the external file to create an output SAS dataset. The input file may be a text file or an Excel spreadsheet. See the documents listed above for more details about the SAS software version you are using. Some of these data access methods (the DBMS=modes) require SAS/ACCESS Interface to PC Files software to function. You must have SAS/ACCESS Interface to PC Files licensed before you can import files directly from some versions Microsoft Excel workbooks. Some features relating to Microsoft Excel 2007, Excel 2010, and Excel 2013 when using Microsoft Windows, LINUX, and UNIX operating systems may not be available in SAS versions prior to the third maintenance release of SAS 9.2. Because the number of SAS, Excel, and operating system versions is large, I once again refer you to the SAS documentation to help you figure out what you have installed.

If you suspect that your SAS and Excel software may have different bit configurations (32 or 64 bit), contact your IT Department.

The DBMS identifiers listed in Table 3.4.1 are relative to the file formats that Microsoft Excel can read or write. The SAS documentation lists other DBMS identifiers that the PROC IMPORT can read. See the SAS documentation for your version of SAS for other options to read file formats available. Different versions of SAS may not be able to read to all of the versions of Excel.

**Table 3.4.1: DBMS Formats Available for Input.**

| DBMS Identifier | SAS/ACCESS Interface to PC Files Required | General Description of the DBMS Output File |
|---|---|---|
| CSV | N | Text file with a comma delimiter |
| TAB | N | Text file with a tab delimiter |
| DLM | N | Text file with a user-defined delimiter |
| EXCEL | Y | Excel workbook (2003 xls – 2013 xlsx) |
| EXCELCS | Y | Excel workbook (2003 xls – 2007 xlsx) using the SAS PC Files Server |
| EXCEL4 | Y | Excel workbook using PROC DBLOAD |
| EXCEL5 | Y | Excel workbook using PROC DBLOAD |
| XLS | Y | Excel workbook using file formats prior to Excel 2007 except Excel 4 and Excel 5 |
| XLSX | Y | Excel workbook using file formats 2007, 2010, and 2013 |

Table 3.4.2 lists some information about the input methods available when reading Excel worksheets. Some of these methods have limitations that are smaller than the full capabilities of the Excel version that created them. These restrictions are as a result of using the Microsoft JET or ACE engines to access the Excel workbooks.

**Table 3.4.2: DBMS Input Methods of Accessing Excel Files.**

| Utility | DBMS Model | Excel Version | Comments |
|---|---|---|---|
| EXCEL | LIBNAME statement | 5, 95, 97, 2000, 2002, 2003, 2007, 2010, 2013 | This DBMS option will use the LIBNAME statement. Depending upon your version of SAS and Excel, access may be limited to the first 65,535 rows and 255 columns. |
| EXCELCS | SAS PC Files Server | 5, 95, 97, 2000, 2002, 2003, 2007, 2010, 2013 | This DBMS option will use the SAS PC Files Server. Depending upon your version of SAS and Excel, access may be limited to the first 65,535 rows and 255 columns. |

| Utility | DBMS Model | Excel Version | Comments |
|---|---|---|---|
| EXCEL4 or EXCEL5 | DBLOAD procedure | 4, 5, 95 | This is supported only on the Microsoft Windows operating systems and is for SAS 6 compatibility. |
| XLS | XLS format | 97, 2000, 2002, 2003 | Some versions of SAS may not support the Chinese, Japanese, or Korean DBCS character sets. |
| XLSX | XLSX format | 2007, 2010, and later formats | Some versions of SAS may not support the Chinese, Japanese, or Korean DBCS character sets or *.xlsb Excel files. |

## 3.5 Overview of the Examples

The examples in this chapter will cover several but not all of the DBMS options used with PROC IMPORT. I like to group the input processing for PROC IMPORT into general categories within the DBMS options. Furthermore, I feel I must place a caveat onto these groupings because both SAS and Microsoft Excel are mature products that have changed over time. While these categories are generally accurate, your SAS version, Excel version, and computer hardware may not support every DBMS option, and each DBMS option might operate slightly differently depending upon what software you have installed. So make sure you verify what is available to you by looking in the SAS manual that relates to your environment.

- An example retained for backward compatibility with files in the Excel 4 and Excel 5 formats.
- Text file output options like CSV, TAB and DLM do not require SAS/ACCESS Interface to PC Files because the methods read text files.
- Options that read directly from a formatted Excel file.
- LIBNAME options that both use and do not use the SAS PC Files Server.

The options that generate text files will show one example and explain the differences that make the other options work.

## 3.6 List of Examples

Table 3.6.1 is a general description of the functions included in the examples shown in this chapter. Some of the examples here have minor overlaps in the features to show how they interact when additional features are included.

**Table 3.6.1: List of Examples for PROC IMPORT.**

| Example Number | General Description |
|---|---|
| 3.1 | **PROC IMPORT Using the DBMS=EXCEL4 or EXCEL5 Option.** This example is included for backward compatibility with Excel formats Excel 4 and Excel 5, although I would consider it rare to find a computer using this Microsoft Excel software today. The example shows how to read to these old Excel formats. |
| 3.2 | **PROC IMPORT Using the DBMS=DLM Option.** This example shows how to use a delimiter to separate input values and read the header row of the input file as data. This example is equivalent to DBMS=CSV and DBMS=TAB but allows you to provide your own delimiter. |
| 3.3 | **PROC IMPORT Using the DBMS=EXCEL Option.** The three parts of this example all read Excel workbooks that do not need the PC Files |

| Example Number | General Description |
|---|---|
| | Server to be processed. The main point of these code routines is to show how to read parts of worksheets within one workbook, and to change variable names and labels as the data is read from Excel into a SAS dataset. |
| 3.4 | **PROC IMPORT Using the DBMS=EXCELCS Option.** This example shows code that was executed on a 64-bit operating system using a 64-bit copy of SAS 9.3 and a 32-bit copy of Microsoft Excel. Since this computer operating system and SAS use a 64-bit configuration but Excel uses a 32-bit configuration, PROC IMPORT requires the use of the SAS PC Files Server. The "CS" part of DBMS=EXCELCS annotates this feature is in use. |
| 3.5 | **PROC IMPORT Using the DBMS=XLS or XLSX to Select Columns.** This example reads an Excel worksheet with no column headers (variable names) in the output Excel worksheet. It also demonstrates that PROC IMPORT will read an Excel sheet name with spaces. |
| 3.6 | **PROC IMPORT Using the DBMS=XLS or XLSX to Select Rows.** Reading Excel data from selected rows of an Excel worksheet. |
| 3.7 | **PROC IMPORT Using the DBMS=XLS or XLSX to Select Excel Ranges.** This example shows you how to use PROC IMPORT to read a range of cells from an Excel worksheet. |

## Example 3.1 PROC IMPORT Using the DBMS=EXCEL4 or EXCEL5 Option

The SAS IMPORT procedure maintains the backward compatibility features required to process Excel workbooks in the Excel 4 and Excel 5 formats. This example shows how to write Excel files in those formats. For Excel 4 workbooks the sheet name is the same as the file name (without the .xls) and there is only one sheet in the workbook. For Excel 5 formatted workbooks, the sheet name is "Sheet1".

```
* SAS code to import data from an Excel4 file.;
* there is only one sheet in Excel4 files;
PROC IMPORT
   DATAFILE='C:\My_Files\shoes_to_Excel_4_file.xls'
   DBMS=EXCEL4
   OUT=shoes_from_Excel_4
   REPLACE;
RUN;

* SAS code to import data from an Excel 5 file.;
PROC IMPORT
   DATAFILE='C:\My_Files\shoes_to_Excel_5_file.xls'
   DBMS=EXCEL5
   OUT=shoes_from_Excel_5
   REPLACE;
RUN;
```

## Example 3.2 PROC IMPORT Using the DBMS=DLM Option

Using PROC IMPORT to read delimited files in Base SAS invokes the External File Interface (EFI), and the following code reads in a delimited file with commas as the delimiter from the external file named Shoes.csv in directory c:\My_files. This example uses the DBMS=DLM option with the DELIMITER=',' option to select a comma for the delimiter. In addition, it uses the DATAROW=1 and GETNAMES=NO options. These options cause the input SAS file to make the first row from the *.csv file appear as data in the SAS file.

**NOTE:** In Example 2.2 in Chapter 2, the code for PROC EXPORT used the PUTNAMES=NO option to write the `'c:\My_Files\Shoes.csv'` output file with no variable names in the first row of the file.

The output log listing below shows the External File Interface SAS code created by the "Generated SAS Datastep" when the PROC IMPORT step above ran. Notice that the input *.csv file did not have a row of headers associated with the data. So, SAS assigned variable names to the input variables (VAR1 to VAR7).

```
PROC   IMPORT
   DATAFILE='c:\My_Files\Shoes.txt'
   DBMS=DLM
     OUT=shoes
     REPLACE;

   DELIMITER=',';
   DATAROW=1;
   GETNAMES=NO;
   GUESSINGROWS=400;
RUN;
```

**Output 3.1: Listing of the External File Interface Code Generated.**

```
1
2        PROC   IMPORT
3           DATAFILE='c:\My_Files\Shoes.txt'
4           DBMS=DLM
5           OUT=shoes
6           REPLACE;
7
8           DELIMITER=',';
9           DATAROW=1;
10          GETNAMES=NO;
11          GUESSINGROWS=400;
12       RUN;

13       /*******************************************************************
14        *   PRODUCT:        SAS
15        *   VERSION:        9.4
16        *   CREATOR:        External File Interface
17        *   DATE:           17FEB14
18        *   DESC:           Generated SAS Datastep Code
19        *   TEMPLATE SOURCE: (None Specified.)
20        ********************************************************************/
21          data WORK.SHOES   ;
22          %let _EFIERR_ = 0; /* set the ERROR detection macro variable */
23          infile 'c:\My_Files\Shoes.txt' delimiter = ',' MISSOVER DSD lrecl=32767 ;
24             informat VAR1 $25. ;
25             informat VAR2 $14. ;
26             informat VAR3 $12. ;
27             informat VAR4 best32. ;
28             informat VAR5 $12. ;
29             informat VAR6 $12. ;
30             informat VAR7 $9. ;
31             format VAR1 $25. ;
32             format VAR2 $14. ;
33             format VAR3 $12. ;
34             format VAR4 best12. ;
35             format VAR5 $12. ;
36             format VAR6 $12. ;
37             format VAR7 $9. ;
38          input
```

```
39                    VAR1 $
40                    VAR2 $
41                    VAR3 $
42                    VAR4
43                    VAR5 $
44                    VAR6 $
45                    VAR7 $
46         ;
47         if _ERROR_ then call symputx('_EFIERR_',1);  /* set ERROR detection
macro variable */
48         run;

NOTE: The infile 'c:\My_Files\Shoes.txt' is:
      Filename=c:\My_Files\Shoes.txt,
      RECFM=V,LRECL=32767,File Size (bytes)=24901,
      Last Modified=17Feb2014:15:55:41,
      Create Time=17Feb2014:16:14:58

NOTE: 395 records were read from the infile 'c:\My_Files\Shoes.txt'.
      The minimum record length was 37.
      The maximum record length was 85.
NOTE: The data set WORK.SHOES has 395 observations and 7 variables.
NOTE: DATA statement used (Total process time):
      real time           0.07 seconds
      cpu time            0.03 seconds

395 rows created in WORK.SHOES from c:\My_Files\Shoes.txt.

NOTE: WORK.SHOES data set was successfully created.
NOTE: The data set WORK.SHOES has 395 observations and 7 variables.
NOTE: PROCEDURE IMPORT used (Total process time):
      real time           0.53 seconds
      cpu time            0.14 seconds
```

For SAS 6.12 and above, the External File Interface writes out "Generated SAS Datastep Code" that could be captured and used elsewhere. The DELIMITER= statement is active only when DBMS=DLM, and this tells PROC IMPORT what character separates the data values within the input file. When DBMS= has a value of CSV or TAB, SAS assumes a delimiter of a comma or Tab character, respectively. The fact that the file name was "Shoes.txt" caused the **"file-format-specific-statement"** DELIMITER=DLM to identify the input file as a text file with values separated by commas not the default of spaces for *.txt files.

## Example 3.3 PROC IMPORT Using the DBMS=EXCEL Option

### Example 3.3 – Part 1

The code in parts 1, 2, and 3 of Example 2.3 in Chapter 2 showed how to create an Excel workbook with different numbers of worksheets. The example shows how to create worksheet names with mixed-case letters in the name. However, this method will not write an Excel worksheet with a blank in the sheet name. The following code will read the Excel file and produce a SAS dataset called "Shoes" in the Work directory. Notice that the RANGE= value for the spreadsheet name was in capital letters and ended in a Dollar sign "$". The spreadsheet name in the "RANGE=" statement did not need to be in uppercase letters.

```
PROC IMPORT
   DATAFILE='c:\My_Files\Shoes.xls'
   DBMS=EXCEL
   OUT=shoes
   REPLACE;
   RANGE='SHOES$'n;
RUN;
```

### Example 3.3 – Part 2.

If we want only part of the input Excel file, there are several ways to go about getting just what we want. The following code brings in only a few cells from the input Excel file. Here, we will also suppress the request to pull the variable names from the first row of the input data, since we are pulling data from the middle of the Excel file.

```
PROC IMPORT
   DATAFILE='c:\My_Files\Shoes.xls'
   DBMS=EXCEL
   OUT=shoes
   REPLACE;
   GETNAMES=NO;
   RANGE='shoes$C2:F4'n;
RUN;
```

This SAS code does that job. The added command "GETNAMES=NO" and the modification of the "RANGE=" operand are the key parts of this SAS code. The SAS output file looks something like the following:

#### Figure 3.1: SAS Output from Reading the Excel Range Using Absolute Addressing of Excel Cells.

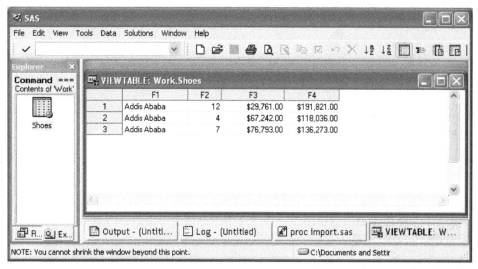

Only 12 cells were read from the Excel worksheet called "SHOES" and the SAS variable names were converted to F1, F2, F3, and F4 because the GETNAMES=NO statement suppressed reading any variable names. The "RANGE=" worksheet name value was in lowercase and included the location of the Excel cells to read into the SAS dataset.

### Example 3.3 – Part 3

Users of Excel Workbooks have the option of creating subsets of cells in a worksheet that can be called by name; these areas are called Named-Ranges. Figure 3.2 below shows one of these named ranges called "small_range". The range name was created while running Excel with the workbook Shoes.xls open.

**Figure 3.2: An Excel 2013 Worksheet with a Named Range Called "small_range" Highlighted.**

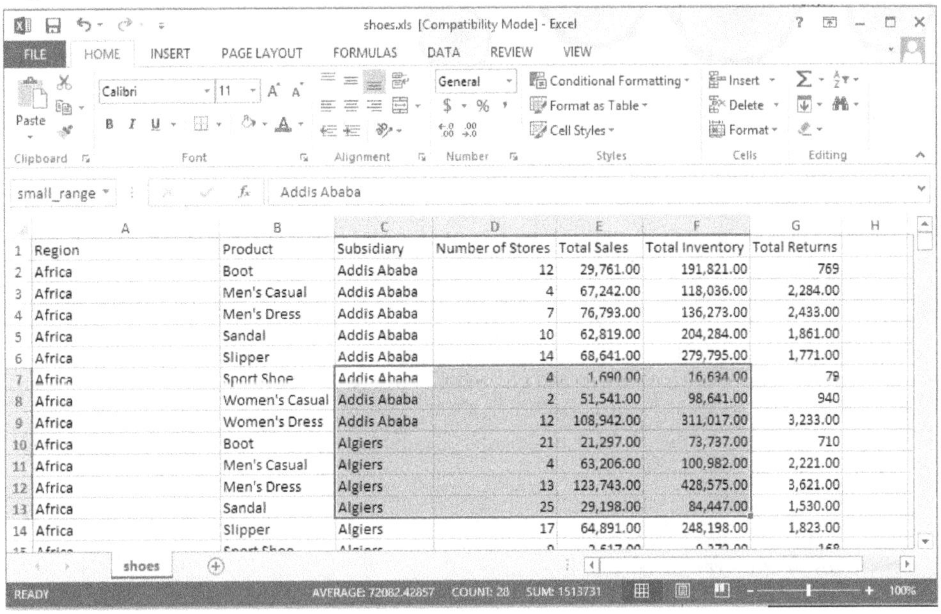

| | A | B | C | D | E | F | G |
|---|---|---|---|---|---|---|---|
| 1 | Region | Product | Subsidiary | Number of Stores | Total Sales | Total Inventory | Total Returns |
| 2 | Africa | Boot | Addis Ababa | 12 | 29,761.00 | 191,821.00 | 769 |
| 3 | Africa | Men's Casual | Addis Ababa | 4 | 67,242.00 | 118,036.00 | 2,284.00 |
| 4 | Africa | Men's Dress | Addis Ababa | 7 | 76,793.00 | 136,273.00 | 2,433.00 |
| 5 | Africa | Sandal | Addis Ababa | 10 | 62,819.00 | 204,284.00 | 1,861.00 |
| 6 | Africa | Slipper | Addis Ababa | 14 | 68,641.00 | 279,795.00 | 1,771.00 |
| 7 | Africa | Sport Shoe | Addis Ababa | 4 | 1,690.00 | 16,634.00 | 79 |
| 8 | Africa | Women's Casual | Addis Ababa | 2 | 51,541.00 | 98,641.00 | 940 |
| 9 | Africa | Women's Dress | Addis Ababa | 12 | 108,942.00 | 311,017.00 | 3,233.00 |
| 10 | Africa | Boot | Algiers | 21 | 21,297.00 | 73,737.00 | 710 |
| 11 | Africa | Men's Casual | Algiers | 4 | 63,206.00 | 100,982.00 | 2,221.00 |
| 12 | Africa | Men's Dress | Algiers | 13 | 123,743.00 | 428,575.00 | 3,621.00 |
| 13 | Africa | Sandal | Algiers | 25 | 29,198.00 | 84,447.00 | 1,530.00 |
| 14 | Africa | Slipper | Algiers | 17 | 64,891.00 | 248,198.00 | 1,823.00 |

The SAS code below shows how to read the data from the Excel named-range called "small_range" into a SAS dataset. Because the GETNAMES=NO option is used, the variable names F1, F2, F3, and F4 that SAS generated are relatively vague variable names; this example will address a way to correct that issue. The DBDSOPTS= option allows you to use other SAS dataset options to change the output SAS dataset while it is being created. The SAS RENAME= dataset option was used here to change the variable names from F1, F2, … to more descriptive variable names. This is done in one pass over the data and makes the output file more useful when PROC IMPORT finishes. You do not need to make another pass over the data to rename the variables. The PROC DATASETS code adds LABEL values to the SAS dataset. The DBMS=EXCEL form of PROC IMPORT does not allow variable labels to be modified on input of the data; therefore, other code is needed to change the variable labels.

```
PROC IMPORT
   DATAFILE='c:\My_Files\Shoes.xls'
   DBMS=EXCEL
   OUT=shoes
   REPLACE;
   GETNAMES=NO;
   DBDSOPTS='RENAME=(F1=Subsidiary F2=Stores F3=Sales F4=Inventory)';
   RANGE=small_range;
RUN;
PROC DATASETS LIBRARY=work NOLIST;
   MODIFY shoes;
      LABEL Subsidiary = "Subsidiary"
            Stores     = "Stores"
            Sales      = "Sales"
            Inventory  = "Inventory";
QUIT;
```

**Output 3.1: Listing of the PROC IMPORT Code generated and the PROC DATASETS Listing.**

```
1
2
3     PROC IMPORT
4         DATAFILE='c:\My_Files\Shoes.xls'
5         DBMS=EXCEL
6         OUT=shoes
7         REPLACE;
8         GETNAMES=NO;
9         DBDSOPTS='RENAME=(F1=Subsidiary F2=Stores F3=Sales F4=Inventory)';
10        RANGE=small_range;
11    RUN;

NOTE: WORK.SHOES data set was successfully created.
NOTE: The data set WORK.SHOES has 7 observations and 4 variables.
NOTE: PROCEDURE IMPORT used (Total process time):
      real time           0.17 seconds
      cpu time            0.06 seconds


12    PROC DATASETS LIBRARY=work NOLIST;
NOTE: Writing HTML Body file: sashtml.htm
13        MODIFY shoes;
14            LABEL Subsidiary = "Subsidiary"
15                  Stores     = "Stores"
16                  Sales      = "Sales"
17                  Inventory  = "Inventory";
18    QUIT;

NOTE: MODIFY was successful for WORK.SHOES.DATA.
NOTE: PROCEDURE DATASETS used (Total process time):
      real time           0.25 seconds
      cpu time            0.15 seconds
```

**Figure 3.3: The SAS Dataset Created by the Code Above.**

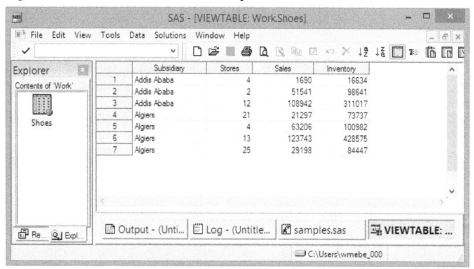

## Example 3.4 PROC IMPORT Using the DBMS=EXCELCS Option

This example is similar to Example 3.2, but the code was executed on a Windows 64-bit configuration. The 64-bit operating system requires the use of the PC Files Server to execute any PROC IMPORT code where DBMS=EXCELCS. The SAS code for Part 1 reads the full Excel worksheet. The difference in the code is the use of the DBMS=EXCELCS option. Note that in most cases the "named-constants" are used as part of the syntax of the RANGE= option; the "named-constants" are not required when a range-name is used with the RANGE= statement.

### Example 3.4 – Part 1
The following SAS code reads a full worksheet from an Excel file on a 64-bit computer; the DBMS=EXCELCS option uses the SAS PC Files Server to access and read the input Excel 32-bit workbook.

```
PROC IMPORT
   DATAFILE='c:\My_Files\Shoes.xlsb'
   DBMS=EXCELCS
   OUT=shoes
   REPLACE;
   RANGE='SHOES$'n;
RUN;
```

### Example 3.4 – Part 2
The following segment of SAS code, while syntactically correct, reads the first row of data as variable names and produces unpredictable results because GETNAMES= is not supported when DBMS=EXCELCS. This code is intended to read three rows of data from the input Excel file. However, the first row is interpreted as SAS variable names.

**NOTE:** The RANGE= value includes Excel cell references, which may not produce your desired output because the GETNAMES= statement is not supported when using the DBMS=EXCELCS option. I suggest that you use the DBMS=XLSX option instead, as shown in Example 3.5. This example shows what happens if you do not use the DBMS=XLSX statement.

```
/* this code does not work */
PROC IMPORT
   DATAFILE='c:\My_Files\Shoes.xlsb'
   DBMS=EXCELCS
   OUT=shoes
   REPLACE;
   RANGE='shoes$C2:F4'n;
RUN;
```

Figure 3.4 shows the output SAS dataset generated by the PROC IMPORT code from above. The intended result was to read three data rows into the SAS dataset. However, the first row was read and translated into variable names.

### Figure 3.4: The SAS Dataset Created by the Code Above.

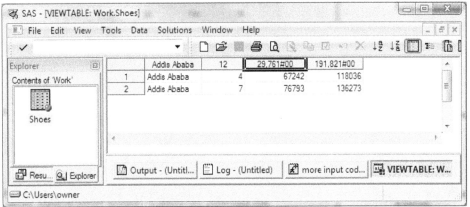

## Example 3.5 PROC IMPORT Using the DBMS=XLS or XLSX to Select Columns

When using the DBMS=XLS option of PROC IMPORT with the ENDCOL and STARTCOL statements, the output SAS dataset is restricted to only the columns requested. This works like a KEEP statement, except the columns have to be contiguous. The input file is the SASHELP.SHOES dataset as exported to an Excel file. This example imports columns 2, 3, and 4 (Product, Subsidiary, and Number of Stores).

**NOTE:** There is a comment in the SAS log about a name change for the variable named "Number of Stores" because this text value has spaces embedded in the value. The value shown in Figure 3.5a for column 3 (Number of Stores) is the label applied to the variable named "Number_of_Stores". Also, ENDCOL= was placed before STARTCOL= to show the statement order is not important. The output SAS dataset has data from three rows and five columns of the input Excel worksheet.

```
PROC IMPORT
   DATAFILE='c:\My_Excel_Files\Shoes.xls'
   DBMS=XLS
   OUT=shoes
   REPLACE;
   ENDCOL="4";
   STARTCOL="2";
RUN;
```

The system output log for Example 3.5 shows the name change of the variable "Number of Stores." The log also verifies that only three columns were output to the SAS dataset from Excel.

```
1    PROC IMPORT
2       DATAFILE='c:\My_Excel_Files\Shoes.xls'
3       DBMS=XLS
4       OUT=shoes
5       REPLACE;
6       ENDCOL="4";
7       STARTCOL="2";
8    RUN;

NOTE:    Variable Name Change.  Number of Stores -> Number_of_Stores
NOTE: The import data set has 395 observations and 3 variables.
NOTE: WORK.SHOES data set was successfully created.
NOTE: PROCEDURE IMPORT used (Total process time):
      real time           0.03 seconds
      cpu time            0.04 seconds
```

### SAS output dataset:

In Figure 3.5a, the SAS dataset label shown for the variable Number_of_Stores has two spaces; however, the actual variable name does not have any spaces embedded.

**Figure 3.5a: The SAS Dataset Created by the Code Above.**

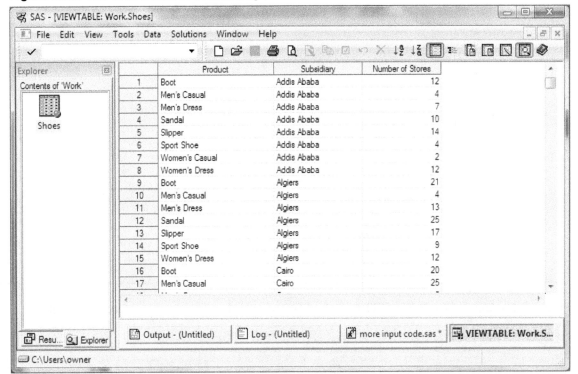

## Example 3.6 PROC IMPORT Using the DBMS=XLS or XLSX to Select Rows

This example uses the PROC IMPORT option pairs STARTROW= / ENDROW= and STARTCOL= / ENDCOL= to show you how you can select a range of cells from an Excel worksheet without creating a named range in an Excel workbook. When the NAMEROW=, GETNAMES=, and RANGE= statements are added to the mix, you can pick names for your variable from inside the Excel file without needing a second pass over the dataset or the need to use PROC DATASETS. The text values with spaces embedded in the value have had an underscore added to replace the space in the variable name. Also, ENDROW= was placed before STARTROW= to show the statement order in not important. The output SAS dataset has data from three columns and five rows of the input Excel worksheet.

```
PROC IMPORT
   DATAFILE='c:\My_Files\Shoes.xls'
   DBMS=XLS
   OUT=shoes
   REPLACE;
   ENDCOL="4";    /* a quoted string is required */
   STARTCOL="2";  /* a quoted string is required */
   ENDROW=10;     /* numeric value is required   */
   STARTROW=6;    /* numeric value is required   */
   NAMEROW=1;
   GETNAMES=NO;
RUN;
```

**Output Log of Code Above**

```
1
2
3      PROC IMPORT
4          DATAFILE='c:\My_Files\Shoes.xls'
5          DBMS=XLS
6          OUT=shoes
7          REPLACE;
8          ENDCOL="4";     /* a quoted string is required */
9          STARTCOL="2";   /* a quoted string is required */
10         ENDROW=10;      /* numeric value is required   */
11         STARTROW=6;     /* numeric value is required   */
12         NAMEROW=1;
13         GETNAMES=NO;
14     RUN;
NOTE:      Variable Name Change.  Number of Stores -> Number_of_Stores
NOTE: The import data set has 5 observations and 3 variables.
NOTE: WORK.SHOES data set was successfully created.
NOTE: PROCEDURE IMPORT used (Total process time):
      real time           0.06 seconds
      cpu time            0.01 seconds
```

**Figure 3.5b: Using PROC IMPORT to Select Rows and Headers from an Excel Worksheet.**

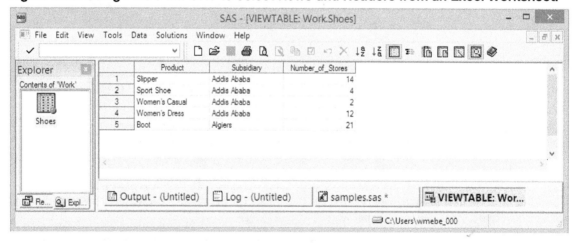

## Example 3.7 PROC IMPORT Using the DBMS=XLS or XLSX to Select Excel Ranges

This example was executed on a computer running 64-bit Windows 8.1 Professional on 64-bit hardware with SAS 9.4 and 32-bit Excel 2013 installed. The DBMS option XLSX provides an alternative method to reading a small group of cells from an Excel spreadsheet. However, this method does not always provide reliable variable names when GETNAMES=YES. GETNAMES=YES looks for variable names in the first row of input cells. Here, GETNAMES=NO is used to turn off the search for variable names in the Excel file. The RANGE='shoes$C2:F4'n command selects only 12 cells from the Excel file.

```
PROC IMPORT
    DATAFILE='c:\My_Files\Shoes.xlsx'
    DBMS=XLSX
    OUT=shoes
    REPLACE;
    GETNAMES=NO;
    RANGE='shoes$C2:F4'n;
RUN;
```

```
1
2
3      PROC IMPORT
4          DATAFILE='c:\My_Files\Shoes.xlsx'
5          DBMS=XLSX
6          OUT=shoes
7          REPLACE;
8          GETNAMES=NO;
9          RANGE='shoes$C2:F4'n;
10     RUN;

NOTE: The import data set has 3 observations and 4 variables.
NOTE: WORK.SHOES data set was successfully created.
NOTE: PROCEDURE IMPORT used (Total process time):
      real time           0.03 seconds
      cpu time            0.01 seconds
```

**Figure 3.6: SAS Output When Using PROC IMPORT to Select a Group of Cells from an Excel Spreadsheet.**

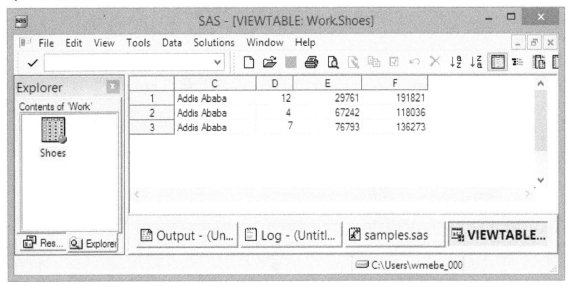

## 3.7 Conclusion

I have shown several methods of reading data and variable names from Excel workbooks. But, there are far too many other combinations of options available for me to present an exhaustive list. This chapter showed features of PROC IMPORT. Some of the important items to take away from this chapter are that the Microsoft Excel JET and ACE engines have limitations. These limitations will occasionally affect the amount of data you can extract from your Excel files. There may be times when you are required to fall back to the tried-and-true delimited file formats to transfer your data to and from Excel. I suggest that you refer to *SAS/ACCESS Interface to PC Files: Reference* for the version of SAS that you have installed. These documents have SAS version-specific descriptions of the syntax and features available for the SAS Import Wizard and PROC IMPORT.

# Chapter 4: Using the SAS LIBNAME to Process Excel Files

| | |
|---|---|
| 4.1 Introduction | 47 |
| 4.2 Purpose | 48 |
| 4.3 Excel-Specific Features of the SAS LIBNAME Statement | 48 |
| 4.4 Syntax of the SAS LIBNAME Statement | 49 |
| 4.5 LIBNAME Statement ENGINE CONNECTION OPTION Descriptions | 50 |
|     4.5.1 HEADER Option to Read Variable Names | 50 |
|     4.5.2 MIXED Option to Select Data Types | 50 |
|     4.5.3 PATH Option to Define Physical File Locations | 51 |
|     4.5.4 VERSION Option to Identify Excel File Version | 52 |
|     4.5.5 PROMPT Option to Interactively Assign a Libref | 52 |
|     4.5.6 Other Common SAS PC File LIBNAME Options | 53 |
| 4.6 Excel-Specific Dataset Options | 53 |
| 4.7 UNIX, LINUX, and 64-Bit Windows Connection Options | 54 |
| 4.8 Overview of the Examples | 55 |
|     4.8 List of Examples | 55 |
| 4.9 Examples | 56 |
|     Example 4.1 Using the Engine Connection HEADER Option | 56 |
|     Example 4.2 Using the Engine Connection MIXED Option | 57 |
|     Example 4.3 Using the Engine Connection PATH Option | 58 |
|     Example 4.4 Using the Engine Connection VERSION Option | 58 |
|     Example 4.5 Using Named Literals with the LIBNAME Statement | 59 |
|     Example 4.6 Using PROC CONTENTS to Examine an Excel Workbook | 60 |
|     Example 4.7 Using Dataset Options to Process Date and Time Values | 62 |
|     Example 4.8 Using Dataset Options to Process Variable Type Conversions | 63 |
|     Example 4.9 Processing on 64-Bit Operating Systems | 64 |
| 4.10 Conclusion | 65 |

## 4.1 Introduction

SAS LIBNAME statements are used everywhere in SAS, and they are the most common way for SAS programmers to access SAS data files. They are used to read SAS data files and write SAS data files. SAS data comes in. SAS data goes out. But, is that all there is? Well, for SAS users without SAS/ACCESS Interface to PC Files installed that may be true. But users who want to read Excel files with Base SAS 9 and above and who have SAS/ACCESS Interface to PC Files installed have other options. The LIBNAME statement will give them direct access to some files that are not SAS files. This tool will allow users to directly read and write data files formatted for Microsoft Windows applications. Many people are not aware that SAS/ACCESS Interface to PC Files will allow SAS programmers to access an Excel spreadsheet the same way as any other SAS file. There are, of course, some restrictions, but there are also a lot of options that help remove some of the bumps in the road. Most programmers use LIBNAME statements only to access SAS data files. When external data is required, the data is converted to a text file and input through a FILENAME connection of some sort. Or, PROC IMPORT or PROC EXPORT is used to process external files. The most popular of these applications (and the focus of this book) is moving data between SAS files and Microsoft Excel. The LIBNAME statement allows the user to define an Excel file in SAS terms. This gives the programmer access to LIBNAME and dataset options to control how the Excel file is defined, accessed and, yes, even how the data will be formatted. Not all of the data formats of Excel can be accessed, but the LIBNAME statement gives the SAS user more control over the data.

### Complexities That Relate to SAS, Excel, and Your Hardware

Before we get started on the basics about the LIBNAME statement as it relates to SAS, Excel, and your hardware, I need to make a few comments. To paraphrase the beginning of Dickens' novel *A Tale of Two Cities*: it is the best of times, it is the worst of times, as far as accessing Excel data from SAS goes. Why would I, as an author who has spent the last several years working on this project, feel that way? The increase in options relating to hardware (32-bit and 64-bit); the availability of 32-bit and 64-bit SAS 8.2, 9.0, 9.1, 9.2, 9.3, to 9.4; and the changes in the Excel version formats from Excel 2003, to Excel 2007 and the addition of 2010 (32-bit and 64-bit), and 2013 (32-bit and 64-bit) present a very large number of combinations. The examples I present here may not work on your configuration.

**NOTE: IMPORTANT RESTRICTION.** In the past, multiple copies of some SAS software versions were able to reside on one computer, but the SAS PC Files Server (pcfservice.exe) for SAS 9.2 and SAS 9.3 have the same name, run as the same service, listen for commands on different computer I/O ports, and are NOT compatible. So, even though you can install both SAS 9.2 and SAS 9.3 on the same computer only one version of SAS will be able to read and write to Excel files using the SAS PC Files Server.

## 4.2 Purpose

SAS is a language of defaults; the runtime system looks to see how a variable is first used. Then, it assigns a default type to the variable. When a programmer types "PROC PRINT; RUN;" the last SAS dataset created is sent to the Output window with all of the observations and all of the variables listed. The SAS language also gives the programmer choices about the code. The ATTRIB command assigns either a numeric or a character data type to a variable. These assignments can change the size of the variable to a small numeric with a limited range of values, or a large character variable bigger than the default. LIBNAME statements are no different; they have options to change the behavior of the input/output processes. This chapter addresses some options available for reading and writing data to and from Excel files using SAS.

## 4.3 Excel-Specific Features of the SAS LIBNAME Statement

### What Is the "LIBNAME" Feature for Excel Files?

The LIBNAME is the SAS door that opens onto Excel data, making it look like SAS data, allowing the SET statement (and dataset options) to directly read and write Excel file data. This allows data to be input or output by a DATA step or SAS procedure. SAS uses either the Microsoft JET or ACE SQL engine to read and write to the Excel files. The LIBNAME statement employs special options to help make the task easier. Of course, options, while being nice to have, often imply freedoms (or restrictions).

**NOTE:** Excel uses binary formats for Excel file versions Excel 5.0 to Excel 2003. However, Excel 2007, Excel 2010, and Excel 2013 do not produce a native binary formatted file. These versions generate files that are output in a new file format called Open XML. The Open XML files have a file name with the extensions "xlsx", "xlsm", and "xlsb". But, Microsoft Excel 2007, Excel 2010, and Excel 2013 can read and write to other Excel version formats. Changing an Excel 2007 or later file extension to *.zip will allow you to view the XML parts of the Excel files.

### What Is Needed to Use a SAS LIBNAME to Access Excel Files?

The ability to use a LIBNAME to access data has long been restricted to SAS datasets. Beginning with SAS 9, SAS/ACCESS Interface to PC Files can use Excel files as SAS input. Furthermore, SAS files can be used for output to Excel. SAS offers other software besides SAS/ACCESS Interface to PC Files that enables users to process data from Oracle, and MySQL databases among others. These and other SAS/ACCESS interfaces will not be discussed here. If you want to find out if you have SAS/ACCESS to PC Files installed on your system, then execute the SETINIT procedure as shown below.

```
proc setinit;
run;
```

### Find Out What Hardware and Software Configuration Your Computer Uses.

All of the Windows operating systems since Windows XP are available in both 32-bit and 64-bit versions. One good thing is that all versions of Excel 2003 and Excel 2007 are 32-bit applications. You can determine the bit configuration of Excel 2010 by checking FILE>HELP and looking on the right side of the screen under "About Microsoft Excel". Similarly, you can determine the bit configuration of Microsoft Excel 2013 by checking FILE>ACCOUNT>ABOUT EXCEL and looking on the upper right of the pop-up screen "About Microsoft Excel". UNIX, LINIX, and other operating systems have their own methods of detecting the bit configuration in use. This information is important to know when selecting options to use with the LIBNAME statement.

### What Can the LIBNAME Statement Do for You?

When you use a SAS LIBNAME to read Excel data, the data can be read directly from native format binary Excel files. The SAS LIBNAME can read data either from a full Excel worksheet or from a part of an Excel worksheet defined by Excel to be a "named range". Within Excel, a "named range" can be as small as a single cell, or as large as the whole worksheet. That same data can be written out in nearly the same way directly to a native Excel binary file. Named ranges can be written out in similar sized units, but not necessarily in the same location. All "named range" output that is written begins in cell A1. Some assumptions are made using LIBNAME statements on the input side and the output side, but knowing the rules can make things easier. Writing data from a SAS file to Excel is generally a little safer. This is because when a SAS file is written out, all of the data in one variable or Excel column is of the same data type. Reading data from Excel is sometimes more of a challenge. The ability of Excel files to change data formats on the data cell level can still present challenges to SAS programs reading input from Excel files. SAS input routines predict the type of data that is in an Excel worksheet column, but generally by only looking at the first few rows of the column. This tends to produce missing data values for columns where both numeric and character values exist. In some cases, this issue can be overcome if named ranges are defined within the Excel worksheet. But full columns of mixed values present a challenge. The MIXED option described below can help to modify this behavior and make your results more consistent with data of unknown types from the spreadsheets. Also, be aware that some configurations of the Microsoft JET/ACE engines may be limited to 255 columns and 65,535 rows.

## 4.4 Syntax of the SAS LIBNAME Statement

First, let us look at the different syntax options and parameters available for the LIBNAME statement as they relate to accessing Microsoft Excel files. Some general LIBNAME options are discussed in this chapter. The SAS/ACCESS engine connection options change the way that the LIBNAME statement interacts with Excel workbooks. But the SAS/ACCESS LIBNAME options work more directly with the data in the worksheets.

```
LIBNAME libref <engine> <physical-file-name>
<SAS/ACCESS-engine-connection-options>
<SAS/ACCESS-libname-options>;
```

The following engine connection options are commonly used by SAS when opening Excel files. These items will be discussed later below; other, less used options also exist.

- HEADER        Affects the way that the first row or observation is processed.
- MIXED         Affects how columns with both numeric and character data are treated.
- PATH          Allows you to supply a path for an Excel file, this is not always needed.

- PROMPT  Enables SAS to provide help menus to assist with assigning librefs.
- VERSION  Assigns the output Excel version; SAS can determine the input version.

The following SAS/ACCESS LIBNAME options are commonly used with Excel. However, space does not permit an example of each option, so I have selected a few values and provided a short description of each here. Other options are available for different input file types and access methods. The defaults for Excel input files are listed in **BOLD** typeface. Aliases may also exist for these options.

- ACCESS=READONLY  Limits SAS to Read-only mode.
- DBGEN_NAME=**DBMS** or SAS  Specify how data source columns are named.
- DBMAX_TEXT=n (1-32767)  Maximum length of an input character string.*
- DBSASLABEL=**COMPT** or NONE  Select whether to save input column names as a label.
- FILELOCK=YES or **NO**  Specify if the input file is locked to other users.
- SCAN_TEXTSIZE=**YES** or NO  Scan the input file character columns for size of strings.
- SCAN_TIMETYPE=YES or **NO**  Scan Excel variables for time values.
- STRINGDATES=YES OR **NO**  Select to read dates as character strings.
- USEDATETYPE=**YES** or NO  Use the SAS format DATE9. for date values.

*Default=1024.

## 4.5 LIBNAME Statement ENGINE CONNECTION OPTION Descriptions

The next part of the LIBNAME statement we will examine here is the way that SAS can open the Excel files. By examining how these options interact with Excel workbooks, I will show how the options treat the data when it is accessed after the Excel file is open. In the LIBNAME syntax chart listed in section 4.4 we see SAS/ACCESS engine connection options and SAS/ACCESS LIBNAME statement options. These are both little-used features of SAS LIBNAME statements. The example portion of this chapter shows some examples and outputs from applying these options

### 4.5.1 HEADER Option to Read Variable Names

The first to be addressed is the set of SAS/ACCESS engine connection options. See Example 4.1, Using the Engine Connection HEADER Option. The HEADER option affects the reading of the first line of the Excel file. When the option value is set to YES (the default), the first row of data is read to build SAS variable names. The cell values are converted to SAS variable names. When the text values would not generate valid SAS variable names, underscores are inserted into the SAS variable name. Duplicate names are avoided by adding a number to the end of the variable name. This number increases as duplicate columns are found, moving from left to right.

When the **HEADER** option is set to **NO** or when the cell value does not convert to a text field, SAS generates a variable name. Also, when the option is set to **NO,** then the first row is considered data, and variable names are generated as F1, F2, F3, ... , to F$n$ with the $n$ being the number of the last column in the Excel spreadsheet.

### 4.5.2 MIXED Option to Select Data Types

The **MIXED** option is provided to assist in reading fields that are not obviously text or numeric. Excel files can be generated by someone typing data into one cell after another without regard to the type of data that they are entering (character or numeric). SAS has to make a default best guess about what the field contains. When an Excel input column contains both character and numeric data, the intent of this option is to read all of the data in the Excel column as character data. This option is handled by the Microsoft Jet or ACE provider and is available only in Windows for Excel files. This does not work for delimited files; remember, delimited files are read with a FILENAME statement.

However, the **MIXED** engine connection option has dependencies that are beyond the control of SAS software. This small feature of the SAS LIBNAME statement is one of the complex concepts dealing with reading data from Excel into SAS. This complexity comes from the environment that the software has to deal with. In the introduction to this chapter, I discussed the complexities that relate to SAS, Excel, and your hardware. Here is where those complexities converge. The following items make implementing this feature difficult.

1. Hardware (32-bit or 64-bit).
2. Operating System (32-bit or 64-bit).
3. SAS Version (8.2, 9.0, 9.1, 9.2, 9.3, 9.4 - 32-bit or 64-bit).
4. SAS/ACCESS Interface to PC Files Server – for SAS 9.2, 9.3, 9.4, 9.4.1.
5. Microsoft Excel Version: Excel 2003 and Excel 2007 are 32-bit.
6. Microsoft Excel Version: Excel 2010 and Excel 2013 can be either 32-bit or 64-bit.
7. Microsoft JET engine processes file formats prior to Excel 2007.
8. Microsoft ACE engine processes all Excel formats including Excel 2007 and later.
9. Microsoft JET and ACE engines are embedded in the OS and use Windows Registry Keys as inputs.
10. Changing Microsoft Registry Keys is risky and affects all programs that use the JET/ACE engines.
11. By default, the **MIXED** engine connection option examines the first eight rows of a spreadsheet.
12. The SAS GUESSINGROWS option is available only to PROC IMPORT.
13. Registry Values **TypeGuessRows** and **ImportMixedTypes** have different locations in each OS.
14. None of these things stays the same very long.
15. When using the Excel engine, HEADER=YES and MIXED=YES work.
16. If you are using the PC Files engine, neither of these will work.

Items 1 to 13 listed above combine to make it difficult for each configuration to work exactly the same. The object of this MIXED option is to convert columns with a mix of numeric and character data into character data so that a SAS programmer can examine the data within a program and make the adjustments relative to the expected data and the data read from the file.

The **MIXED** option works, but by default checks only the first 8 rows of an Excel worksheet (without registry key changes). The first row is typically considered to be a header row and, if all numeric values are returned in rows 2 through 7, then the variable is considered to be numeric. In this case, any character values are returned as missing values. All the worksheet columns are tested, and SAS variables are all created based upon the results of the JET/ACE engine tests. Also, if the first different row is beyond the limit of the guessing row parameter, then data of the opposite type is returned as missing. When this MIXED option is used, then the Excel file is read in import mode, and no updates are allowed to the file.

When the **MIXED** option has a value of **NO** (the default value), a specified number of rows of the Excel column are searched. A guess is returned about the data type of the input Excel data field. The default is to search the first eight rows of an Excel spreadsheet. Examples later in this chapter will show you how to use this option and some alternative methods.

**NOTE:** This option may cause the wrong variable length to be assigned to the field. The Windows Registry settings **TypeGuessRows** and **ImportMixedTypes** control the behavior of this option, and require administrative privileges to modify. I do not recommend changing these settings. Improperly edited changes to the Windows Registry can cause your computer to stop working. These settings are described in the SAS documentation.

### 4.5.3 PATH Option to Define Physical File Locations

The LIBNAME statement will at times allow you to enter a "`<physical-file-name>`". You can enter this value on the LIBNAME statement when the file type can be determined by the file name. However, when your computer, operating system, and application software are not all the same bit configuration, it is more difficult to determine how to access an Excel file using the SAS LIBNAME statement. Therefore, you need to give SAS some more information about your environment or files that you are using. The engine connection option **PATH=** (in conjunction with the <engine> option of the LIBNAME statement) allows SAS to select access methods that are appropriate for your configuration.

All of this may seem a little confusing because the "`<physical-file-name>`" is listed separately as part of the LIBNAME statement, and the PATH option syntax as shown in Example 4.3, Using the Engine Connection PATH Option, also includes a data path and file name. If the PATH option is used, "**PATH=**" should immediately precede the "`<physical-file-name>`" and the LIBNAME statement can include only one "`<physical-file-name>`". More important, when you use the PATH option you must also include an engine option on the LIBNAME statement.

Some engine names used to access Excel workbooks in conjunction with the PATH option are:
- EXCELCS
- PCFILES

The PCFILES engine resolves bit configuration differences through the SAS PC Files Server, which is a background windows service that can be configured to start when your computer starts. Example 4.3, Using the Engine Connection PATH Option, will show you how to use this option.

## 4.5.4 VERSION Option to Identify Excel File Version

The **VERSION** option is not required on input because the Microsoft JET/ACE provider engines can determine the format of the input file. However, the output format can be chosen to be one of the three output Excel file formats (Excel 5, Excel 97 – 2003, and Excel 2007+). The default version is '97'. The actual option codes are listed here '2007','2003', '2002', '2000', '97', '95', '5' (the quotes are valid but not required.) NOTE – 2007 is valid only for SAS 9.2 Mod 2 and later. The Excel 2010 and Excel 2013 formats are compatible with the file format for Excel 2007.

## 4.5.5 PROMPT Option to Interactively Assign a Libref

This option can be used to interactively assign a file to a libref. There are several options available for this command. I would not like to assign a libref to a file using interactive prompts every time that I use the file. But it turns out that this option does have a useful feature. It will provide access to an initialization string of characters that is useful when using the ODBC data sources. Both an Excel file and an Oracle database or other databases can be accessed this way. When you use the PROMPT method to connect to an ODBC data source, you can issue the following command to display the initialization string on the log.

```
%PUT %SUPERQ (SYSDBMSG)
```

Because each string is different, no example will be shown here. Also, because of the advanced nature of this process, it should be used only when the exact syntax is known for all of the connection options required for the ODBC data source being accessed.

The options available are the following.
- Yes
- No
- Required
- Noprompt
- Prompt
- UDL

**NOTE:** When you use **PROMPT=NO** or **PROMPT=NOPROMPT**, a "`<physical-file-name>`" or **PATH=** for file identification is required.

A "UDL" is a Microsoft Datalink File, which is not discussed here.

## 4.5.6 Other Common SAS PC File LIBNAME Options

The final part of the syntax is for LIBNAME options. They, along with dataset options, modify how the data elements are processed. The programmer can set the file to be Read-only, increase the default size of character strings, change how dates are processed, and make several other file-level changes. The options above applied to the files in their entirety, and the options are set for all data within the file. This next set of options is not as absolute in the uniform application, and can affect the file or the variables within the file.

- ACCESS=READONLY　　　　　Limits SAS file access to Read-only mode.
- DBGEN_NAME=DBMS or SAS　　Specify how data source columns are named.
- DBMAX_TEXT=n (1-32767)　　　Maximum length of an input character string.*
- DBSASLABEL=COMPT or NONE　Select whether to save input column names as a label.
- FILELOCK=YES or NO　　　　　Specify if the input file is locked to other users.
- SCAN_TEXTSIZE=YES or NO　　Scan the input file character columns for size of strings.
- SCAN_TIMETYPE=YES or NO　　Scan Excel variables for time values.
- STRINGDATES=YES OR NO　　　Select to read dates as character strings.
- USE_DATETYPE=YES or NO　　　Use the SAS format DATE9. for date values.

*Default=1024.

## 4.6 Excel-Specific Dataset Options

Another important feature of reading data with a LIBNAME statement is that you have control of dataset level options. With typical native SAS formatted datasets used with a LIBNAME statement, you are able to modify what you allow into and out of the dataset when it is opened. While these options are not directly part of the LIBNAME statement, I like to consider them a feature of using a LIBNAME statement.

The following are dataset options available to Excel files.

| Option | Description |
|---|---|
| DBCREATE_TABLE_OPTS | Appends specific SQL code to the CREATE TABLE command. |
| DBENCODING | Defines a different character set to be used in accessing the Excel file. |
| DBFORCE | Forces insertion of character strings larger than defined length into the file. |
| DBGEN_NAME | Determines generation of SAS dataset variable names. |
| DBLABEL | Chooses to use SAS variable labels or names as the output column names. |
| DBMAX_TEXT | Defines the maximum length of a character variable. |
| DBSASLABEL | Chooses to use the source column names as SAS variable labels. |
| DBSASTYPE | Specifies a data type that overrides input Excel default variable data type. |
| DBTYPE | Specifies a data type that overrides output default SAS variable data type. |
| INSERT_SQL | Determines the SQL insert method used to add new rows. |
| READBUFF | Determines how many rows are read into the input buffer for processing. |
| SASDATEFMT | Defines specific date formatting by column when processing dates. Similar formats must be available in SAS and Excel. |

*54 Exchanging Data between SAS and Microsoft Excel*

The following are general dataset options that can also be applied to excel files.

| | |
|---|---|
| DROP | Excludes variables from being written or read to/from an input or output file. |
| FIRSTOBS | Specifies the first record read from an input Excel file. |
| IN | Creates a flag to indicate if a variable is from a specific input file. |
| KEEP | Includes variables being written or read to/from an input or output file. |
| RENAME | Renames variables when either reading or writing a file. |
| WHERE | Selects observations for input or output based upon user-defined conditions. |

## 4.7 UNIX, LINUX, and 64-Bit Windows Connection Options

This section of the chapter is still dealing with the LIBNAME statement, but here the hardware and software changes. We begin to focus a little more directly upon the 64-bit options and options that are not Windows options. Also, we will look at some of the features that extend the access of the LIBNAME statement beyond the scope of a single computer into the realm of multiple computer environments. As was mentioned earlier, the complexities introduced by multiple bit configurations and software versions has made the job of maintaining ease of access a considerable task. I only have enough space here to introduce a few simple concepts about interconnecting the many systems that exist.

SAS/ACCESS Interface to PC Files comes with what I call a translator. SAS calls it the SAS PC Files Server. This program is normally started when the computer is re-booted and runs in the background waiting for a command from SAS to interact with a PC file. The PC Files Server connection options are available on UNIX, Linux, and 64-bit Windows Server systems. These descriptions, when used on computer that is not a Windows computer, may be slightly different from the option descriptions used on 32-bit PC systems that are based on Windows. When the options contain characters that are not valid in SAS names, then the values must be enclosed in quotation marks. Environment variables and system options may be substituted for some of these connection options. The syntax for specifying options is **OPTIONNAME=value**.

**NOTE:** To access PC files from UNIX, Linux, and 64-bit Windows systems, an active network connection must exist from these systems to a PC-based system. The 64-bit Windows operating system is in fact a PC-based system. However, the Excel files are accessed via the "PCFILES" server which may be either a 32-bit or 64-bit interface. The "PCFILES" server must be accessible, and a Windows 64-bit PC may not require the SERVER or PORT options.

This connection can be created using a LIBNAME statement similar to the following for these systems:

**LIBNAME** libref **pcfiles** <connection-options> <libname-options>;

### 4.7.1 Accessing Excel files on LINUX, UNIX, and 64-Bit Windows Systems

Contact your system administrator for the correct server name and port number because these can be changed locally. An example would look something like the following LIBNAME statements. "PCFILES" is the SAS/ACCESS engine name required for accessing files on a PC or PC network from Linux, UNIX, or a 64-bit Windows system, and must be running on the server identified in the "SERVER" option. If you are using 64-bit SAS and the PC Files server on the same computer, the SERVER and PORT options are not required.

**LIBNAME** myxls **PCFILES SERVER**=xxxx **PORT**=8621 **PATH**="c:\demo.xls"; * default port V9.2;
**LIBNAME** myxls **PCFILES SERVER**=xxxx **PORT**=9621 **PATH**="c:\demo.xls"; * default port V9.3;

## 4.8 Overview of the Examples

### 4.8 List of Examples

**Figure 4.8.1: List of Examples for Using LIBNAME Options.**

| Example Number | General Description |
|---|---|
| 4.1 | **Using the Engine Connection HEADER Option.** This example shows how the first row of an excel spreadsheet is treated when the HEADER=NO option is used alone. |
| 4.2 | **Using the Engine Connection MIXED Option.** This example shows how the MIXED=YES option can convert all of the data columns into character values. But, be careful. This example takes advantage of the HEADER=NO option, too. When all columns have a character value in the first row, all columns will be character when read from the file. Be ready to rename your variables. |
| 4.3 | **Using the Engine Connection PATH Option.** This example shows two output Excel files being processed with a PATH= option to write the spreadsheets. |
| 4.4 | **Using the Engine Connection VERSION Option.** The code in this example shows how to write data to three different versions of Excel output workbooks. |
| 4.5 | **Using Named Literals with the LIBNAME Statement.** The use of Named Literals is explained in this example. These constants allow the use of some special characters when reading or writing to an Excel spreadsheet. The values are used to represent the sheet names and allow some latitude when referencing Excel worksheets. The code first writes a new worksheet into either a new or existing workbook. The code then reads the same file into a SAS dataset. |
| 4.6 | **Using PROC CONTENTS to Examine an Excel Workbook.** The code shown here produces a listing of the variables in the Excel workbook. Both full worksheets and named ranges are displayed when the output of the PROC CONTENTS is displayed. |
| 4.7 | **Using Dataset Options to Process Date and Time Values.** Dates and time values are reformatted directly from Excel as the data is being read into a SAS dataset. This can eliminate programming effort and computer time to process date fields. But, always look to see how the data is stored if you need the converted data values. |
| 4.8 | **Using Dataset Options to Process Date Variable Type Conversions.** The DBSASTYPE option will allow you to convert an Excel column of data values into SAS variables with specific formats. This example converts two dates into variables with different formats. |
| 4.9 | **Processing on 64-bit Operating Systems.** This is an introduction to the use of the PATH= option and the PCFILES engine as they relate to 32-bit and 64-bit software on a 64-bit computer. |

## 4.9 Examples

### Example 4.1 Using the Engine Connection HEADER Option

This example relates how the first row of Excel data is treated. If the Excel file contains descriptive names of the data and is useful to be converted to a SAS variable name, then the default value of "**HEADER=YES**" should be used. The SAS programmer can also code this option directly on the LIBNAME statement. The output is shown in Figure 4.9.1.

```
LIBNAME xls_data EXCEL 'C:\My_excel_files\my_excel_file.xls' HEADER=NO;
DATA Excel_data;
   Set xls_data.'sheet1'n;
Run;
```

**Figure 4.9.1: Reading Excel Data with the HEADER=NO Option Turned On.**

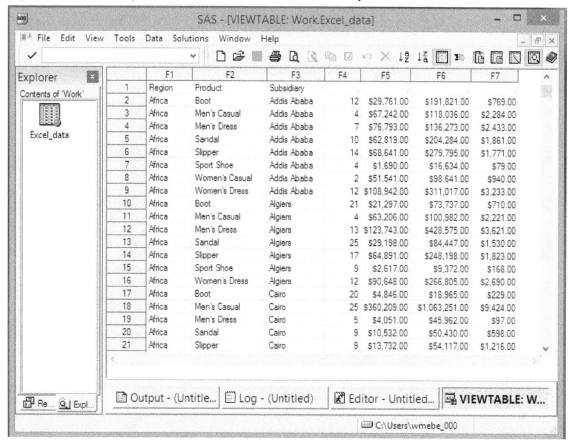

The **HEADER=NO** option suppresses using row one of the data in the Excel worksheet for the variable names in the SAS output dataset. The SAS variable names generated are F1, F2, F3, F4, F5, F6, and F7 for this file. The SAS dataset created is shown here. When we look at the first observation of the SAS dataset, we expect to see the names of the variables as data values of observation number one. However, the first row of the SAS dataset contains data values Region, Product, Subsidiary, ".", ".", ".", and ".". This occurs because, when reading the data from Excel, an estimate is made about what type of data is in any given column. Typically, the first eight rows of the Excel worksheet are scanned and the data is checked to see if it has more character or numeric values. This testing determines whether or not the column is assigned to use a character or numeric informat when the data is transferred to the SAS dataset. Therefore, the character values in row 1, columns F4 to F7, of the Excel worksheet fail to be converted to numeric values and are assigned a missing numeric value. This would be true for any character value in columns 4 through 7. Conversely, any numeric value in column 1 to 3 will be converted to a character missing value.

**NOTE:** There is a way to change the system-wide default for the GUESSINGROWS option, but it involves modifying the SAS or WINDOWS system REGISTRY values.

### Example 4.2 Using the Engine Connection MIXED Option

This example uses the **MIXED=YES** engine connection option to capture all data in the Excel worksheet as character data values. This eliminates missing values in fields with mixed character and numeric values in individual Excel cells. But it also places the burden of converting numeric data on the program that uses the SAS dataset that was created. The code in Figure 4.9.2 will run using 32-bit SAS V9.2 on a 32-bit operating system or 32-bit SAS V9.3 on a 32-bit operating system.

```
LIBNAME xls_data EXCEL 'C:\My_excel_files\my_excel_file.xls'
HEADER=NO MIXED=YES;
   DATA Excel_data;
      SET xls_data.'sheet1'n;
   RUN;
```

**Figure 4.9.2: Reading Excel Data with the MIXED=YES and the HEADER=NO Options Turned On.**

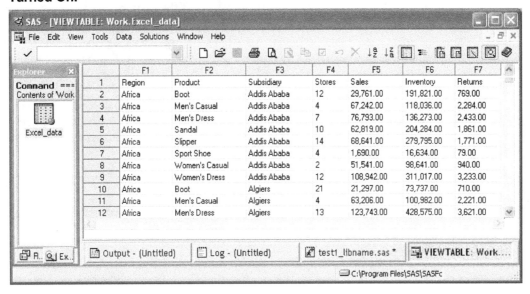

The data shown in Figure 4.9.2 is an example of the **MIXED** option with a value of **YES,** combined with **HEADER=NO option.** The combined use of these two options changed the values in row 1 from missing values to the text from the worksheet for columns F4 through F7. The **HEADER=NO** option changes only the top row of the SAS file. In addition, note that the **MIXED=YES** option affected all of the data in all of the columns by left justifying it. That means that all of the data in the worksheet is read as character data. This trick works only if the first row has character headers in every column.

As an aside, and not shown here, notice that the SAS variable names are all in a pattern: "F1", "F2", etc. An alternative to hand-coding a rename for all of the variables is to read only one row from the file and process the SAS file of headers with PROC CONTENTS and SQL to create a macro variable that can be used to create a rename command.

## Example 4.3 Using the Engine Connection PATH Option

This example shows a simple application of the **PATH**= option. The SAS code shown also uses the SAS Excel engine.

**NOTE:** Starting with SAS 9.2 TS Level 2M3, varying levels of support for reading the newer versions of Excel have become available with each new release of SAS. Of course, the issues of hardware, software, and bit configurations may affect each computer differently.

```
LIBNAME xls_data EXCEL path='C:\My_excel_files\my_excel_file.xls' ;
   DATA xls_data.'sheet1'n;
      Set sashelp.shoes;
   Run;

LIBNAME xls_data EXCEL path='C:\My_excel_files\my_excel_file.xlsb' ;
   DATA xls_data.'sheet1'n;
      Set sashelp.shoes;
   Run;
```

## Example 4.4 Using the Engine Connection VERSION Option

The following example outputs three files, and each file will be written in a different Excel format. This example shows how the **VERSION**= option is used to write the output files. These examples run on 32-bit SAS 9.2 on a 32-bit operating system and 32-bit SAS 9.3 on a 32-bit operating system. The SAS code shown also uses the SAS Excel engine.

**NOTE:** Starting with SAS 9.2 TS Level 2M3, varying levels of support for reading the newer versions of Excel have become available with each new release of SAS. Of course, the issues of hardware, software, and bit configurations may impact each computer differently. Workbooks with the Excel extension *.xlsm may not be valid for this type of processing.

```
LIBNAME xls_V95 EXCEL 'C:\My_excel_files\my_excel_file_V95.xls'
version=95;
LIBNAME xls_V97 EXCEL 'C:\My_excel_files\my_excel_file_V97.xls'   version=97;
LIBNAME xls_V07 EXCEL 'C:\My_excel_files\my_excel_file_V07.xlsb'
version=2007;
DATA  xls_V95.'shoes'n
   xls_V97.'shoes'n
   xls_V07.'shoes'n;
Set sashelp.shoes;
Run;
```

```
36   LIBNAME xls_V95 EXCEL 'C:\My_excel_files\my_excel_file_V95.xls'
version=95;
NOTE: Libref XLS_V95 was successfully assigned as follows:
      Engine:         EXCEL
      Physical Name: C:\My_excel_files\my_excel_file_V95.xls
37   LIBNAME xls_V97 EXCEL 'C:\My_excel_files\my_excel_file_V97.xls'
version=97;
NOTE: Libref XLS_V97 was successfully assigned as follows:
      Engine:         EXCEL
      Physical Name: C:\My_excel_files\my_excel_file_V97.xls
38   LIBNAME xls_V07 EXCEL 'C:\My_excel_files\my_excel_file_V07.xlsb'
version=2007;
NOTE: Libref XLS_V07 was successfully assigned as follows:
      Engine:         EXCEL
      Physical Name: C:\My_excel_files\my_excel_file_V07.xlsb
39   DATA  xls_V95.'shoes'n
40       xls_V97.'shoes'n
41       xls_V07.'shoes'n;
```

```
42    Set sashelp.shoes;
43    Run;

NOTE: SAS variable labels, formats, and lengths are not written to DBMS
tables.
NOTE: SAS variable labels, formats, and lengths are not written to DBMS
tables.
NOTE: SAS variable labels, formats, and lengths are not written to DBMS
tables.
NOTE: There were 395 observations read from the data set SASHELP.SHOES.
NOTE: The data set XLS_V95.shoes has 395 observations and 7 variables.
NOTE: The data set XLS_V97.shoes has 395 observations and 7 variables.
NOTE: The data set XLS_V07.shoes has 395 observations and 7 variables.
NOTE: DATA statement used (Total process time):
      real time           0.06 seconds
      cpu time            0.04 seconds
```

## Example 4.5 Using Named Literals with the LIBNAME Statement

Now let's put some data into an Excel file. We will use what is called a named literal. This example creates a new Excel workbook with one worksheet. When we create the new worksheet, we use the name "Sheet1" to create the workbook and worksheet. When I open the SAS Explorer window, I see two items in the list of objects that SAS can view. They are "Sheet1" and "Sheet1$". See Figure 4.9.3 for an example. "Sheet1" is treated as a "Named Range". "Sheet1$" refers to all of the data on the worksheet. Either name can be selected within a DATA step SET statement or procedure call. Either sheet icon in the Explorer window should also be able to be selected and viewed by double-clicking on the icon and opening the SAS "Viewtable" window. If you had defined any named ranges in any worksheets of the Excel workbook, those named ranges would also be displayed in the SAS Explorer window. You can also view those Excel cells. Remember, when you look at these items from the Explorer window you are looking at the Excel data--not a SAS dataset. The code below shows one way to create an Excel worksheet and one way to read the same full Excel worksheet.

```
LIBNAME xls_data EXCEL 'C:\My_excel_files\my_excel_file.xls';
DATA xls_data.'sheet1'n;
    SET sashelp.shoes;
RUN;
DATA test;
   set xls_data.'sheet1$'n;
RUN;
```

To show you how complex this processing is I want to point out the following cases that you can use when you create an Excel worksheet. Three of these commands do the same thing. Can you guess which one does not work? The same conditions exist when using a SET statement to read the Excel worksheet.

**DATA Step Code**                    **Status**

- DATA xls_data.sheet1;              Valid SAS Command
- DATA xls_data.sheet1$;             Invalid SAS Command (unquoted dollar sign)
- DATA xls_data.'sheet1'n;           Valid SAS Command
- DATA xls_data.'sheet1$'n;          Valid SAS Command

### Figure 4.9.3: SAS Enhanced Editor and Explorer Windows Showing Excel Worksheet References.

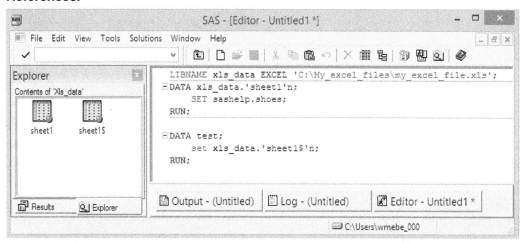

The next step, after verifying there are no errors on the log, is to look at the Excel file Figure 4.9.4 to see what was generated. Notice that the SAS variable names were used as the Excel column headers in row 1. Columns with numbers are right-justified, and columns with text values are left-justified. Multiple SAS DATA steps can write to the same Excel workbook defined by one LIBNAME statement. However, each DATA step or procedure must write to a separate worksheet. The LIBNAME statement does not support a REPLACE option.

### Figure 4.9.4: Data Written to an Excel Worksheet.

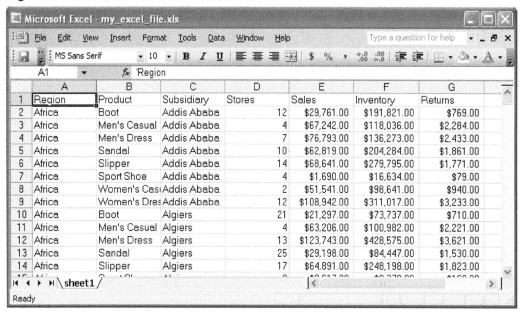

Figure 4.9.4 Excel file contents after the cells were widened to show the full dollar amounts in the columns. Also notice that the sheet name is "sheet1" not "Sheet1" because the named constant generates a case-sensitive sheet name.

## Example 4.6 Using PROC CONTENTS to Examine an Excel Workbook

Figure 4.9.5a shows the SAS code to display the contents of the Excel file we created. First, we notice that the Excel file has one worksheet, but the PROC CONTENTS output below shows two data tables, sheet1 and sheet1$. Next we ask "Why are there two sheets?" The answer is this: Excel has a feature called a named-range that is a subset of the data on an Excel worksheet. SAS accounts for this by assigning a name to the subset of used cells in the worksheet and the full worksheet. Both of names listed in the PROC CONTENTS listing point to the same data. If the workbook had a named range created by Excel, the named range would also show up in the PROC CONTENTS listing as a table. Full worksheet names end in a $ when viewed as a SAS file, while the named ranges do not have a trailing $.

```
LIBNAME xls_data EXCEL 'C:\My_excel_files\my_excel_file.xls';
   proc contents data=xls_data._all_;
   run;
```

**Figure 4.9.5a: Contents of the Created Excel File.**

```
                       The CONTENTS Procedure
                             Directory

             Libref          XLS_DATA
             Engine          EXCEL
             Physical Name   C:\My_excel_files\my_excel_file.xls
             User            Admin
                                             DBMS
                                  Member    Member
                      #   Name    Type      Type
                      1   sheet1  DATA      TABLE
                      2   sheet1$ DATA      TABLE

                       The CONTENTS Procedure

      Data Set Name          XLS_DATA.sheet1   Observations          .
      Member Type            DATA              Variables             7
      Engine                 EXCEL             Indexes               0
      Created                .                 Observation Length    0
      Last Modified          .                 Deleted Observations  0
      Protection                               Compressed           NO
      Data Set Type                            Sorted               NO
      Label
      Data Representation    Default
      Encoding               Default

             Alphabetic List of Variables and Attributes

       #   Variable    Type   Len   Format      Informat    Label
       6   Inventory   Num    8     DOLLAR21.2  DOLLAR21.2  Inventory
       2   Product     Char   14    $14.        $14.        Product
       1   Region      Char   25    $25.        $25.        Region
       7   Returns     Num    8     DOLLAR21.2  DOLLAR21.2  Returns
       5   Sales       Num    8     DOLLAR21.2  DOLLAR21.2  Sales
       4   Stores      Num    8                             Stores
       3   Subsidiary  Char   12    $12.        $12.        Subsidiary

                       The CONTENTS Procedure

      Data Set Name          XLS_DATA.'sheet1$'n  Observations       .
      Member Type            DATA                 Variables          7
      Engine                 EXCEL                Indexes            0
      Created                .                    Observation Length 0
      Last Modified          .                    Deleted Observations 0
      Protection                                  Compressed        NO
      Data Set Type                               Sorted            NO
      Label
      Data Representation    Default
      Encoding               Default
```

```
                 Alphabetic List of Variables and Attributes

        #    Variable       Type      Len     Format         Informat       Label
        6    Inventory      Num         8     DOLLAR21.2     DOLLAR21.2     Inventory
        2    Product        Char       14     $14.           $14.           Product
        1    Region         Char       25     $25.           $25.           Region
        7    Returns        Num         8     DOLLAR21.2     DOLLAR21.2     Returns
        5    Sales          Num         8     DOLLAR21.2     DOLLAR21.2     Sales
        4    Stores         Num         8                                   Stores
        3    Subsidiary     Char       12     $12.           $12.           Subsidiary
```

## Example 4.7 Using Dataset Options to Process Date and Time Values

In the next SAS code segment, the same Excel worksheet is input twice. I wanted to give an example with date and date/time processing options, but SAS has limited data in the SASHELP library with dates in the file. Therefore, I created a small file that uses dates. This file will be available for download from the Author website, http://support.sas.com/publishing/authors/benjamin.html. The file has a product name, amount, and three dates with time segments attached. It is in a CSV format and can be loaded directly into an Excel file. I do not show the full file in Figure 4.9.5b, but the examples here will give you enough detail to understand what is going on in the processing. The first code line assigns a libref xls_data so that we can read the data from the sales_data.xlsx file.

There are two DATA steps that follow and read the same Excel file to produce different results. The SAS dataset Sales_data1 was created with default options applied to all of the date fields. So, all of the date fields were read in and are displayed with a DATE9. format. In the second SAS dataset, fields are changed upon input using the SAS dataset option SASDATEFMT as shown in the list below.

| Field Name | Processing Option Applied |
|---|---|
| • Product | Default as a charcter string. |
| • Amount | Default as a numeric currency value. |
| • Date_Ordered | Default as a DATE9. Value. |
| • Date_Shipped | SASDATEFMT option converted the date/time value to a TIME8. value. |
| • Date_Delivered | SASDATEFMT option converted the Excel date to a DATETIME21.2. value |

The SAS and Excel systems both recognize dates, but each system has its own opinion about how to store and represent dates. The reason for the difference is that SAS and Excel use different base dates for their date reference point. SAS uses a base date of 01 Jan 1960, and Excel uses the Dublin Julian Day (30 DEC 1899) as the base date.

```
LIBNAME xls_data EXCEL 'C:\My_excel_files\sales_data.xlsx';

DATA Sales_data1;
    SET xls_data.'Sales_data$'n;
Run;

DATA Sales_data2;
    SET xls_data.'Sales_data$'n
        (SASDATEFMT=(Date_Shipped=TIME8. Date_Delivered=DATETIME21.2));
Run;
```

**Figure 4.9.5b: SAS View Table of the Sales_data1 File Read from Excel.**

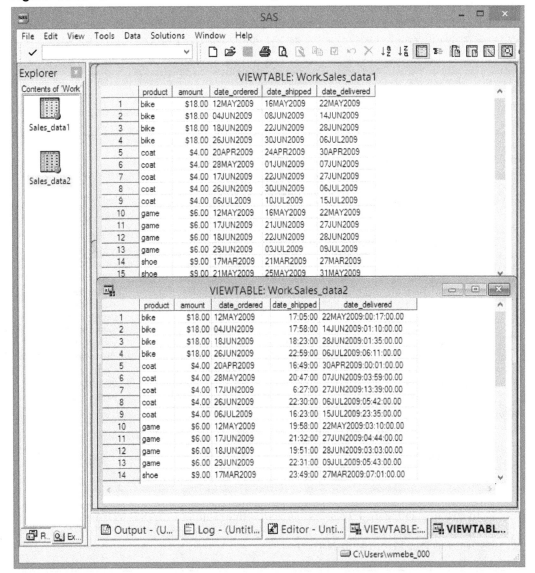

## Example 4.8 Using Dataset Options to Process Variable Type Conversions

The next segment of SAS code reads all five fields from the Sales_Data Excel file, and did nothing with the first three fields (Product, Amount, and Date_Ordered). The **DBSASTYPE=** dataset option took control of formatting the dates when processing the Date_Shipped and Date_Delivered fields.

| Field Name | Processing Option Applied |
|---|---|
| • Product | Default as a character string. |
| • Amount | Default as a numeric currency value. |
| • Date_Ordered | Default as a DATE9. value. |
| • Date_Shipped | DBSASTYPE option converted the date to a 22-byte character value. |
| • Date_Delivered | DBSASTYPE option converted the Excel date value to a real number. |

The SAS documentation for your version will point out that the SAS date format and date informat must be equivalent. This restricts the number and type of changes you can make on input from an Excel file. But, you now have control over what changes are made. The following code is similar to Example 4.7, but it changes the dates to different formats using the dataset option **DBSASTYPE=**.

```
LIBNAME xls_data EXCEL 'C:\My_excel_files\sales_data.xls';
DATA Sales_data3;
     SET xls_data.'Sales_data$'n
        (DBSASTYPE=(Date_Shipped='CHAR22.' Date_Delivered=NUMERIC));
Run;
```

**Figure 4.9.6: SAS View Table of the Sales_data1 File Read from Excel.**

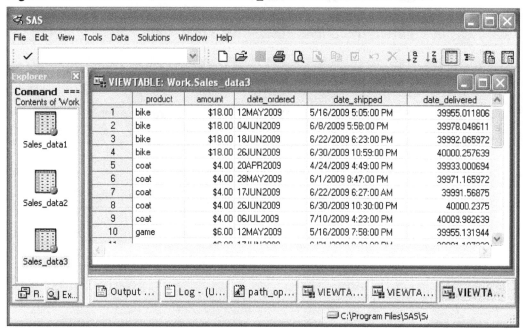

The Date_Delivered field in Figure 4.9.6 shows numeric values similar to SAS DATE-time values except the values are much larger than a SAS DATE numeric value. As mentioned above, the Excel base date is 30 DEC 1899, which is long before 01 JAN 1960.

## Example 4.9 Processing on 64-Bit Operating Systems

This example demonstrates the syntax for accessing Excel files when running SAS on a Windows 64-bit operating system. Similar syntax is required for using UNIX or LINUX operating systems. SAS/ACCESS Interface to PC Files contains a program described as the SAS PC Files Server (pcfservice.exe). This program runs as a service and can be configured to start automatically when your computer starts running. This service program runs on a Windows operating system. It can be either a PC or a server, but it must run on a Windows computer. Therefore, when UNIX and LINUX access Excel files, SAS must first log on to the Windows system before they can access an Excel workbook. The PC Files Server must be running on a Microsoft Windows platform that SAS can communicate with (i.e., send execution commands) for SAS to process Excel data. The SAS PC Files Server program was written to communicate across the 32-bit / 64-bit boundary. It was written before Excel 2010 became available as both a 32-bit and 64-bit application.

Remember, Excel 2007 and earlier are all 32-bit applications. Some of the options for UNIX and LINUX are not displayed here because they are site-specific (**CONNECT_STRING, DSN, PORT, PASSWORD, PATH, PORT, SERVER, SERVERUSER, and SSPI**).

```
LIBNAME xls_V97   PCFILES PATH='C:\My_excel_files\my_excel_file_V97.xls' ;
LIBNAME xls_V07x  PCFILES PATH='C:\My_excel_files\my_excel_file_V07.xlsx';
LIBNAME xls_V07b  PCFILES PATH='C:\My_excel_files\my_excel_file_V07.xlsb';
   DATA xls_V97.'sheet1'n
        xls_V07x.'sheet1'n    /* writes a *.xlsb formatted file not *.xlsx */
        xls_V07b.'sheet1'n;
        Set sashelp.shoes;
   RUN;
```

## 4.10 Conclusion

The SAS LIBNAME statement is a very powerful tool, with many options that are opened up by having SAS /ACCESS Interface to PC Files available. Since the release of SAS 9.2 TS Level 2M3, SAS users again have access to read and write all versions of Microsoft Excel on a 32-bit Windows system and also on the UNIX, LINUX, Windows 64-bit, and Windows 64-bit server systems. In addition, the Excel application has been upgraded to run in a 64-bit mode. The connection options that process an Excel workbook on a Windows 32-bit or 64-bit OS are critical to accessing Excel data. Not shown here: how to read and write Excel files using PROC SQL. That is reserved for Chapter 10. The connection options from Chapter 10 allow PROC SQL to find and access Excel files. This chapter looked only at the options that apply directly to Microsoft Excel files on a PC. SAS/ACCESS Interface to PC Files will also process Microsoft Access files. The LIBNAME and DATASET options described here enhance access to data files that are not SAS files. The DATASET options are available in DATA steps, in SAS procedure executions, and in PROC SQL processing, thus pushing the power of the LIBNAME engine far beyond the Import/Export engines of Base SAS. Below is a summary of the options explained. However, it is not an exhaustive list of all options available.

Engine Connection Options

- HEADER option     controls processing of the first line of Excel worksheets.
- MIXED option     controls conversion of input Excel data into character variables.
- PATH option     assists in locating files for processing.
- VERSION option     controls the output version of an Excel workbook.

Examples Using Selected Dataset Options

- Using Named Literals with the LIBNAME statement.
- Using PROC CONTENTS to Examine an Excel workbook.
- Using Dataset Options to Process Date and Time Values.
- Using Dataset Options to Process Variable Type Conversions.
- Processing on 64-bit Operating Systems.

# Chapter 5: SAS Enterprise Guide Methods and Examples

5.1 Introduction .................................................................................................. 67
5.2 Purpose ....................................................................................................... 68
5.3 Typical Methods to Access Excel from SAS Enterprise Guide ....................... 68
5.4 Overview of the Examples ........................................................................... 68
5.5 List of Examples .......................................................................................... 68
5.6 Examples ..................................................................................................... 69
    Example 5.1 Using the Export Method with Enterprise Guide ............................ 69
    Example 5.2 Using the "Send To" Method ........................................................ 71
    Example 5.3 Using the "Send To" Method to Output a Graph or Report ............ 71
    Example 5.4 Using the "Export" Method to Output a Graph or Report .............. 75
    Example 5.5 Using "Open" or "Import" Toolbar Options to Read Excel Workbooks ... 77
    Example 5.6 Using the "Import Data" Toolbar Option to Read a Range of Cells ... 80
5.7 Conclusion .................................................................................................. 84

## 5.1 Introduction

This chapter is an introduction to moving data between SAS Enterprise Guide and Microsoft Excel. SAS Enterprise Guide has a lot of features that process and analyze data. Here we are interested in a few processes that present data to and read data from Excel, and that are available when using SAS Enterprise Guide. Some of these methods are found on the File tab of the Enterprise Guide toolbar and are handled by software "wizard" tools. Many of the examples in this chapter offer step-by-step instructions with pictorial explanations of the steps needed to perform the tasks. This chapter assumes a working knowledge of SAS Enterprise Guide 4.3, 5.1, or 6.1 and shows examples of the following methods of moving data between SAS and Excel.

    Send data to    -    (Output to Excel)
    Export data to    -    (Output to Excel)
    Export a graph    -    (Output to Excel or Word)
    Open a spreadsheet    -    (Input from Excel)
    Import a spreadsheet    -    (Input from Excel)
    Import a range of cells    -    (Input from Excel)

SAS Enterprise Guide 5.1 was the first release to be available in both 32-bit and 64-bit versions. As with all of the SAS and Excel products, the bit compatibility will need to be watched to make sure the products work together. SAS Enterprise Guide establishes a Microsoft client/server method of writing data to Excel files even when executing on one computer. The Microsoft product DDE is not available to SAS Enterprise Guide. Some of the examples in this chapter rely on the SAS Add-In for Microsoft Office, which is part of the SAS Enterprise BI Server and SAS Enterprise Miner.

68 *Exchanging Data between SAS and Microsoft Excel*

**NOTE:** The examples shown in this chapter use SAS Enterprise Guide 6.1. Prior versions of SAS Enterprise Guide may have slight different interfaces. While the "Import", "Export", and "Send To" options also appear in several places within the SAS Enterprise Guide menus, each of the three options performs similarly regardless of how you start the option.

## 5.2 Purpose

This chapter describes methods of sharing data between SAS and Microsoft Excel. I have explained most of the methods previously; SAS Enterprise Guide just packages them into a wizard to standardize the processing. The overall bit configuration of the hardware and software may influence the figures and wizard options presented. This book does not deal with features of SAS Enterprise Guide other than those that permit the movement of data between Excel and SAS Enterprise Guide. SAS Enterprise Guide can use some advanced data access methods like OLEDB (Object Linking and Embedding, Database) and ODBC (Open Database Connectivity). These will not be discussed in this chapter.

## 5.3 Typical Methods to Access Excel from SAS Enterprise Guide

Common methods of accessing Excel data from within SAS Enterprise Guide include importing and exporting data from text files or Excel files. Most of the examples in this chapter deal directly with those methods. In addition to Microsoft Excel workbooks, the SAS Enterprise Guide wizards can process other types of data files like Microsoft Access database files, *.csv, *.tab, *.txt, *.htm, and *.html text files.

## 5.4 Overview of the Examples

The examples in this chapter were generated with SAS Enterprise Guide 6.1 on a system where the SAS Add-In for Microsoft Office has been installed for Microsoft Word, Excel, Power Point, and Outlook. Some of the features described here require this software. However, some older versions of SAS Enterprise Guide are able to transfer data and graphs to Microsoft Excel and Word without the SAS Add-In for Microsoft Office. While you were able to transfer data and graphs in much earlier versions of SAS Enterprise Guide, this ability was not well publicized. These features can be found by searching the tool or option menus of SAS Enterprise Guide for the Word (🗎) and Excel (🗎) icons and activating them. There are no examples in this chapter using these icons, but they are mentioned here to increase your awareness of the features present in older versions of SAS Enterprise Guide.

## 5.5 List of Examples

**Figure 5.5.1: List of Examples for Enterprise Guide.**

| Example Number | General Description |
|---|---|
| 5.1 | **Using the Export Method with Enterprise Guide.** This example shows you how to open a SAS dataset within SAS Enterprise guide and then Export the data to an Excel workbook. This example uses the "Save" option. |
| 5.2 | **Using the "Send To" Method.** This example shows how to export the data to an Excel workbook using the "Send To" option. |
| 5.3 | **Using the "Send To" Method to Output a Graph or Report.** This example shows how to export a graph or report to an Excel workbook using the "Send To" option. |
| 5.4 | **Using the "Export" Method to Output a Graph or Report.** This example shows how to export a graph or report to an Excel workbook using the "Export" method. |

| Example Number | General Description |
|---|---|
| 5.5 | **Using "Open" or "Import" Toolbar Options to Read Excel Workbooks.** This example shows how to import Excel data into SAS Enterprise Guide. |
| 5.6 | **Using the "Import Data" Toolbar Option to Read a Range of Cells.** This example shows how to import a range of Excel data cells into SAS Enterprise Guide. |

## 5.6 Examples

### Example 5.1 Using the Export Method with Enterprise Guide

This example has the following steps. Note that the "Export" option button is in three locations.

1. Select a file to output.
2. Select the "Export" option on the File tab.
3. Select the "Export <DataSetName>" option, here shown as "Export SHOES".
4. On the EXPORT window select the output file location and data format, your version of SAS Enterprise Guide may allow you to choose "Microsoft Excel 97-2003 Workbooks (.xls)" or "Microsoft Excel Workbooks (.xlsx)", or one of several text formats that Excel can read.
5. Press the "SAVE" button, close the SAS file, and open the Excel file to view the results.

After opening SAS Enterprise Guide and selecting a new project, you have many ways to open a SAS file to export the data to Excel. Figure 5.1 shows one simple method of selecting the SASHELP.shoes dataset using the SAS toolbar options: FILE▶OPEN▶DATA▶SERVERS▶LOCAL▶SASHELP▶SHOES

**Figure 5.1: SAS Enterprise Guide After Opening a SAS Dataset.**

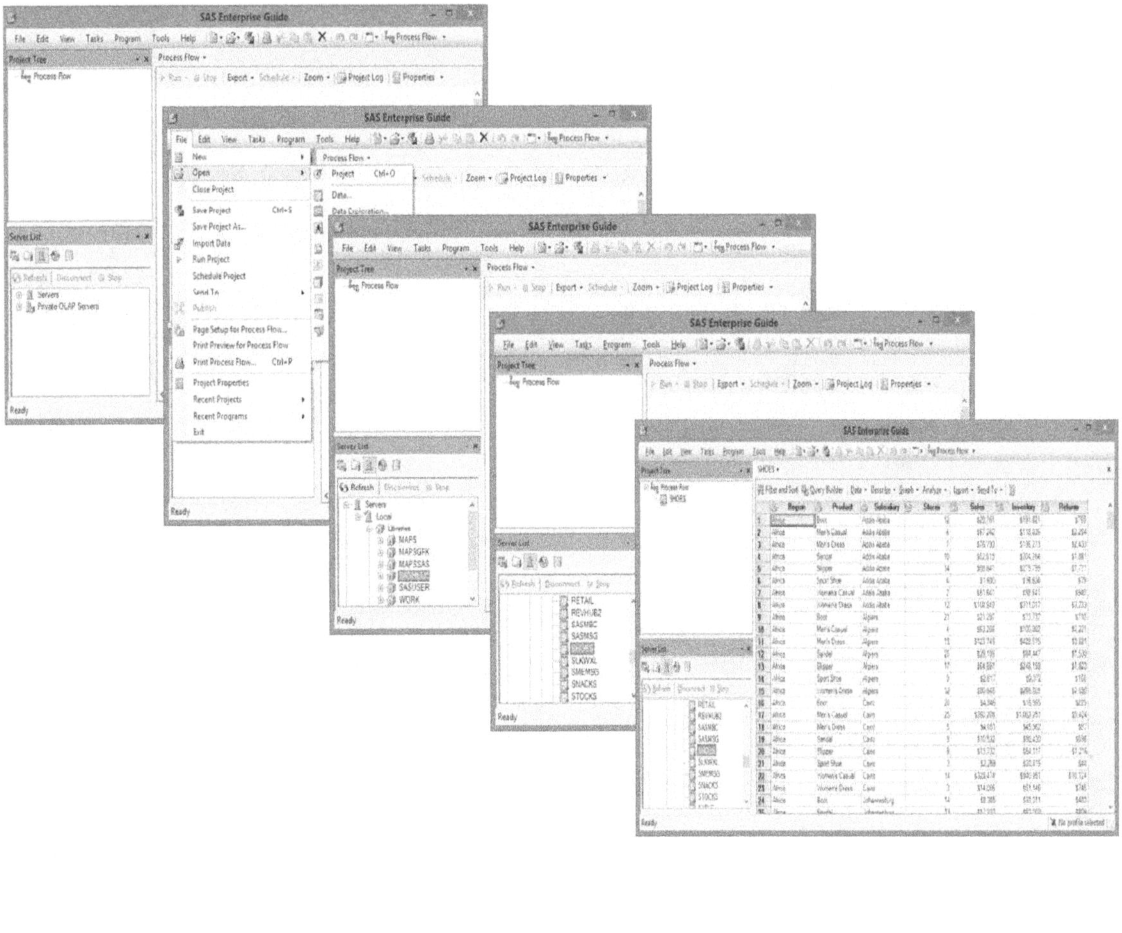

This document does not describe output options available to the Microsoft Exchange Server. SAS documentation indicates that with the Microsoft Exchange Server, Excel 2000, 2003, and 2007 can be output. (Excel 2007, Excel 2010, and Excel 2013 use the same output format). Enterprise Guide 5.1 for 64-bit SAS has the ability to write to the *.xlsx files along with some text formats that Excel can read. It worked with Excel 2013. The exact list of output file formats available will depend upon the number and type of SAS/ACCESS routines you have installed.

Four other output formats (shown in Figure 5.2) can be chosen instead of the *.xls or *.xlsx format. These options create text files, but Microsoft Excel is able to process these file formats and create an Excel workbook. These formats can also be input into Excel 2007, Excel 2010, or Excel 2013. They also can be input into Excel using methods described in Chapter 2.

**Figure 5.2: SAS Enterprise Guide "SAVE" Dialogue Window.**

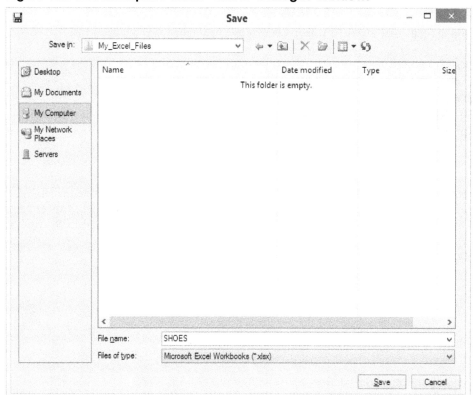

This is a list of other output file formats that Excel is able to read. Your system configuration may be able to create other formats.

- Microsoft Excel 97-2003 workbooks (.xls)
- Microsoft Excel workbooks (.xlsx)
- Text files (Comma delimited) (*.csv)
- Text files (*.txt)
- Text files (Tab delimited) (*.tab)
- Text files (Space delimited) (*.txt)

## Example 5.2 Using the "Send To" Method

This example has the following steps. Note that the "Send To" option button is in three locations.

1. Select a file to output. Any SAS dataset will do. See Example 5.1.
2. Select the "Send To" option on the File tab.
3. Select the "Microsoft Excel" option displayed on the pop-up menu.
4. Press the Microsoft Excel button to send the input SAS file directly to Microsoft Excel and to save the workbook using Excel save options. This option opens the Microsoft Excel version installed on your computer and allows you to save the output in any Excel format you have available including Excel 2007, 2010, and 2013. Be aware that this method of sending data to Excel may directly open Excel with the data displayed.

### Figure 5.3: SAS Enterprise Guide "SEND TO" Menus.

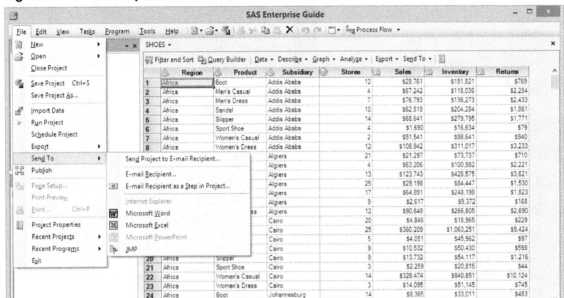

## Example 5.3 Using the "Send To" Method to Output a Graph or Report

This section assumes that you have generated a graph or report using SAS Enterprise Guide 4.3 or later. SAS Add-In for Microsoft Office is required for this example of how to send data to Excel, Word, or Power Point.

SAS Enterprise Guide 4.3 or later can send graphs and reports to Microsoft programs like Word, Excel, and Power Point. This example will focus on Excel. Figure 5.4a is an example of a graph created with the bar graph wizard. Note that the Enterprise Guide project tree also shows that a report has been generated. This report simply includes a title and a footnote. Yours will be much nicer.

**Figure 5.4a: A Graph Generated Using the Enterprise Guide Bar Graph Wizard.**

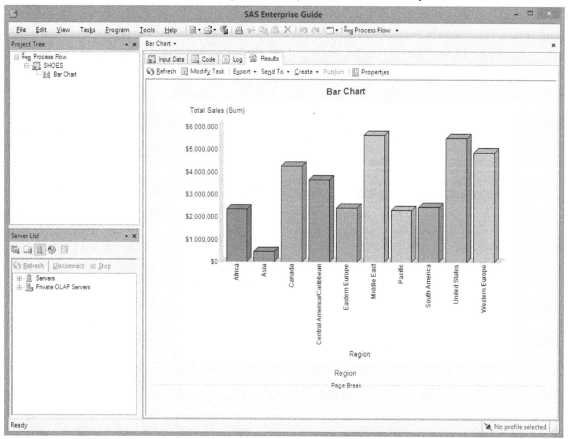

Note that the Enterprise Guide "Send To" option can be located in three places. This option allows you to send the currently selected (active) file to the requested output destination you choose with the "Send To" option menu.

1. In the Enterprise Guide toolbar under the "File" option.
2. When a graph or report is displayed in the toolbar as the "Send To" option.
3. In the project tree window by right-clicking on the "Report" you want to save. This option may not be available in Enterprise Guide 5.1. See the documentation for your installation.

**Figure 5.4b: Sending a Graph Generated Using Enterprise Guide Bar Graph Wizard to Excel.**

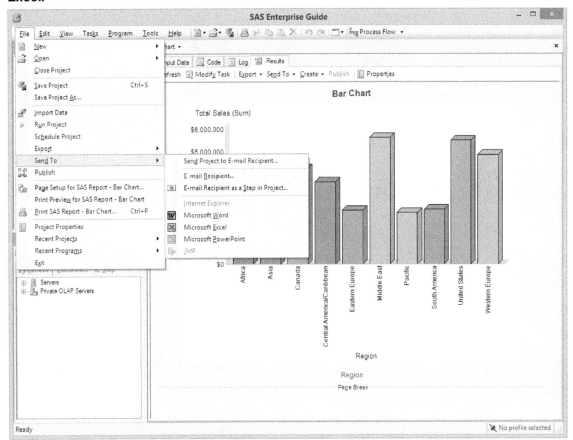

In Figure 5.4b, the "Send To" option in the Bar Chart toolbar was selected to send the graph to Excel. Different messages may appear between the time you select the Excel button (▣) and when the workbook opens.

If there are no Excel workbooks open when you are sending the graph to Excel, then SAS Enterprise Guide will send the graph to "Book$n$" in your default "Documents" folder. If "Book1" does not exist, it will create "Book1". Otherwise, "$n$" will increment to the next available number. When a Microsoft Excel workbook is open, you are given a choice of creating a new workbook (in your documents folder), sending the graph to an open workbook, or opening a workbook to receive the output. The end result can be seen in Figure 5.5. Depending upon your system default settings, your version of SAS Enterprise Guide might even open the Excel workbook that is created.

**Figure 5.5: The Resulting Graph in an Excel Workbook.**

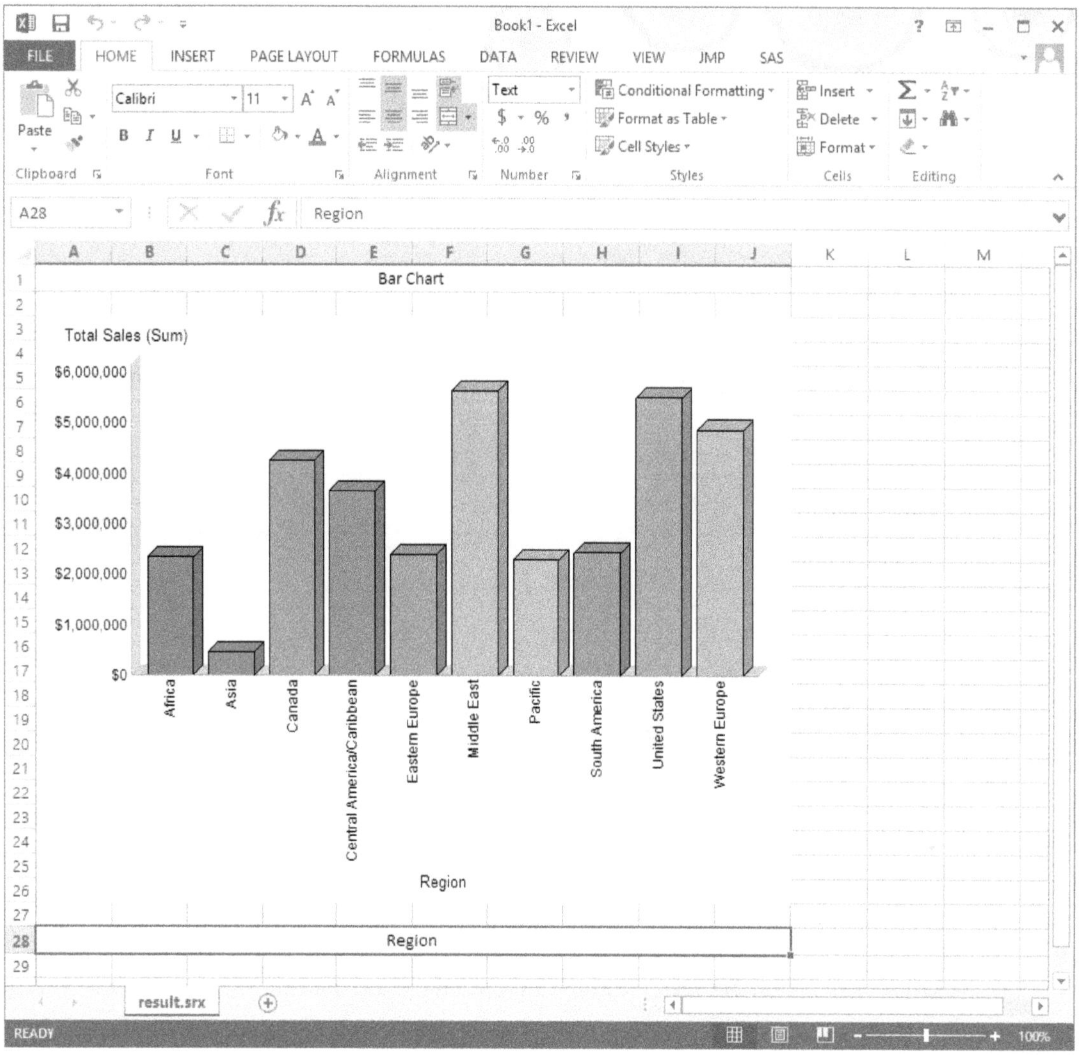

Also note that the graph is an "ActiveX" enabled graph. While it is outside of the scope of this book to explain those features. An ActiveX-enabled graph can be manipulated by selecting parts of the graph, right clicking, and selecting from a menu to modify the graph. The image in Figure 5.6 shows the result of right clicking on the graph and selecting the "Pie Chart" option. Another option is "Copy", which will allow the graph to be placed into the Windows "Copy Buffer" so that you can "Paste" it wherever it is needed.

**Figure 5.6: The Result of Changing an ActiveX Bar Chart to a Pie Chart in an Excel Workbook.**

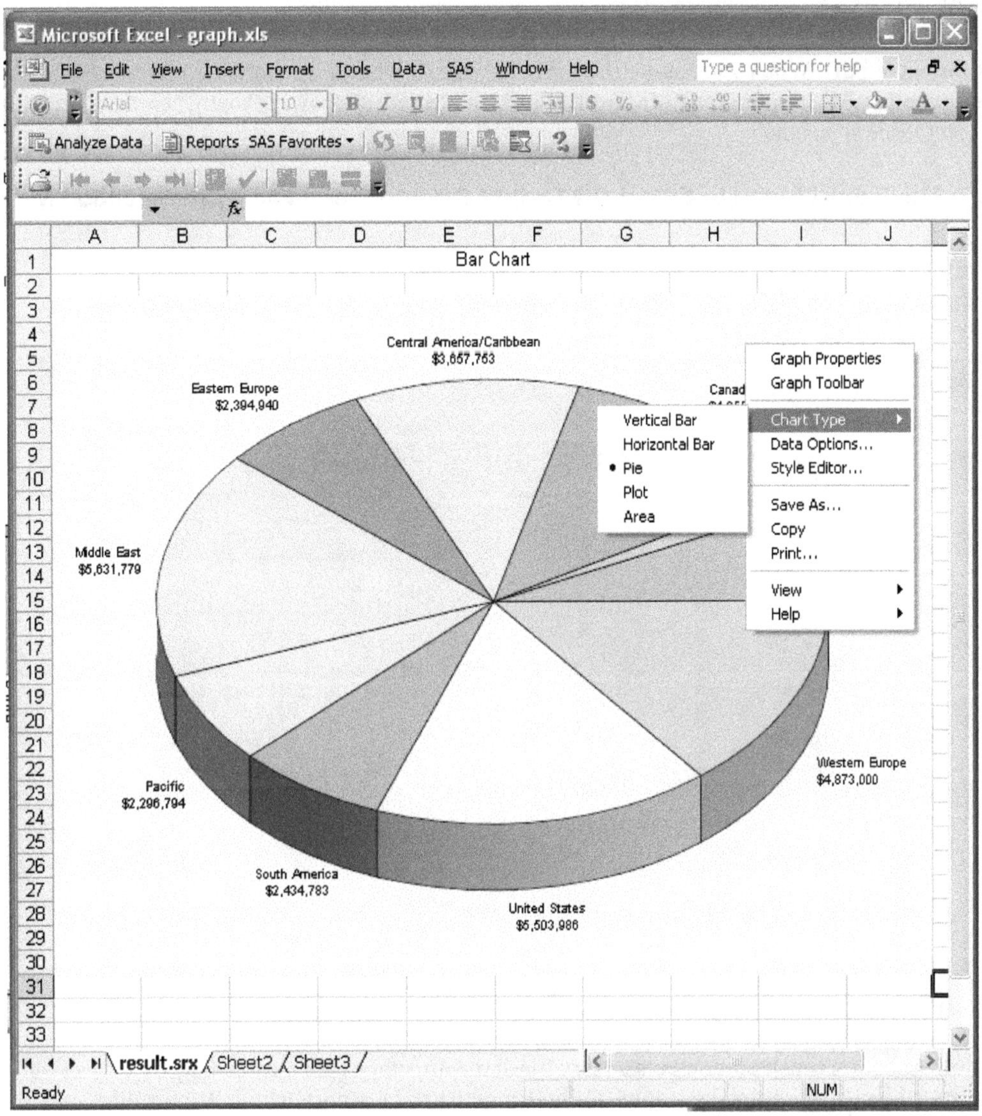

The Excel version shown in Figure 5.6 is an older version of Excel than the image in Figure 5.5. I used the older version because I was able to show the ActiveX command menus in this image. A SAS Enterprise Guide report can be processed in the same manner as the graph above when you have the SAS Add-In for Microsoft Office installed.

## Example 5.4 Using the "Export" Method to Output a Graph or Report

This example shows how SAS can create a graph or report file to export to Microsoft Excel. SAS Add-In for Microsoft Office must be installed. This section also assumes that you have generated a graph or report using SAS Enterprise Guide 4.3 or later. The next chapter will deal directly with the features of SAS Add-In for Microsoft Office.

Choosing the "Export" option (See Figure 5.7) on the graph or report toolbar or the File tab of the Enterprise Guide toolbar will open the Export window where you select the output directory to which you write your files. The SAS Enterprise Guide Export option can generate different types of output files, including SAS report files (*.srx), HTML files (*.htm, and *.html) XML files (*.xml), and portable document format (*.pdf). When you choose the *.srx file type, SAS will generate a SAS report file. This file can be opened by Microsoft Excel, Microsoft Word, and Microsoft Power Point when the SAS Add-In for Microsoft Office is installed.

**Figure 5.7: The Export Menu to Select the Output Directory and File Type Selection Menu.**

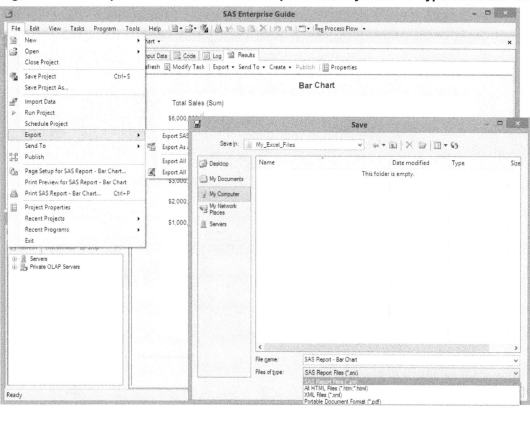

When the Export option finishes, it outputs a *.srx file and a directory with a cascading style sheet for each object exported. Figure 5.8 shows the outputs for both a graph and a report output. A cascading style sheet is a file or a series of commands that tells the program opening the *.srx file how to format and color each part of the file being opened. In this case, the cascading style sheet will describe our bar chart generated by SAS Enterprise Guide. Because not all computers are configured the same way, the cascading style sheet will usually allow for several fonts to be acceptable when displaying the output graph. This is why the same input file may look different on two different computers.

**Figure 5.8: The Directories and Files Output by the EXPORT Option for *.srx Files.**

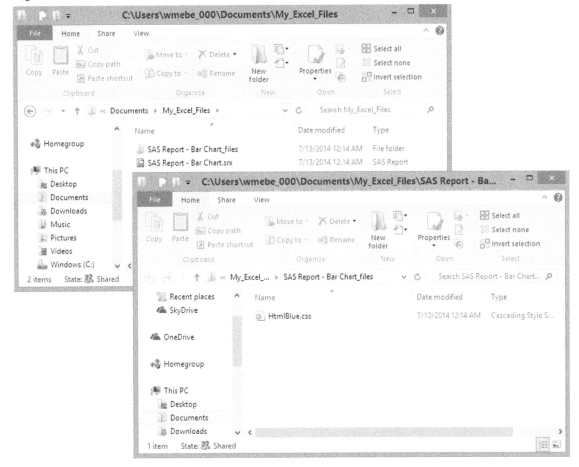

## Example 5.5 Using "Open" or "Import" Toolbar Options to Read Excel Workbooks

This example shows how to read an Excel workbook using the Open or Import options on the SAS Enterprise Guide toolbar. The example has the following steps. Note that the "Open" and "Import" option buttons are in two places.

1. Select either the "Import" or "Open" option on the File tab.
2. Select a file using the "Open" file dialogue box.
3. Follow the instructions on the "Import Data" screens that are presented.
4. Press the "Finish" button to read the data into a SAS dataset.

**NOTE:** Once again, the user is cautioned that this process uses the Microsoft JET or ACE engine to read data from Excel and is limited to 255 columns and 65,535 rows of data.

**78** *Exchanging Data between SAS and Microsoft Excel*

Both the Open and Import options will begin the Import Wizard. When you select either of the toolbar tab sequences shown in Figure 5.9 "FILE▶OPEN▶DATA▶MY_COMPUTER" or "FILE▶IMPORT DATA▶MY_COMPUTER", a window similar to Figure 5.10 will appear.

### Figure 5.9: The File Tab Menu Showing the "Open" and "Import" Options.

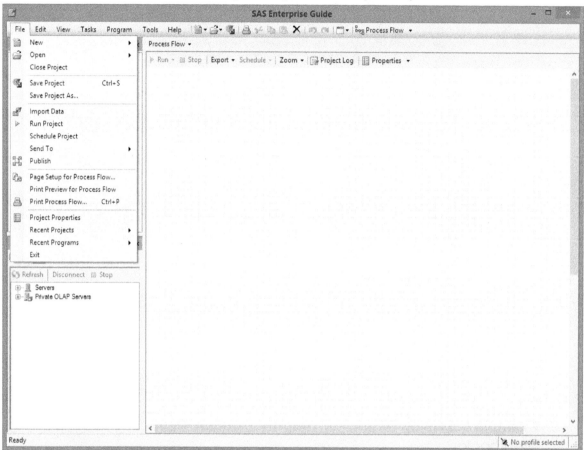

Chapter 5: SAS Enterprise Guide Methods and Examples   79

**Figure 5.10: Screens Showing the Input Directory and Input File Types.**

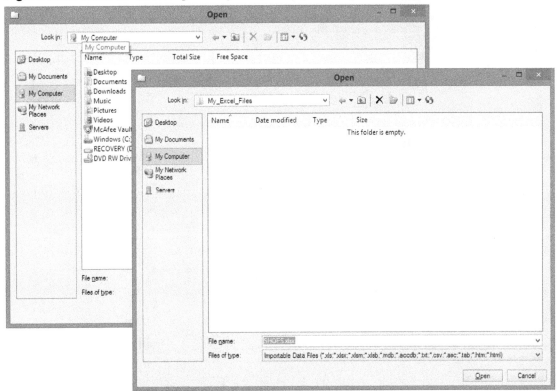

**Figure 5.11: Four Screens Showing the SAS Wizard to Read and Select Data from an Input Excel File.**

Figure 5.11 shows the screens that comprise the Import wizard, and allow you to read data into SAS Enterprise Guide, as shown in Figure 5.12.

**Figure 5.12: The SAS Dataset Generated Using the Import Wizard.**

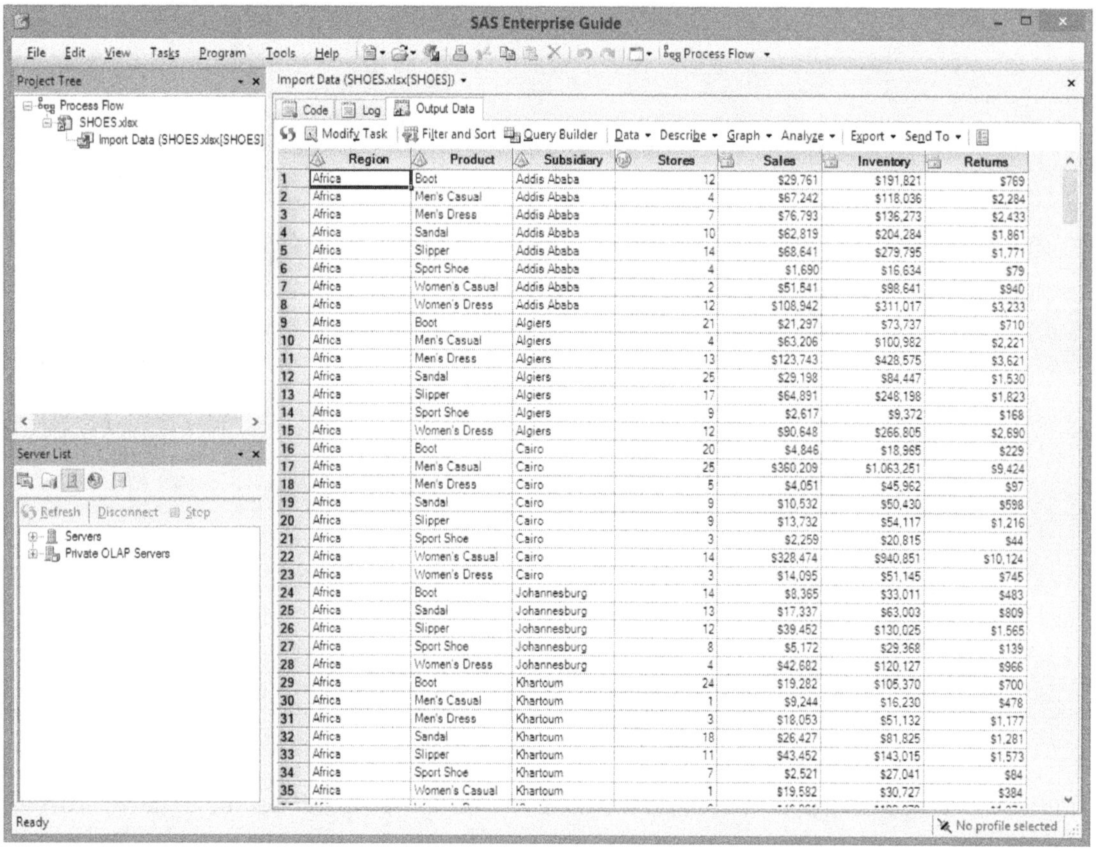


| | Region | Product | Subsidiary | Stores | Sales | Inventory | Returns |
|---|---|---|---|---|---|---|---|
| 1 | Africa | Boot | Addis Ababa | 12 | $29,761 | $191,821 | $769 |
| 2 | Africa | Men's Casual | Addis Ababa | 4 | $67,242 | $118,036 | $2,284 |
| 3 | Africa | Men's Dress | Addis Ababa | 7 | $76,793 | $136,273 | $2,433 |
| 4 | Africa | Sandal | Addis Ababa | 10 | $62,819 | $204,284 | $1,861 |
| 5 | Africa | Slipper | Addis Ababa | 14 | $68,641 | $279,795 | $1,771 |
| 6 | Africa | Sport Shoe | Addis Ababa | 4 | $1,690 | $16,634 | $79 |
| 7 | Africa | Women's Casual | Addis Ababa | 2 | $51,541 | $98,641 | $940 |
| 8 | Africa | Women's Dress | Addis Ababa | 12 | $108,942 | $311,017 | $3,233 |
| 9 | Africa | Boot | Algiers | 21 | $21,297 | $73,737 | $710 |
| 10 | Africa | Men's Casual | Algiers | 4 | $63,206 | $100,982 | $2,221 |
| 11 | Africa | Men's Dress | Algiers | 13 | $123,743 | $428,575 | $3,621 |
| 12 | Africa | Sandal | Algiers | 25 | $29,198 | $84,447 | $1,530 |
| 13 | Africa | Slipper | Algiers | 17 | $64,891 | $248,198 | $1,823 |
| 14 | Africa | Sport Shoe | Algiers | 9 | $2,617 | $9,372 | $168 |
| 15 | Africa | Women's Dress | Algiers | 12 | $90,648 | $266,805 | $2,690 |
| 16 | Africa | Boot | Cairo | 20 | $4,846 | $18,965 | $229 |
| 17 | Africa | Men's Casual | Cairo | 25 | $360,209 | $1,063,251 | $9,424 |
| 18 | Africa | Men's Dress | Cairo | 5 | $4,051 | $45,962 | $97 |
| 19 | Africa | Sandal | Cairo | 9 | $10,532 | $50,430 | $598 |
| 20 | Africa | Slipper | Cairo | 9 | $13,732 | $54,117 | $1,216 |
| 21 | Africa | Sport Shoe | Cairo | 3 | $2,259 | $20,815 | $44 |
| 22 | Africa | Women's Casual | Cairo | 14 | $328,474 | $940,851 | $10,124 |
| 23 | Africa | Women's Dress | Cairo | 3 | $14,095 | $51,145 | $745 |
| 24 | Africa | Boot | Johannesburg | 14 | $8,365 | $33,011 | $483 |
| 25 | Africa | Sandal | Johannesburg | 13 | $17,337 | $63,003 | $809 |
| 26 | Africa | Slipper | Johannesburg | 12 | $39,452 | $130,025 | $1,565 |
| 27 | Africa | Sport Shoe | Johannesburg | 8 | $5,172 | $29,368 | $139 |
| 28 | Africa | Women's Dress | Johannesburg | 4 | $42,682 | $120,127 | $966 |
| 29 | Africa | Boot | Khartoum | 24 | $19,282 | $105,370 | $700 |
| 30 | Africa | Men's Casual | Khartoum | 1 | $9,244 | $16,230 | $478 |
| 31 | Africa | Men's Dress | Khartoum | 3 | $18,053 | $51,132 | $1,177 |
| 32 | Africa | Sandal | Khartoum | 18 | $26,427 | $81,825 | $1,281 |
| 33 | Africa | Slipper | Khartoum | 11 | $43,452 | $143,015 | $1,573 |
| 34 | Africa | Sport Shoe | Khartoum | 7 | $2,521 | $27,041 | $84 |
| 35 | Africa | Women's Casual | Khartoum | 1 | $19,582 | $30,727 | $384 |

## Example 5.6 Using the "Import Data" Toolbar Option to Read a Range of Cells

While this example is similar to Example 5.5, it is worth showing because of the differences in the input and output values. An Excel file has embedded information about the nature of the file. This tool will analyze the input file and provide information about the file structure. This example shows that the Import Data tool can recognize a named range within an Excel file and extract only those cells from the spreadsheet. In addition, you can choose to keep or drop fields within the range before reading the data into a SAS dataset. Figure 5.13 shows how to use a predetermined named range as input. Excel named ranges can be generated using the "FORMULA" tab and the "Name Manager" option shown in Figure 5.13. The highlighted selection is entitled Addis_Ababa and is part of the "Shoes" worksheet that starts with the upper left cell (C2) and includes all of the cells in the rows and columns that are between C2 to G9. This is a group of 40 cells.

**Figure 5.13: Excel Workbook Showing a Named Range called Addis_Ababa.**

| | A | B | C | D | E | F | G |
|---|---|---|---|---|---|---|---|
| 1 | Region | Product | Subsidiary | Number of Stores | Total Sales | Total Inventory | Total Returns |
| 2 | Africa | Boot | Addis Ababa | 12 | 29,761.00 | 191,821.00 | 769 |
| 3 | Africa | Men's Casual | Addis Ababa | 4 | 67,242.00 | 118,036.00 | 2,284.00 |
| 4 | Africa | Men's Dress | Addis Ababa | 7 | 76,793.00 | 136,273.00 | 2,433.00 |
| 5 | Africa | Sandal | Addis Ababa | 10 | 62,819.00 | 204,284.00 | 1,861.00 |
| 6 | Africa | Slipper | Addis Ababa | 14 | 68,641.00 | 279,795.00 | 1,771.00 |
| 7 | Africa | Sport Shoe | Addis Ababa | 4 | 1,690.00 | 16,634.00 | 79 |
| 8 | Africa | Women's Casual | Addis Ababa | 2 | 51,541.00 | 98,641.00 | 940 |
| 9 | Africa | Women's Dress | Addis Ababa | 12 | 108,942.00 | 311,017.00 | 3,233.00 |
| 10 | Africa | Boot | Algiers | 21 | 21,297.00 | 73,737.00 | 710 |
| 11 | Africa | Men's Casual | Algiers | 4 | 63,206.00 | 100,982.00 | 2,221.00 |
| 12 | Africa | Men's Dress | Algiers | 13 | 123,743.00 | 428,575.00 | 3,621.00 |
| 13 | Africa | Sandal | Algiers | 25 | 29,198.00 | 84,447.00 | 1,530.00 |
| 14 | Africa | Slipper | Algiers | 17 | 64,891.00 | 248,198.00 | 1,823.00 |
| 15 | Africa | Sport Shoe | Algiers | 9 | 2,617.00 | 9,372.00 | 168 |
| 16 | Africa | Women's Dress | Algiers | 12 | 90,648.00 | 266,805.00 | 2,690.00 |
| 17 | Africa | Boot | Cairo | 20 | 4,846.00 | 18,965.00 | 229 |

Although Example 5.5 is the basis of this example, there are several differences are between it and this example. The first difference occurs within the Import Wizard when showing screen 2 of 4 of the SAS Import Wizard (Figure 5.14). Notice that the "Use a predetermined named range" button is selected and that the "First row of range contains field names" checkbox is not selected. Some versions of SAS Enterprise Guide may also show a names range that would appear something like "SHOES$(SHOES$A1:G396)". This occurs because of the way that older versions of SAS opened Excel workbooks and assigned a named range to all active cells of a worksheet. For you to be able to select data from the Excel workbook, the data cells you wish to import must be within a named range.

**Figure 5.14: SAS Import Wizard Screen 2 of 4.**

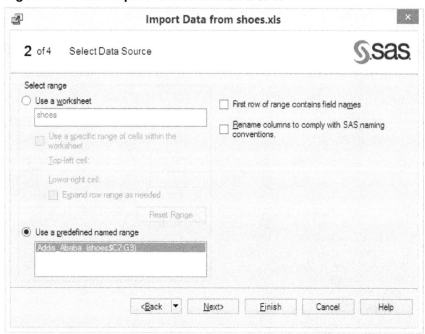

**Figure 5.15: SAS Import Wizard Screen 3 of 4.**

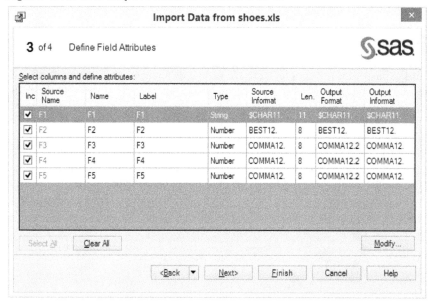

Screen 3 of the Import Data task allows you to select or reject fields, and modify any of the attributes that were found when the tool analyzed the input file. The check mark in the "Inc" column will include the variable in the output file. As noted above, notice that Figure 5.14 does not have the checkbox "First row of range contains field names" checked. The Import Data Wizard detected the absence of field names and supplied the names F1, F2, F3, F4, and F5. When you highlight any row and press the "Modify" button, Figure 5.16 will be displayed and you can make changes to the attributes of the variable. That will be left as an exercise for you to play with. The original field names were "Subsidiary", "Stores", "Sales", "Inventory", and "Returns". Figure 5.16 shows the unchanged values.

**Figure 5.16: Screen to Modify the Field Attributes of Input Variables.**

The rest of the import process is the same as Example 5.5. In Figure 5.17, the imported data is displayed without modifying any of the variable names or labels.

**Figure 5.17: SAS Enterprise Guide Resulting Data File.**

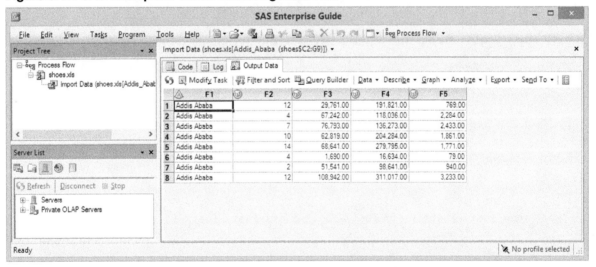

## 5.7 Conclusion

Within this chapter I have shown how SAS Enterprise guide is able to import and export data to and from Microsoft Excel using these processes.

| | | |
|---|---|---|
| Send data to | - | (Output to Excel) |
| Export data to | - | (Output to Excel) |
| Export a graph | - | (Output to Excel or Word) |
| Open a Spreadsheet | - | (Input from Excel) |
| Import a Spreadsheet | - | (Input from Excel) |
| Import a range of cells | - | (Input from Excel) |

All SAS Enterprise Guide versions are excellent tools that give the SAS user many ways to process data with both graphical interfaces and SAS program code. This chapter just touched upon a few of the methods that allow SAS Enterprise Guide to interact with Microsoft Office products. These are the kinds of tools that allow a team of SAS programmers to easily provide managers with data in Excel format.

# Chapter 6: Using JMP to Share Data with Excel

| | |
|---|---|
| 6.1 Introduction | 85 |
| 6.2 Purpose | 85 |
| 6.3 Methods of Sharing Data between JMP and Excel | 86 |
| 6.4 List of Examples | 87 |
| 6.5 Examples | 87 |
|     Example 6.1 Within Excel, Set the JMP Preferences for Loading Excel Data | 87 |
|     Example 6.2 Reading Data from Excel to JMP | 88 |
|     Example 6.3 Writing Data from JMP to Excel | 89 |
| 6.6 Conclusion | 90 |

## 6.1 Introduction

JMP is a SAS product that works like a spreadsheet. I apologize to all JMP users who believe differently, but from my vantage point as a SAS and Excel user, that is my first impression. Since JMP is an independent product from SAS that seems to run with or without both Excel and SAS, that is what I see. But, just because something *"works like"* something else, that does not mean that the two are always and forever just alike. I am impressed by JMP and, after having started my career using punched cards, a sorter, hard-wired plug boards, and paper tape, it takes a lot to impress me. There are a lot of good books written about JMP; the only thing I have to add here is how the tab on the ribbon in Excel helps get data into JMP. While any data that is in an Excel spreadsheet will work just fine, we will build upon the data files built in a previous chapter about using the SAS Add-In for Microsoft Office. We will do that because there is a direct path from SAS to Excel to JMP. It is also done all from within Excel, as though you are using one tool.

## 6.2 Purpose

Once again, like many other chapters in this book, the object is not to do anything with the data, but just to find and move the data. We will assume that we have already spent hours (well, a few seconds anyway) moving the "Shoes" data from a SAS dataset into Excel. Figure 6.1a shows the result of all that hard work. The data from the SASHELP.SHOES dataset has been copied into a Microsoft Excel workbook, and filters have been added to the Excel columns. Figure 6.1b shows the expanded "JMP" tab on the Excel ribbon. This chapter will examine the contents of the JMP tab.

**Figure 6.1a: Excel Workbook of SHOES Data with Excel Filters Applied.**

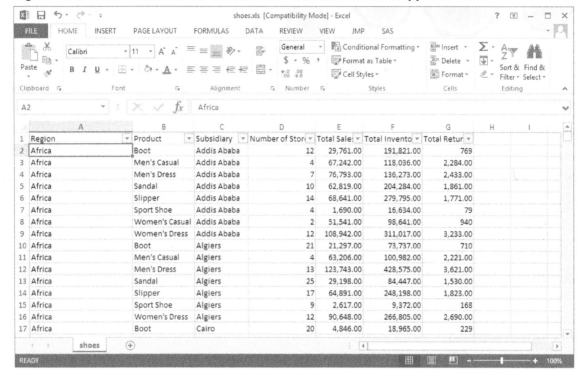

## 6.3 Methods of Sharing Data between JMP and Excel

The following table shows several methods of moving data between JMP and Excel. This list of options is found by selecting the JMP tab on the Excel ribbon along with a short description of their functions. Other options exist on the tabs, but do not relate to moving data between JMP and Excel.

**Table 6.1: Excel Workbook Options under the JMP Tab That Deal with Excel Data and JMP.**

| Group | Option | Function |
|---|---|---|
| Transfer to JMP | Preferences | Identify Excel data to load into JMP. |
| | Data Table | Load data into a JMP table; this will also start JMP if it is not running. |
| | Graph Builder | Transfer Excel data and launch the Graph Builder platform. |
| | Distribution | Transfer Excel data and launch the Distribution platform. |
| Profile in JMP | Create/Edit Model | Set up preferences in the JMP Profiler with Excel data. |
| | Run Model | Run the JMP Profiler. |

**Figure 6.1b: Excel Workbook with JMP Tab Showing "Transfer to JMP" and "Profile in JMP".**

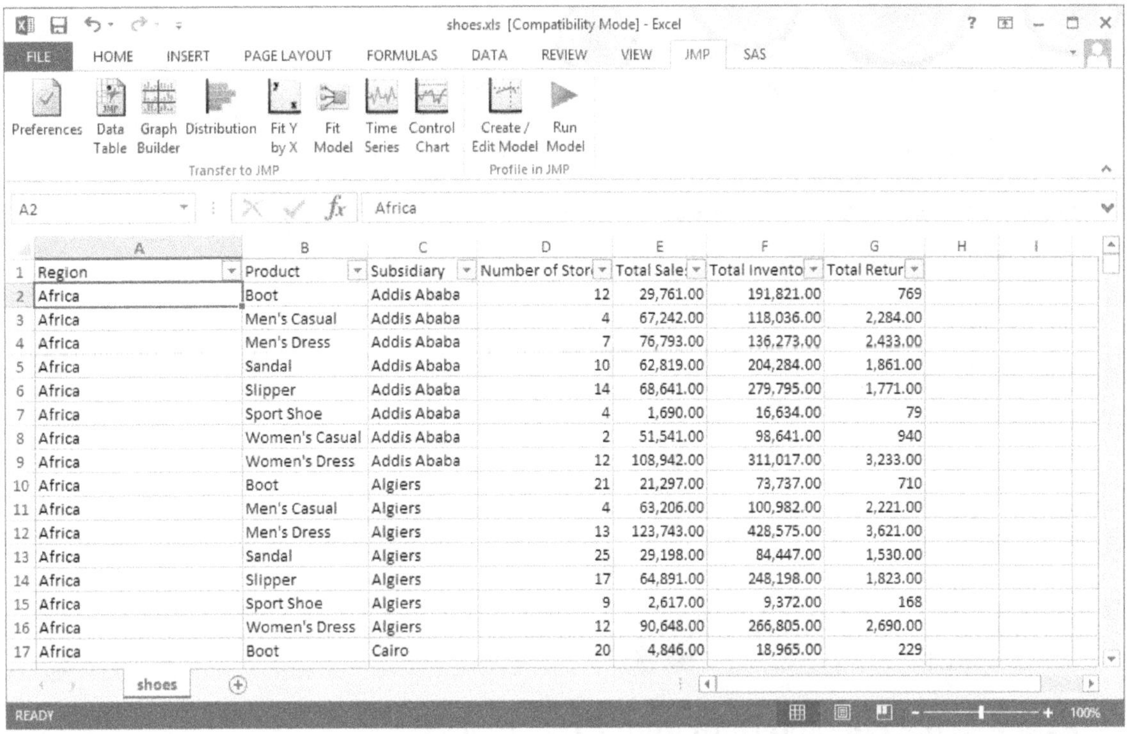

## 6.4 List of Examples

**Figure 6.6.1: List of Examples for JMP.**

| Example Number | General Description |
| --- | --- |
| 6.1 | **Within Excel, Set the JMP Preferences for Loading Excel Data.** This example shows you how to set up your preferences that describe how you want to read data into JMP from Excel workbooks. You can select to process the first row of data as headers or data, and whether or not to transfer hidden rows and columns. |
| 6.2 | **Reading Data from Excel to JMP.** This example shows how to read data from Excel into JMP. |
| 6.3 | **Writing Data from JMP to Excel.** This example shows how to write data from JMP to Excel. |

## 6.5 Examples

### Example 6.1 Within Excel, Set the JMP Preferences for Loading Excel Data

After data is loaded into an Excel spreadsheet, the JMP Preferences pop-up menu gives you the following choices about how to select your data for JMP, make these choices before copying any data to JMP.

**Preferences:**
- Select the output JMP table name.
- Choose whether or not to use the first rows of Excel cells as column names.
- Choose how many rows to use as column names.

- Turn on or off the transfer of hidden rows.
- Turn on or off the transfer of hidden columns.

Figure 6.2 shows the setting options of the JMP Preferences window. Your actual display may differ from these settings. Use this window to set the preferences listed above. The data table name "My_JMP_Shoes_data" is the user-supplied output JMP data table name. As listed above, there are four other options that you can set when you transfer data from Excel to JMP.

**Figure 6.2: Excel JMP Preferences Window Showing User-Supplied Values.**

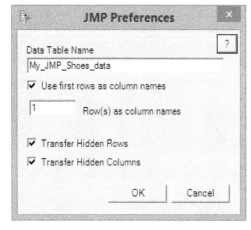

## Example 6.2 Reading Data from Excel to JMP

This process also occurs while using Excel and the JMP tab on Windows. Next, use the Data Table menu option to transfer data to JMP. Selecting that menu option produces the following when using JMP 11. The JMP program starts, and the JMP Home window (Figure 6.3) appears along with the JMP Data window that displays the imported data file (Figure 6.4).

**Figure 6.3: JMP Home Window.**

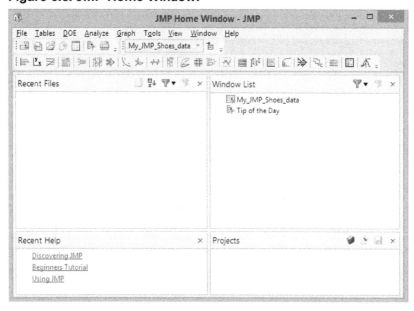

Figure 6.4 is also a JMP window; it looks similar to the Excel windows with data columns and rows, but has many extra options and features not available in Excel. Once again, this book is about moving data, not using the data.

**Figure 6.4: JMP Data Window.**

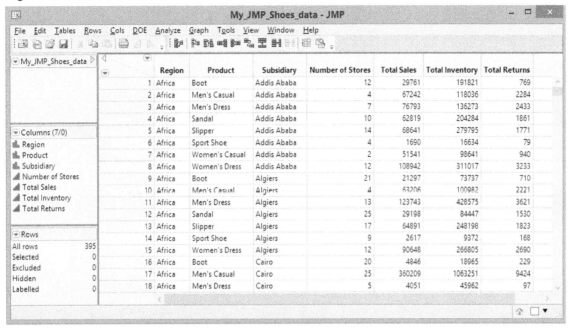

## Example 6.3 Writing Data from JMP to Excel

This example will demonstrate how to send data from JMP to Excel (and other file types we have seen before). Figure 6.4 shows the JMP data window for JMP 11.1 from Example 6.2. The "File" tab has a "SEND" option and, when it is opened, the "Save JMP File As" menu opens and looks similar to Figure 6.5. The output files can be created as Excel files, comma-separated value files (CSV), tab-delimited files (TSV), and text files (TXT). Other SAS output formats are also available; and you can select the older Excel *.xls format for Excel 97-2003 formats.

**Figure 6.5: JMP Save Data Window.**

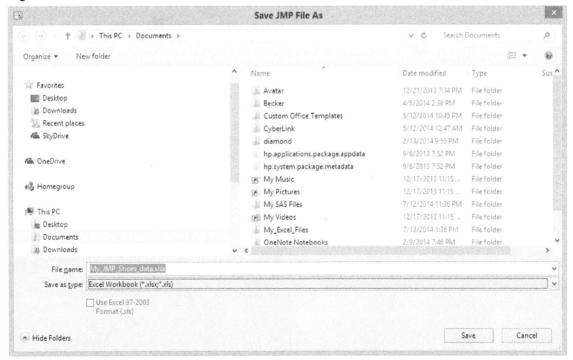

## 6.6 Conclusion

JMP may also open up four windows: a home window, a JMP log window, a "Tip of the Day" window, and a data window. The number of windows that open when you select an option may vary if you have already seen the tip of the day or if you have more JMP data files open. Once the data is in the JMP data window, there are several options that relate to sending it somewhere else. You can save the file and any changes made; you can also save the file as something else. Or, if you have the proper metadata connections established, the data can be sent to any of the metadata connections. No JMP application or metadata examples are shown here because the object of this book is moving the data, not how you use the data.

While many JMP features require a metadata connection, it is possible to use JMP without that connection because you are able to access the data. That is true either because it is output by SAS to an Excel file or because JMP can read a file formatted by SAS. The object of this chapter is to show that real SAS dataset files can be processed through Excel to hand off the data to JMP.

# Chapter 7: SAS Add-In for Microsoft Office (Excel)

7.1 Introduction..................................................................................................... 91
7.2 Purpose........................................................................................................... 91
7.3 Methods of Sharing Data Using SAS Add-In for Microsoft Office.................... 92
7.4 List of Examples.............................................................................................. 94
7.5 Examples......................................................................................................... 94
    Example 7.1 Open a SAS Dataset Using SAS Add-In for Microsoft Office.....................94
    Example 7.2 Open a SAS Report Dataset (*.srx) Using SAS Add-In for Microsoft Office..................................................................................................................99
7.6 Conclusion ..................................................................................................... 105

## 7.1 Introduction

The software architecture of the Microsoft products incorporate something called a Component Object Model (COM) software package. The SAS Add-In for Microsoft Office is a COM package. Once installed into Microsoft Excel and other Microsoft products, the SAS Add-In for Microsoft Office gives you access to SAS features directly from Excel. The SAS Add-In for Microsoft Office is part of SAS Enterprise Guide, SAS Enterprise Business Intelligence Server, and SAS Enterprise Miner. These are separately available SAS products.

The ability to use Excel to access SAS data will be described here. My system has these products installed, so I will show you some of the methods that Excel can use to access SAS data using the SAS Add-In for Microsoft Office. This software extends the functionality of Microsoft Office products using the add-in by allowing direct access to SAS datasets from within the Microsoft products.

Here we will demonstrate how you can identify and use SAS data from within Excel to perform limited Read and Write functions from Excel with SAS Add-In for Microsoft Office 6.1. Specifically, SAS Add-In for Microsoft Office can function and process SAS datasets without using a metadata server, as long as SAS Enterprise Business Intelligence Server and SAS Enterprise Miner are installed.

## 7.2 Purpose

This chapter will introduce you to the features of SAS Add-In for Microsoft Office as it relates to reading and writing SAS data to and from Excel workbooks. Once the data is in the workbook or out of the workbook, then my task of showing you how to move the data is complete.

**Table 7.1: SAS Add-In for Microsoft Office Task Availability, by Product and Task.**

| Product | Task | Process SAS Data | Run SAS Tasks | Run SAS Reports | Share SAS Reports |
|---|---|---|---|---|---|
| Excel | | Yes | Yes | Yes | No |
| Word | | No | Yes | Yes | No |
| PowerPoint | | No | Yes | Yes | No |
| Outlook | | No | No | No | Yes |

## 7.3 Methods of Sharing Data Using SAS Add-In for Microsoft Office

The very first interaction with SAS Add-In for Microsoft Products that your computer, but probably not you, will have is when your SAS software is installed. During deployment of SAS software, one of the questions asked is this: Into which Microsoft products should I install the SAS Add-In for Microsoft Office? The default answer for this question is to install the add-in into Microsoft Excel, Microsoft Word, and Microsoft PowerPoint. In shown in Figure 7.1, the option to install the add-in into Microsoft Outlook is also selected. Since your system administrator set up your machine, this may not be the case on your computer.

**Figure 7.1: SAS V9.3 Deployment Wizard Showing Options for SAS Add-In for Microsoft Office.**

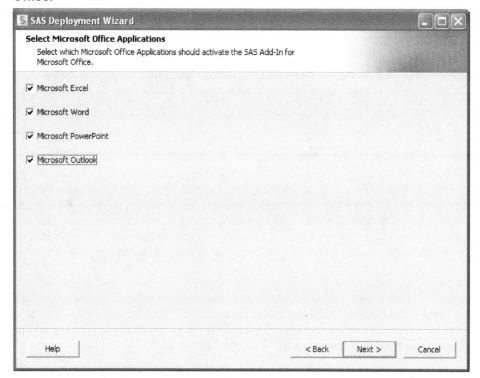

When the SAS Add-In for Microsoft Office is installed, a new tab appears on the toolbar or ribbon of the Microsoft Office products that can use the add-in. If your version of Excel (or Word, PowerPoint, or Outlook) does not have a SAS tab, then stop here. You do not have the SAS Add-In for Microsoft Office installed, and cannot use these features.

When we click on the SAS tab on the Excel toolbar or ribbon, we will see a screen similar to Figure 7.2. This figure was generated using Windows XP Professional with Excel 2010 that had SAS 9.3 and SAS Add-In for Microsoft Office 4.3 installed.

**Figure 7.2: Microsoft Excel with the SAS Tab on the Ribbon for the SAS Add-In for Microsoft Office.**

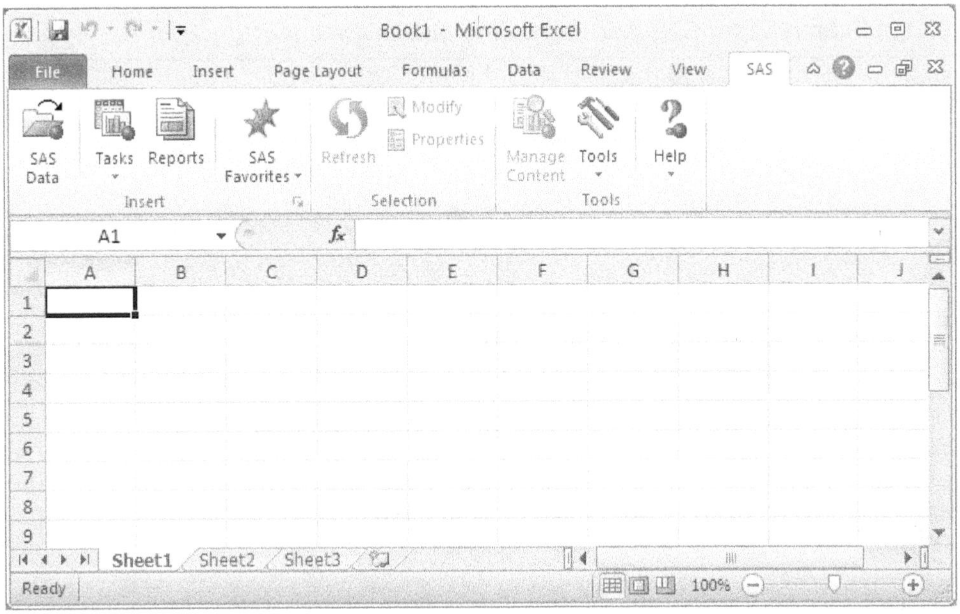

Table 7.2 shows the features available on the SAS tab for Microsoft Excel, but the other Microsoft products (Word, PowerPoint, and Outlook) all have different sets of features available. Microsoft Excel is the only Microsoft product that can open a data source. A data source is either a SAS dataset or an OLAP cube.

**NOTE:** The features described here do not require the activation of SAS Business Intelligence software, but do require a license for SAS Business Intelligence software.

**Table 7.2: SAS Add-In for Microsoft Office (Excel) Tab Options.**

| Group | Option | Function |
|---|---|---|
| Insert | SAS Data | Opens a SAS data source. |
| | Tasks | Performs one of over 75 options including sorting, transposing, and graphing. |
| | Reports | Allows searching the network and executing available *.srx report routines. |
| | Quick Start | Allows for the selection of multiple graphing tasks, automatic charts, and statistical processes. |
| | SAS Favorites | Allows storing and executing favorite report routines in a profile file. |
| Selection | Refresh ** | Re-runs an analysis. |
| | Modify ** | Changes any optional parameters available to the task. |
| | Properties** | Allows review and changing of report properties. |
| Tools | Manage Content** | Allows changing optional parameters and running selected reports. |
| | Tools | Contains several features for managing data, reports, styles, and scheduling reports. |
| | Help | Contains Help information for SAS Add-In for Microsoft Office. |

** Available after a report is executed.

## 7.4 List of Examples

**Table 7.3: List of Examples for SAS Add-In for Microsoft Office.**

| Example Number | General Description |
|---|---|
| 7.1 | Open a SAS Dataset Using SAS Add-In for Microsoft Office. |
| 7.2 | Open a SAS Report Dataset (*.srx) Using SAS Add-In for Microsoft Office. |

## 7.5 Examples

### Example 7.1 Open a SAS Dataset Using SAS Add-In for Microsoft Office

The first thing I do when I get a new piece of software is to figure out how to open a file. So, I will also use that as my first example here. Using Excel 2013 and SAS 9.4, I click on the SAS Data Ribbon option on the SAS tab. The pop-up View SAS Data menu appears, as shown in Figure 7.3 (View SAS Data). The menu allows me to select the input SAS dataset, customize the display, and select the number of records that will be visible. One of the features of the "View SAS Data" pop-up menu is the "Filter & Sort" option. When that is selected, additional screens appear that allow you to select variables, filters, and sort order for the input SAS data. Under the "Number of Records toView" option, be aware that when "All" is selected I would always caution you to verify that the number of records and the number of columns that are actually read into the Excel workbook match the number of rows and columns in the original SAS dataset. This is especially true when reading more than 65,536 rows or 255 columns.

**NOTE:** The features described here do not require the activation of SAS Business Intelligence software, but do require a license for SAS Business Intelligence software.

## Figure 7.3: "View SAS Data" Screen Allows You to Navigate Your System to Pick a SAS File to Process.

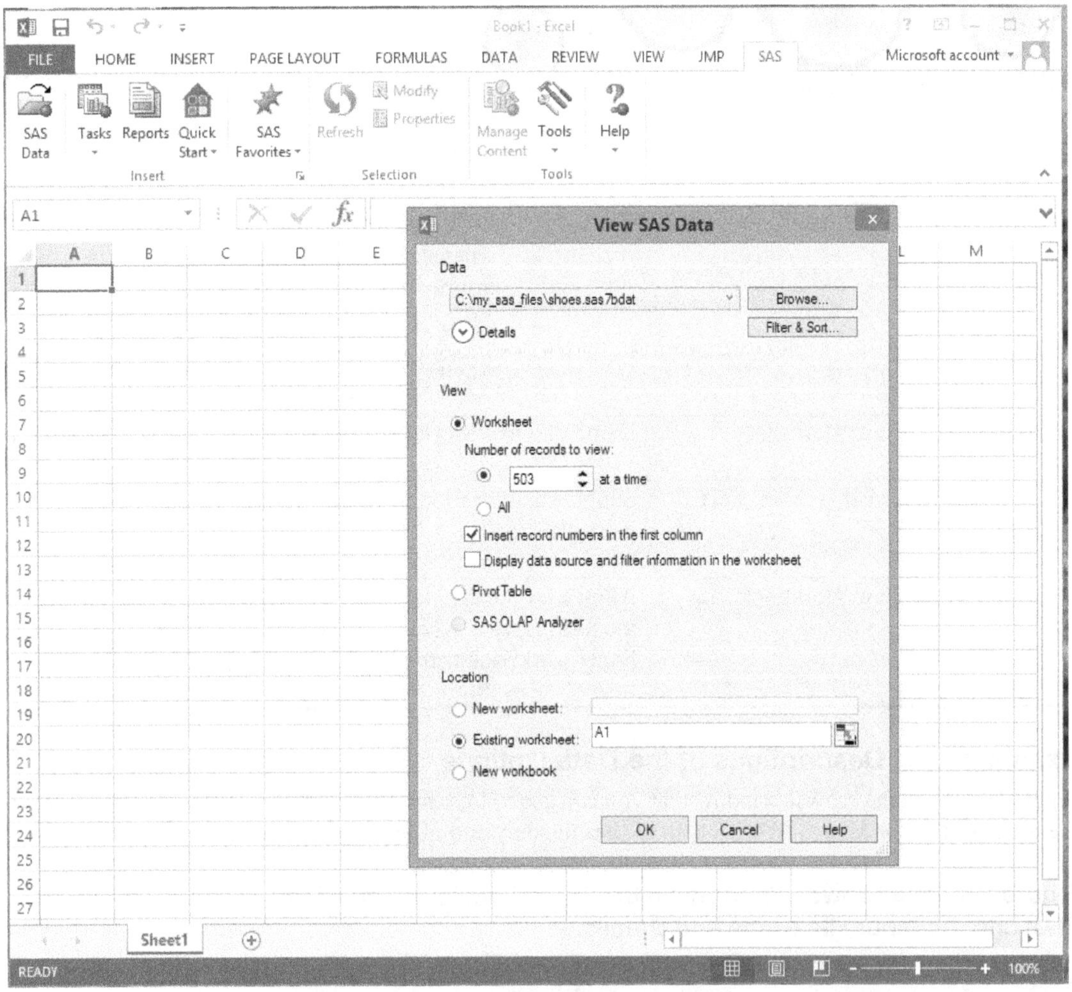

The menu in Figure 7.3 has three sections called Data, View, and Location. Table 7.4 describes these sections.

## Table 7.4: Description of Options Available When Reading SAS Data into Excel.

**Options for Loading a SAS Dataset into Excel Using the SAS Add-In for Microsoft Office**

| Section | Option | Description |
| --- | --- | --- |
| Data | Combo Box | Lists file names that have been used. |
| | Browse | Allows you to search a file to select for input. |
| | Details | Displays or hides the data options selected. |
| | Filter & Sort | Causes a pop-up menu to appear that allows you to<br><br>• select columns to read or exclude.<br>• filter the rows to input.<br>• sort the data before displaying it. |

**Options for Loading a SAS Dataset into Excel Using the SAS Add-In for Microsoft Office**

| Section | Option | Description |
|---------|--------|-------------|
| View | Worksheet | Controls the following:<br>• the number of records to view.<br>• whether or not record numbers are inserted into the Excel worksheet.<br>• whether or not to display data source and filter information in the worksheet.<br>• whether or not to create a pivot table.<br>• whether or not to use the SAS OLAP Analyzer. |
| Location | New Worksheet | Name of a new worksheet to create. |
|  | Existing Worksheet | Supplies the location to place the data into an existing worksheet. |
|  | New Workbook | Allows for the creation of a new workbook, usually named in the form Book*n*, where *n* is the next number not used for an Excel workbook named BOOK. |

### Example 7.1.1 Descriptions of the Data Options

Figure 7.4 shows the browse window that you can use to look for your SAS datasets to call into Excel using the SAS Add-In for Microsoft Office. Excel has the only add-in that allows you read SAS datasets directly.

**Figure 7.4: The Browse Window on the "SAS Data" Feature of the SAS Add-In Tab, Showing the Input File Name and Details.**

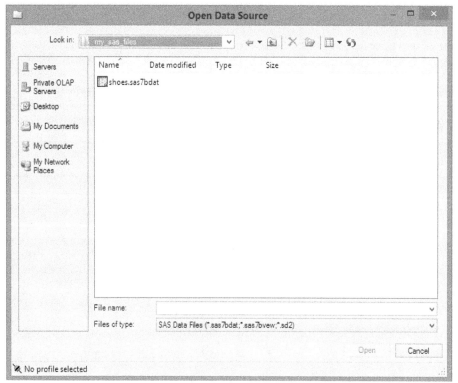

**Figure 7.5: The First Window on the "Filter & Sort" Tab. From Here, Select Variables to Read.**

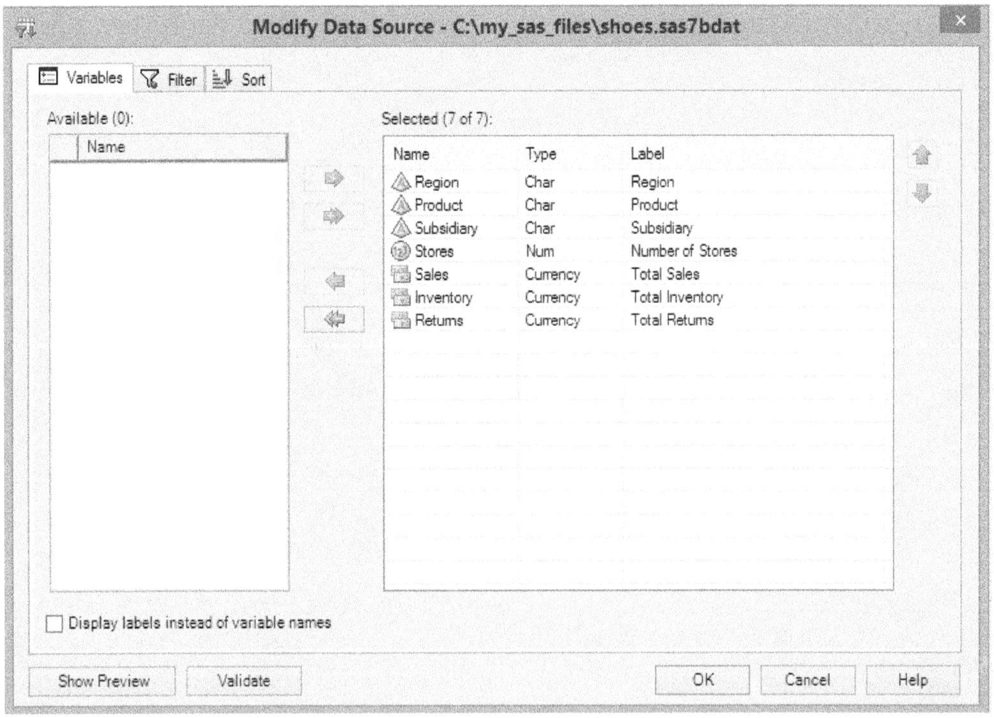

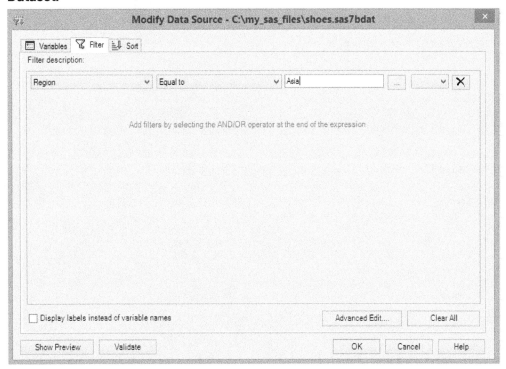

Figure 7.5 shows all of the variable names and types found in the input SAS dataset. All variables are shown here as selected. They had been highlighted and moved over to the right pane by clicking one of the right-pointing arrows. You will be able to select a subset of the variables by highlighting only the variables you wish to move. Figure 7.6 shows how to filter the selection to only the region "Asia".

**Figure 7.6: Filter Option Using SAS Add-In for Microsoft Office (Excel) to Read a SAS Dataset.**

**Figure 7.7: Data Options on the "SAS Data" Feature of the SAS Add-In Tab, Showing the Input File Name and Details.**

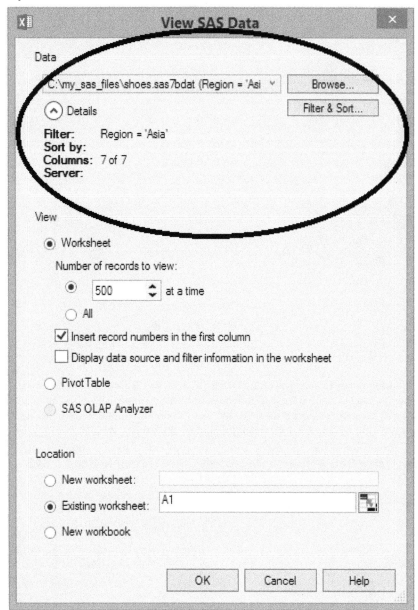

Figure 7.8 shows the result of adding the SAS data to an Excel worksheet with SAS Add-In for Microsoft Office. The columns have the Excel filter option applied and the import routine read only the records where "Region" equals "Asia". You may notice that only the variable names and 14 data rows were imported into the Excel worksheet.

**Figure 7.8: Excel Worksheet Showing Data Read by Selecting OK in Figure 7.3.**

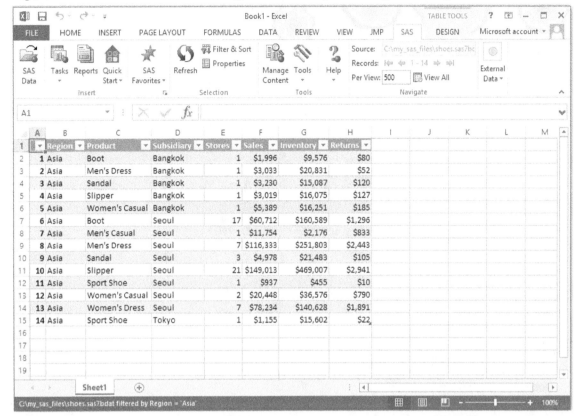

## Example 7.2 Open a SAS Report Dataset (*.srx) Using SAS Add-In for Microsoft Office

This example involves both SAS Enterprise Guide and Excel with the SAS Add-In for Microsoft Office installed. SAS Add-In for Microsoft Office is used to read SAS datasets into an Excel workbook. The most common type of SAS dataset is the familiar table of "Observations" and data "Variables". These look much like the Excel "Rows" and "Columns" of data. The second common, but not so well known, type of SAS data file is the *.srx SAS report file. SAS Enterprise Guide can produce one of these files.

This example has the following general steps.

1. Build a chart from the SASHELP.SHOES dataset with the Bar Chart Wizard. The chart was plotted by Region and Inventory.
2. Export the chart by using the Export SAS Report – Bar Chart option. This will create a file with the .srx extension that the SAS Add-In for Microsoft Office can read into Excel as a bar chart. A window will appear to allow you to save the report file anywhere on your network.
3. Import the output SAS *.srx file into Excel.

### Step 1: Build a bar chart using SAS Enterprise Guide.

Remember that in Example 5.3 we built a bar chart similar to Figure 7.9. This Example shows how to create a *.srx file using SAS Enterprise Guide and read it into Excel using the SAS Add-In for Microsoft Office. Here we will start with the bar chart already created and show how to create a *.srx file.

**Figure 7.9: An Enterprise Guide Bar Chart of the Shoes Dataset Showing Total Sales by Region.**

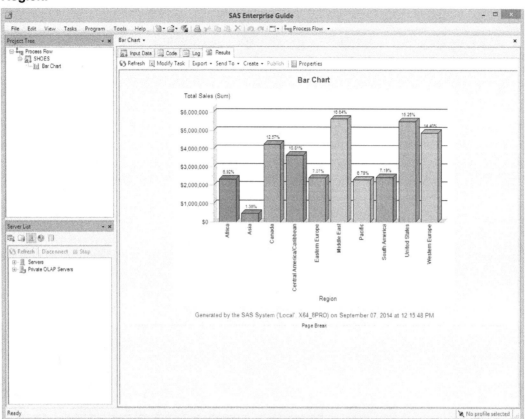

## Step 2: Export the bar chart as a *.srx file from using SAS Enterprise Guide.

Next, we need to create the *.srx file by using the "EXPORT" feature of SAS Enterprise Guide. Figure 7.10 shows an expanded view of the Export tab on the Project toolbar with the two options.

**Figure 7.10: Expanded Area of Figure 7.9 Showing the Export Tab Options from SAS Enterprise Guide.**

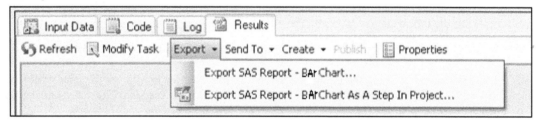

Either option will create a *srx file. The first option will output the displayed bar chart directly into a directory you choose. The output will be a *.srx file and a directory that contains a cascading style sheet. This option will place the information into the directory when you press Save. Figure 7.11 appears when you select "Export SAS Report – Bar Chart". The message that appears when other charts or graphs are exported may be different from this message. Figure 7.11 also shows that no other *.srx files exist in the same directory; that may not always be the case.

**Figure 7.11: Save Screen for Outputting the *.srx Files.**

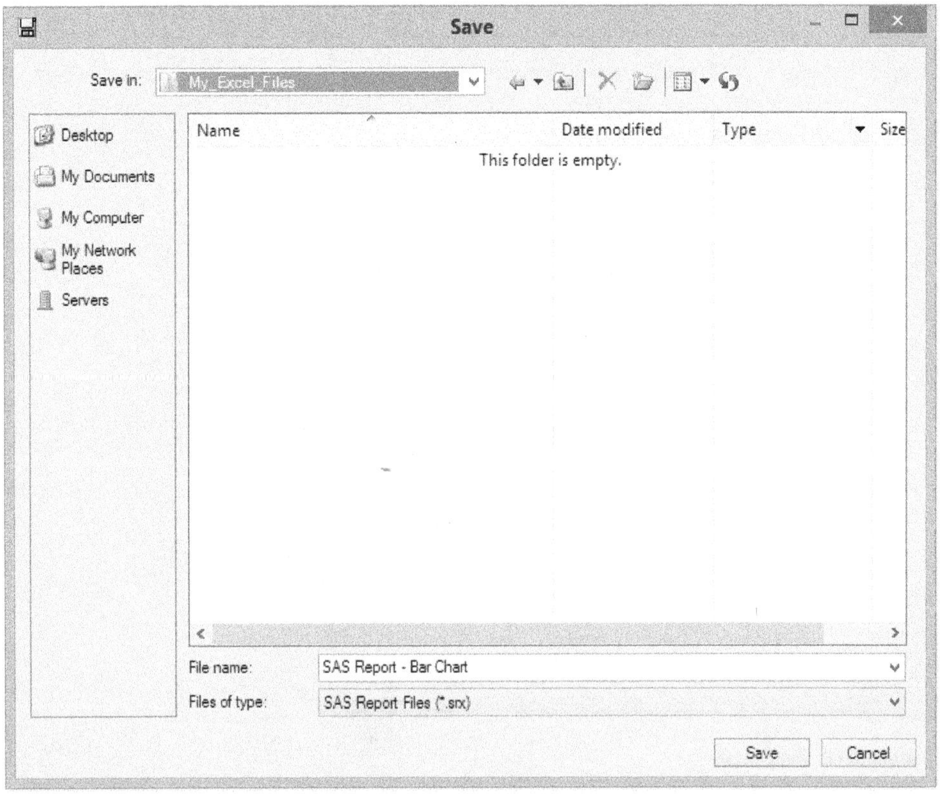

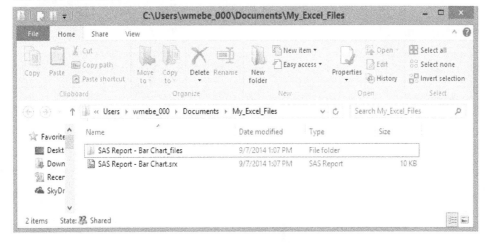

**Figure 7.12: A Windows Explorer View of the Output Directory Used in Figure 7.11-- After the Save Is Done.**

When you select "Export SAS Report – Bar Chart As A Step In Project …" the screens shown in Figure 7.13 appear and allow you to select output options that occur as part of running your report.

**Figure 7.13: Export Windows That Enable You to Generate a *.srx File as Part of Your Report.**

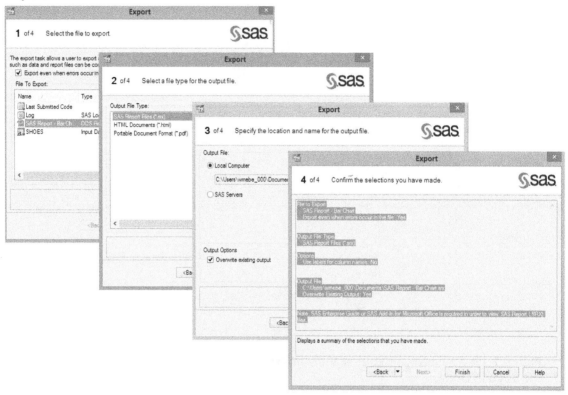

## Step 3: Import the *.srx file into Excel.

After opening Excel, you can read the *.srx file by selecting the **SAS▶Insert▶Reports** tabs and then browsing your system until you locate the report you just created with SAS Enterprise Guide (See Figure 7.14). After you select the report, an Open button appears that allows you to read the graph into Excel. Figure 7.15 shows the result.

**Figure 7.14: Excel SAS Add-In for Microsoft Office "Reports" Window to Open a *.srx Report File.**

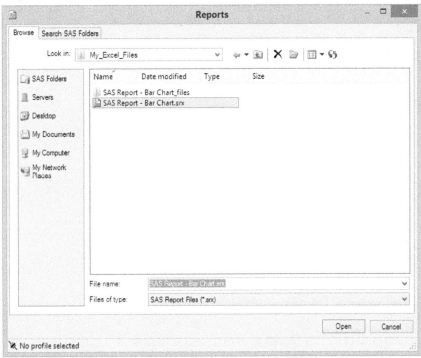

**Figure 7.15: Excel Worksheet with the *.srx SAS Report File Open.**

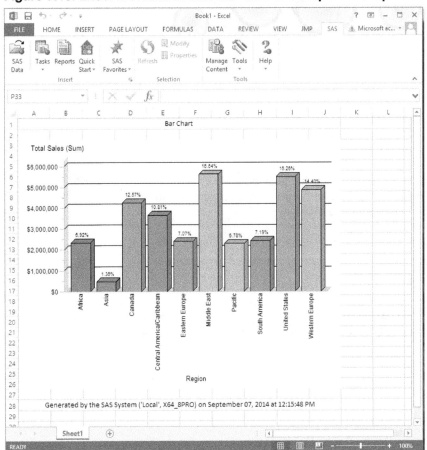

The image shown in Figure 7.15 is the same image that was produced by the SAS Bar Chart Wizard. I allowed the graph to have the default header and trailer and did not make any changes other than making each column of data a different color. This was produced with SAS Enterprise Guide 6.1, SAS 9.4, and Excel 2013. The first time I built this chart, I used Windows XP Pro operating system.

**Figure 7.16: The Properties Window Associated with the Excel Graph from the *.srx SAS Report.**

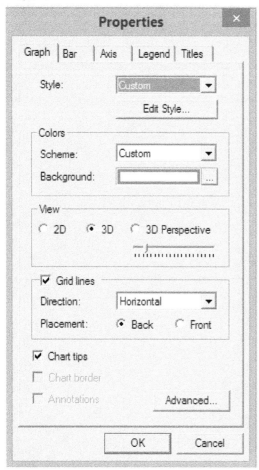

The header and footer may not have exactly the same format in Excel, Word, and PowerPoint as is available in SAS Enterprise Guide. With the proper profile connections, the graph can also be saved into Microsoft Outlook. Now at this point, we have:

- Used Example 5.3 to build a graph in SAS Enterprise Guide.
- Exported a SAS bar chart (or any graph) to a SAS *.srx file.
- Read the *.srx file into an Excel workbook.

An added bonus is that the graph in Excel has Active-X capabilities. Figure 7.17 shows the same data changed to a pie chart using the features available in Excel to make those changes.

**Figure 7.17: Excel Workbook with *.srx File Loaded and Changes Made to Graph Using Excel Options.**

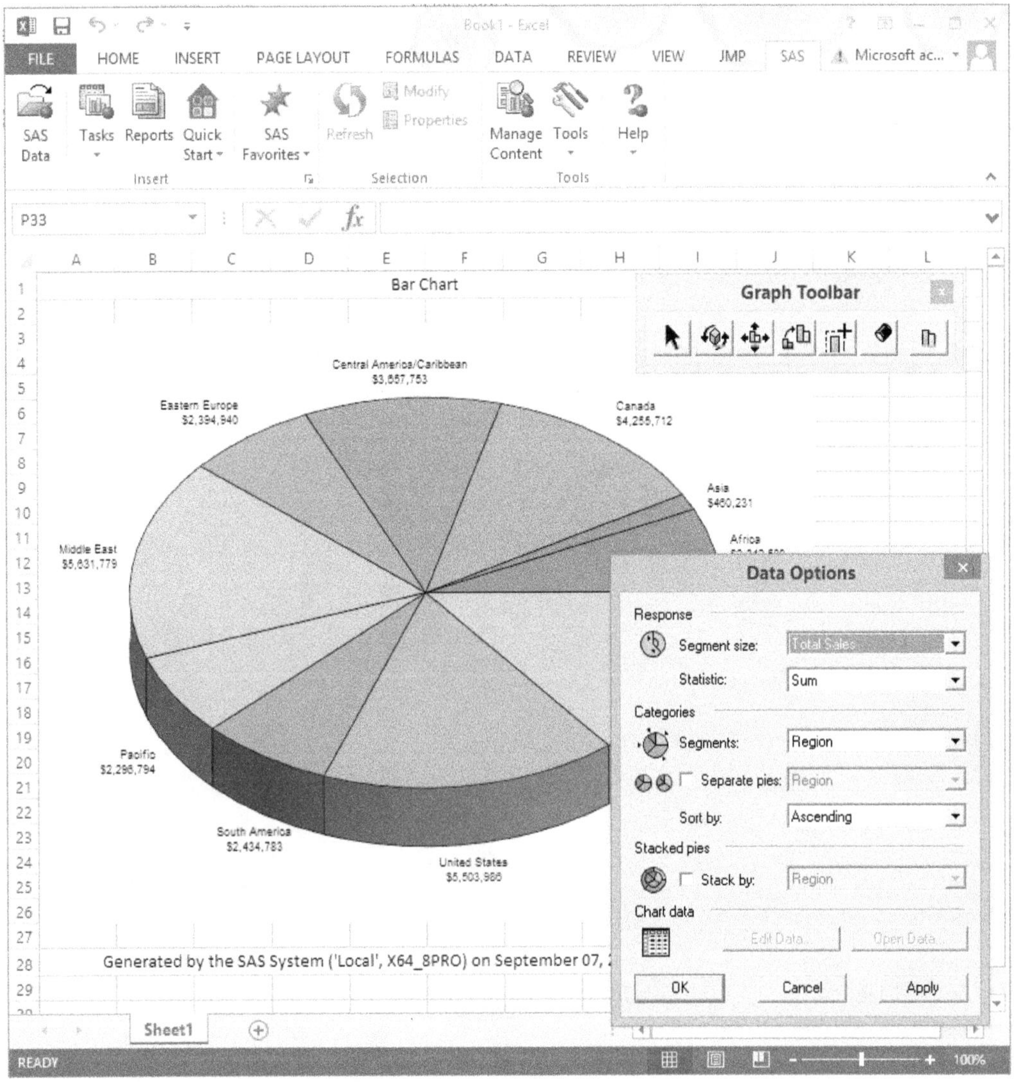

## 7.6 Conclusion

SAS Add-In for Microsoft Office provides an excellent method of transferring tables and graphs to Microsoft products, by enabling Microsoft Office to read the SAS report files (*.srx). The *.srx files can be generated by several of SAS products. SAS Add-In for Microsoft Office is part of SAS Enterprise Guide, SAS Enterprise Business Intelligence Server, and SAS Enterprise Miner. These products are licensed separately. The features described here do not require the activation of all features of SAS Business Intelligence software, but do require a license for SAS Business Intelligence software.

# Chapter 8: Creating Output Files with ODS for Use by Excel

8.1 Introduction .......................................................................................................... 108

8.2 Purpose ................................................................................................................ 108

8.3 An Introduction to SAS Tagset Templates That Create Files for Excel ........ 109
    8.3.1 How to Locate a Tagset Template ................................................................. 110

8.4 Difference Between an ODS Tagset and an ODS Destination ...................... 111

8.5 Syntax of the ODS CSV and CSVALL Output Processes ............................... 111

8.6 CSV and CSVALL Tagset Options .................................................................... 111

8.7 Overview of CSV and CSVALL Examples ....................................................... 113

8.8 CSV and CSVALL Examples to Write *.csv Files ............................................ 113
    Example 8.8.1 Simple CSV and CSVALL File Default Output Differences ............ 113
    Example 8.8.2 CSV and CSVALL Title and Footnote Output Differences ............ 115
    Example 8.8.3 Write Currency Values as Unformatted Numbers ........................ 118
    Example 8.8.4 Change Delimiters When Outputting Data with CSV Tagset ....... 120
    Example 8.8.5 Save Leading Zeroes in Character Fields Sent to Excel .............. 123

8.9 Syntax of ODS MSOFFICE2K Output Processes to Write HTML Files ......... 124

8.10 MSOFFICE2K Tagset Template Options ....................................................... 125

8.11 Overview of MSOFFICE2K Examples ............................................................ 126

8.12 MSOFFICE2K Examples to Write HTML Files .............................................. 126
    Example 8.12.1 Generating an HTML Output File with No Options ................... 126
    Example 8.12.2 Generating an HTML File Using the Summary_Vars Option .... 127

8.13 Syntax of the ODS EXCELXP Tagset Template Output Processes ............ 128

8.14 ODS EXCELXP Tagset Options ...................................................................... 130

8.15 Overview of EXCELXP Examples ................................................................... 132

8.16 EXCELXP Examples to Write XML Files ....................................................... 133
    Example 8.16.1 Generating an XML Output File with No Options ..................... 133
    Example 8.16.2 Adjusting Column Width Using Tagset Template Options ........ 134
    Example 8.16.3 Tagset Option to Hide Columns While Writing the File ............ 135
    Example 8.16.4 Apply an Excel "AUTOFILTER" to Selected Output Columns .... 136
    Example 8.16.5 Using Multiple Options to Produce a "Ready-to-Print" Spreadsheet ............................................................................................................ 137
    Example 8.16.6 Creating a Table of Contents in an Excel Workbook ................ 138
    Example 8.16.7 Methods of Naming Excel Worksheets ..................................... 140
    Example 8.16.8 Splitting One Report onto Multiple Excel Worksheets ............. 141
    Example 8.16.9 Methods of Placing Labels in Excel Worksheet Names ........... 142
    Example 8.16.10 Use SHEET_INTERVAL= BYGROUP to Create Worksheets ..... 143
    Example 8.16.11 Use SHEET_INTERVAL= PROC to Create Worksheets ........... 144
    Example 8.16.12 Build Separate Worksheets with Titles on Each Sheet .......... 146

**8.17 The New ODS Destination EXCEL for Writing Workbooks .......................... 147**

**8.18 Conclusion ................................................................................................. 148**

## 8.1 Introduction

With the release of SAS 7 came the introduction of the Output Delivery System (ODS). This powerful enhancement to SAS software expanded the standardization of SAS output processes. From the user's perspective, a new default system appeared that managed output files but did not intrude on the way things worked already. While at the same time, the new ODS processes provided options that exceeded the status quo and opened new ways to use the output data from SAS procedures. The ODS destinations called OUTPUT and HTML first appeared with this release. SAS 8 soon followed with more options.

Delimited files had long provided a method to transport data between systems, including between SAS and Excel. A common method of generating delimited files was to use the SAS DATA step and FILENAME statement to write to an external text file that could be read and processed by Excel. The data in these files was separated by characters not normally found within the data. The development of ODS provided a platform for new tools that enabled the standardization of output methods for transporting data between software systems like SAS and Excel. One of the first of these methods was the CSV destination, which allowed an automatic output of text files formatted so that Excel could read the file and place data directly into individual cells of a spreadsheet. Other options were soon to follow.

SAS 8.2 provided the first experimental look at the new tagset output feature of the ODS system. With the proper ODS commands your output could be directed into a new processing path that formatted the output data. With this new processing path ODS acquired the ability to store command files that controlled the output formatting. These new command files are called tagsets. This chapter will discuss not only some of these tagset command files, but also the new ODS destination called EXCEL. This destination became available as an experimental ODS destination in SAS 9.4 TS Level 1M1. As with all of the experimental SAS products, the final product may change.

## 8.2 Purpose

The purpose of this chapter is to examine some of the ODS data output methods. As noted in the introduction, HTML was one of the first ODS destinations. While it is not a recommended method, it is still supported and used by the SAS Explorer window. I will give you my interpretation of how I see the ODS-to-Excel-readable-file generation process works. I do not expect you to find this definition in any SAS manual, but here goes.

- Start running SAS.
- Issue an ODS command similar to the following to open the output file.
  ODS (*any SAS ODS Destination or* TAGSETS=) (*any output file name*) (*any optional options*);
- Run any SAS process that creates an output file.
- Issue an ODS command similar to the following to close the output file.
  ODS (*any SAS ODS destination* or TAGSETS) CLOSE;
- Do anything else or end SAS.

Of course, this sample of pseudo code will not execute anywhere and to be useful the "ODS" commands must be provided in the correct SAS syntax for the task you wish to accomplish. The important thing to remember here is that prior to SAS 8.2 and SAS 7 (for the few people who used it) every non-HTML file used to transport data had to be coded in a DATA step. It is important to realize that the ability to process tagsets is a huge step forward for the ability to generate formatted SAS output. SAS software has long had users wanting output that was easy to transport to other systems. Tagsets crossed that gap. This chapter will discuss the following ODS tagsets and destinations because they have a direct relationship with making the transfer of data from SAS to Excel easier for you as a programmer. The tagsets and destinations listed reduce your coding effort from pages of SAS code to lines of code defining the output file, locations, and structures.

**Table 8.1: ODS Options That Create Files That Can Be Opened by Microsoft Excel.**

| ODS Option | Destination or Tagset | Output |
| --- | --- | --- |
| CSV | Both | Create a text file with values separated by a delimiter |
| CSVALL | Both | Create a text file with values separated by a delimiter |
| MSOFFICE2K | Both | Create a text file with HTML formatted data values |
| HTML | Both | Create a text file with HTML formatted data values |
| EXCELXP | Tagset | Create a text file with XML formatted data values |
| EXCEL | Destination | Create a Microsoft Excel file in *.xlsx native format |

The SAS ODS destinations and tagsets listed in Table 8.1 allow SAS to write delimited or formatted files without interrupting the flow of the job, thereby enabling programmers to do more work in less time. Simple commands turn the feature on and off, and most outputs in between are written to the output file. The capabilities of ODS have been steadily increasing ever since the release of SAS 7. The power of tagsets has increased from writing text-based (*.txt, *.csv, *.tab) files to writing Hyper Text Mark-up Language (HTML) files and then to outputting Extensible Mark-up Language (XML)-formatted files. The new experimental ODS destination EXCEL will actually write native Excel 2007 and later formatted files. As with all of the experimental SAS products, the final product options may change.

## 8.3 An Introduction to SAS Tagset Templates That Create Files for Excel

As I mentioned in earlier, tagsets are text files stored for use by the Output Delivery System. Many of the tagsets are delivered with, and are part of, Base SAS. PROC TEMPLATE prepares SAS code to be used as a tagset. The syntax for coding a tagset is slightly different from that of regular SAS code. There are also specific methods in place to store an access tagset templates. Tagsets delivered with your SAS product are usually stored in a location that is Write protected (see Figure 8.1), so the contents cannot be easily overwritten. Other storage levels are easily defined and are used to store changes to original tagsets. The tagset names and the concatenated list of storage directories allow changes to be applied without overwriting the original files. Tagsets can be updated by SAS and are stored on the SAS web site. You can also make changes to a tagset and install the changes on your system.

## 8.3.1 How to Locate a Tagset Template

SAS menus are context driven menus, this means that the command line menus may change depending upon your currently displayed screens. For instance in Figure 8.1 you have to have the "Results" window displayed as the current window before the "View" option will display the menu listing "Tagsets" as the first item. Then the selection of "Sashelp.Tmplmast", "Tagsets", and "ExcelXP" will display the information in Figure 8.2. The production SAS 9.4 TS Level 1M1 version of the ExcelXP tagset template simply defines the name and includes a separate tagset code module. The code in your ExcelXP tagset code file may be different.

**Figure 8.1: How to Locate a Tagset Template.**

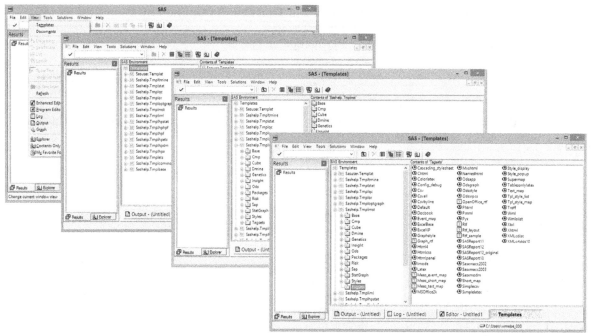

Figure 8.1 shows the path to locating the tagset production code modules, and Figure 8.2 shows the SAS 9.4 TS Level 1M1 contents. This tagset simply defines the ExcelXP name and includes (references) other tagset instructions from the module "tagsets.ExcelBase". The code module "tagsets.ExcelBase" has over 8,500 lines of code that define and implement all of the options and features of the ExcelXP tagset template.

**Figure 8.2: Contents of the ExcelXP Tagset PROC TEMPLATE Code Module.**

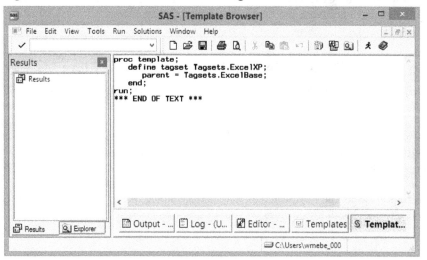

See the SAS Help files on your system for information about using PROC TEMPLATE.

## 8.4 Difference Between an ODS Tagset and an ODS Destination

An ODS tagset template is code that is used by PROC TEMPLATE to perform a task, like write a *.csv file. Base SAS ships with many templates that are available to use. These tagsets can be edited and modified and then stored in a place where you can have access to them. This is discussed later. An ODS destination like the new EXCEL destination can be provided with options at the time you execute the ODS command, but you cannot change the actual code that is used to process the data. Many SAS web pages provide references to and downloadable copies of tagsets.

## 8.5 Syntax of the ODS CSV and CSVALL Output Processes

This section will describe CSV tagset templates, including options, along with the default values. Many of the options will be further explained by an example. Section 8.3 describes how to locate a specific tagset. You can then view the actual code, but it is often better left alone.

The tagset templates "CSV" and "CSVALL" are effectively the same tagset template; the difference is that the "CSVALL" tagset template sets the "BYLINES", "TITLES", "PROC_TITLES", "NOTES" options to "yes". Example 8.8.1 parts "a" and "b" show the actual differences. The "CSV" tagset template is the "Parent" tagset of "CSVALL". Therefore, everything not defined in tagset template CSVALL is defined in the CSV tagset template. I refer to both the CSV and the CSVALL tagset template as the "CSV tagset template" unless an option is specific to only one of them. The CSV tagset templates convert a SAS file to an output text file of values separated by a delimiter. Text files of data separated by a delimiter have long been used to transfer data between software packages.

The options for CSV tagset templates are described below, and in general fall in to several categories based upon what part of the output the option modifies.

- Modify or display SAS outputs for titles, footnotes, bylines, and table headers
- Display of SAS log messages and tagset template "Help" messages
- Numeric variable formatting features
- Character variable formatting features

## 8.6 CSV and CSVALL Tagset Options

Table 8.2 highlights some of the CSV and the CSVALL tagset options, the option names default values and a description. When you are using either the CSV or CSVALL tagset, current information about what is installed on your computer can be printed to your log by adding either "OPTIONS(DOC='Quick') or OPTIONS(DOC='help') to the ODS statement. The tagset options for the CSV and CSVALL tagset templates fall into the following general categories. The asterisk "*" indicates that an option will be shown in an example below.

**Table 8.2: ODS Options for the CSV and CSVALL Tagset Templates.**

| ODS Option | Description |
| --- | --- |
| **Options related to titles and footnotes** | |
| BYLINES | Select if byline records will be put into the output file. |
| PROC_TITLES | Select if procedure title records will be put into the output file. |
| TITLES | Select if title and footnote records will be put into the output file. |
| TABLE_HEADERS | Select if header sections of output tables will be put into the output file. |
| **Options related to data and columns** | |
| CURRENCY_AS_NUMBER | If selected, currency symbols and punctuation are removed from output values, |
| CURRENCY_SYMBOL | Select currency symbol to use, |
| DECIMAL_SEPARATOR | Select decimal separator to use, |
| DELIMITER | Select field delimiter to use, |
| QUOTE_BY_TYPE | Select field types to enclose in quotes, |
| QUOTED_COLUMNS | Select columns to enclose in quotes. |
| PERCENTAGE_AS_NUMBER | Select if percent signs and punctuation are removed from output values. |
| PREPEND_EQUALS | Use with QUOTE_BY_TYPE to prepend an equal sign in front of numeric values. |
| THOUSANDS_SEPARATOR | Select thousands separator to use in numeric fields, usually for use in Europe or countries with different numeric formatting. |
| **Options related to error messages** | |
| NOTES | Select whether or not to suppress SAS system error messages. |
| **Options related to Tagset template documentation** | |
| DOC | Write user Help information about the tagset template options to the system log. |

## 8.7 Overview of CSV and CSVALL Examples

Example 8.8.1 includes ODS commands to execute the CSV and CSVALL features. There is more than one way to write a CSV file using ODS. All six sets of code shown in Example 8.8.1 write the same data to an output file into either a CSV or CSVALL directory with a slightly different name. All six of these methods are part of Base SAS, and they all produce the same output. The code in Figures 8.3 and 8.4 and shown in Example 8.8.1 demonstrates that there are only minor differences in the output files. However, if you need an output file without titles you should use the CSV tagset.

Table 8.3 describes the examples shown in this chapter. Some of the examples here have minor overlaps in the features to show how they interact when additional features are included. Example 8.8.1 will show outputs from both the CSV and CSVALL tagset executions. However, the other examples will focus on the output from only one tagset.

**Table 8.3: List of Examples for ODS CSV and CSVALL Tagsets or Destinations.**

| Example Number | General Description |
| --- | --- |
| 8.8.1 | **Simple CSV and CSVALL File Default Output Differences.** This example compares the default behavior of both the CSV and CSVALL tagset output files. |
| 8.8.2 | **CSV and CSVALL Title and Footnote Output Differences.** This example shows the differences in the output files when more than one ODS output is written to a CSV file, and compares the CSV output to output produced by using CSVALL. Both PROC PRINT and PROC FREQ are used in this example, along with WHERE and BY clauses. |
| 8.8.3 | **Write Currency Values as Unformatted Numbers.** This example shows how to write currency values as numbers without the dollar signs and punctuation. The input file has whole dollar amounts, and therefore the output file displays whole dollar values only. Only a CSVALL example is shown. |
| 8.8.4 | **Change Delimiters When Outputting Data with CSV Tagset.** This example shows how to create a "Tab"-delimited file using the CSV tagset. Any other single character can be used as a delimiter, but try to pick one that does not appear in the data to avoid problems. Only a CSV example is shown. This example also shows the difference between a *.csv file and a *.tab file when opened with Excel. |
| 8.8.5 | **Save Leading Zeroes in Character Fields Sent to Excel.** This example uses the QUOTE_BY_TYPE='Yes' and the PREPEND_EQUALS='Yes' options to preserve leading zeros for variables that are read into an Excel workbook. Only a CSV example is shown. |

## 8.8 CSV and CSVALL Examples to Write *.csv Files

### Example 8.8.1 Simple CSV and CSVALL File Default Output Differences

The examples shown in this section relate to the CSV and CSVALL tagset templates. I will start by showing the differences between the two tagset templates. The CSV tagset template is the "Parent" tagset template, while the CSVALL tagset template is a child template. Essentially, the CSVALL tagset template just changes some defaults, and then uses all of the code in the CSV tagset template. The CSV tagset template is referred to in PROC TEMPLATE as a "PARENT."

**Figure 8.3: Sample ODS Commands to Create a CSV Output File Using CSV Tagset or Destination.**

```
ODS MARKUP BODY='c:\temp\csv\shoes1.csv' TAGSET=CSV;
        PROC PRINT DATA=sashelp.shoes;
        RUN;
ODS MARKUP CLOSE;

*************************************************************;

ODS CSV BODY='c:\temp\csv\shoes2.csv';
        PROC PRINT DATA=sashelp.shoes;
        RUN;
ODS CSV CLOSE;

*************************************************************;

ODS TAGSETS.CSV BODY='c:\temp\csv\shoes3.csv';
        PROC PRINT DATA=sashelp.shoes;
        RUN;
ODS TAGSETS.CSV CLOSE;
```

**Figure 8.4: Sample ODS Commands to Create a CSV Output File Using CSVALL Tagset or Destination.**

```
ODS MARKUP BODY='c:\temp\csvall\shoes1.csv' TAGSET=CSVALL;
        PROC PRINT DATA=sashelp.shoes;
        RUN;
ODS MARKUP CLOSE;

*************************************************************;

ODS CSVALL BODY='c:\temp\csvall\shoes2.csv';
        PROC PRINT DATA=sashelp.shoes;
        RUN;
ODS CSVALL CLOSE;

*************************************************************;

ODS TAGSETS.CSVALL BODY='c:\temp\csvall\shoes3.csv';
        PROC PRINT DATA=sashelp.shoes;
        RUN;
ODS TAGSETS.CSVALL CLOSE;
```

The two output screens shown in Figure 8.5 show the differences between using the ODS CSV and the ODS CSVALL tagset or destination options. All three pairs of code listed in Figures 8.3 and 8.4 produce the same output files. The output file names are the same in Figure 8.5, but the top left image was written using the "TAGSETS.CSV" code and the bottom right image was created using the "TAGSETS.CSVALL" code.

**Figure 8.5: Output Differences Between CSV and CSVALL ODS Output Files.**

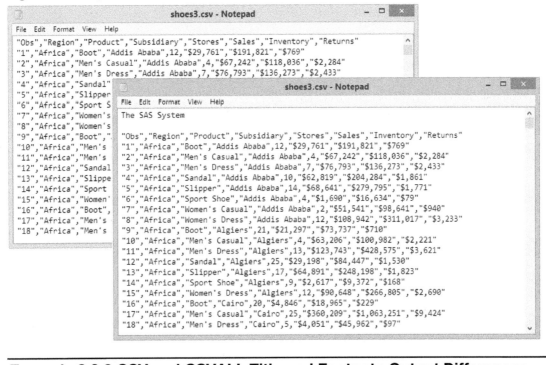

## Example 8.8.2 CSV and CSVALL Title and Footnote Output Differences

In this example, the tagset templates CSV and CSVALL will both be used to show the differences between the default settings of the CSV and CSVALL tagset templates. This example highlights the differences in the output *.csv files. By default, options "BYLINES", "TITLES", "PROC_TITLES", and "NOTES" are set to "yes" in the CSVALL tagset processing. The code in Figure 8.6 processes both the CSV and CSVALL output files. The WHERE= clause in the PROC PRINT code in Figure 8.6 restricts the output, and the "BY" clause groups the data when the output files are created.

**Figure 8.6: SAS Code to Process Both CSV and CSVALL Output.**

```
TITLE1
  "Testing Title options BYLINES - PROC_TITLES - TITLES and FOOTNOTES";

FOOTNOTE1
  "Testing footnote options BYLINES - PROC_TITLES - TITLES and FOOTNOTES";

ODS TAGSETS.CSV    BODY='c:\temp\csv\csv_shoes.csv';
ODS TAGSETS.CSVALL BODY='c:\temp\csvall\csvall_shoes.csv';

PROC PRINT DATA=sashelp.shoes
      (WHERE=(product = 'Boot' and region in('Asia' , 'Canada')));
      BY region;
RUN;
PROC FREQ DATA=sashelp.shoes;
    TABLE product / LIST;
RUN;

ODS TAGSETS.CSV CLOSE;
ODS TAGSETS.CSVALL CLOSE;
```

### Figure 8.7: ODS Output from Tagset CSV as a Comma-Separated Value Text File.

Figure 8.7 show ODS output from a CSV tagset template showing a standard output with no options set. This output was produced using SAS 9.4.

### Figure 8.8: ODS Output from Tagset CSVALL as a Comma-Separated Value Text File.

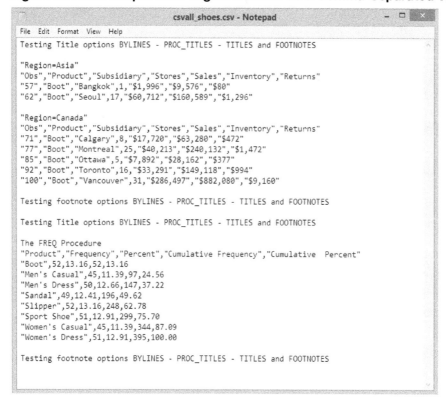

Figure 8.8 shows ODS output from CSVALL tagset template showing standard output with options "BYLINES", "TITLES", "PROC_TITLES", "NOTES" set to "yes" in the CSVALL processing by default. This output was produced using SAS 9.4.

**Figure 8.9: ODS Output from Tagset CSV as Read into an Excel Workbook.**

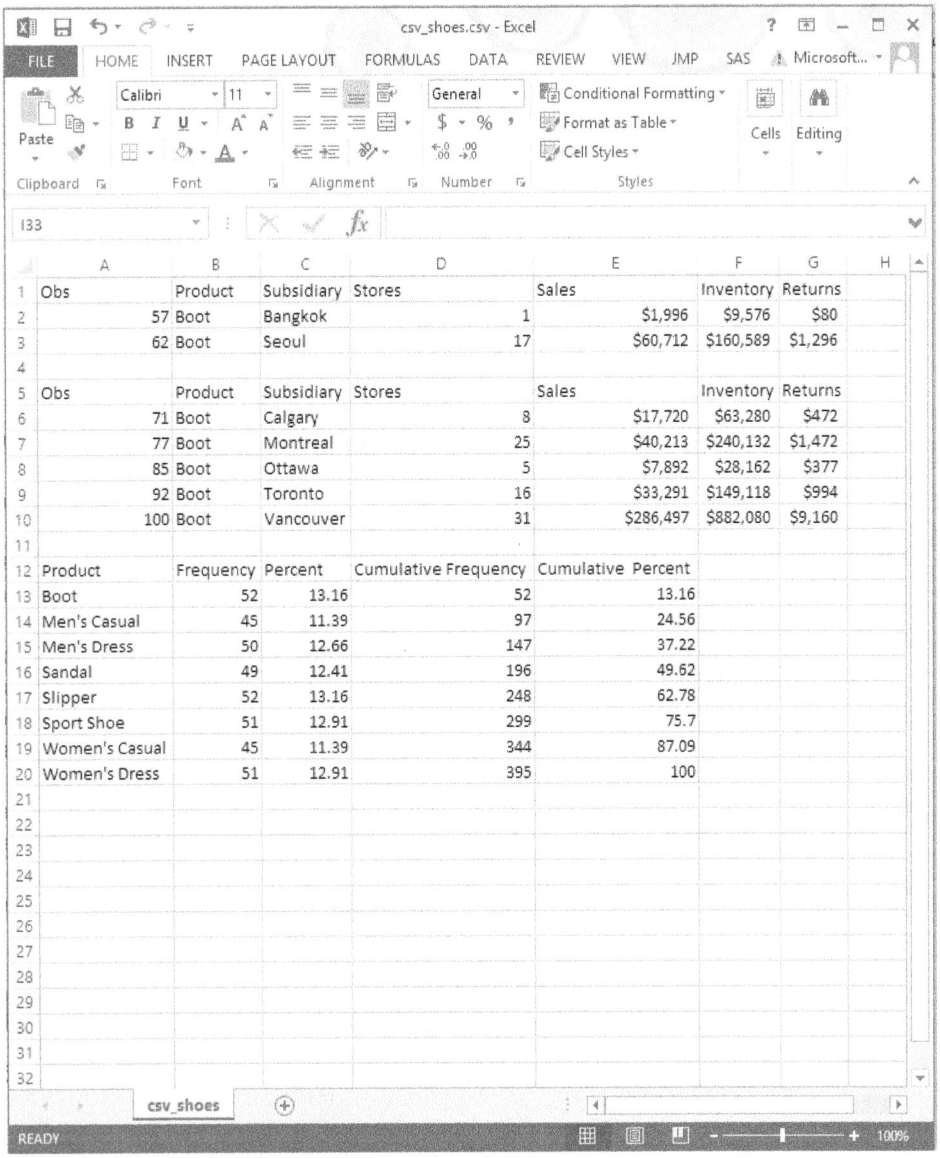

| | A | B | C | D | E | F | G |
|---|---|---|---|---|---|---|---|
| 1 | Obs | Product | Subsidiary | Stores | Sales | Inventory | Returns |
| 2 | 57 | Boot | Bangkok | 1 | $1,996 | $9,576 | $80 |
| 3 | 62 | Boot | Seoul | 17 | $60,712 | $160,589 | $1,296 |
| 4 | | | | | | | |
| 5 | Obs | Product | Subsidiary | Stores | Sales | Inventory | Returns |
| 6 | 71 | Boot | Calgary | 8 | $17,720 | $63,280 | $472 |
| 7 | 77 | Boot | Montreal | 25 | $40,213 | $240,132 | $1,472 |
| 8 | 85 | Boot | Ottawa | 5 | $7,892 | $28,162 | $377 |
| 9 | 92 | Boot | Toronto | 16 | $33,291 | $149,118 | $994 |
| 10 | 100 | Boot | Vancouver | 31 | $286,497 | $882,080 | $9,160 |
| 11 | | | | | | | |
| 12 | Product | Frequency | Percent | Cumulative Frequency | Cumulative Percent | | |
| 13 | Boot | 52 | 13.16 | 52 | 13.16 | | |
| 14 | Men's Casual | 45 | 11.39 | 97 | 24.56 | | |
| 15 | Men's Dress | 50 | 12.66 | 147 | 37.22 | | |
| 16 | Sandal | 49 | 12.41 | 196 | 49.62 | | |
| 17 | Slipper | 52 | 13.16 | 248 | 62.78 | | |
| 18 | Sport Shoe | 51 | 12.91 | 299 | 75.7 | | |
| 19 | Women's Casual | 45 | 11.39 | 344 | 87.09 | | |
| 20 | Women's Dress | 51 | 12.91 | 395 | 100 | | |

### Figure 8.10: ODS Output from Tagset CSVALL as Read into an Excel Workbook.

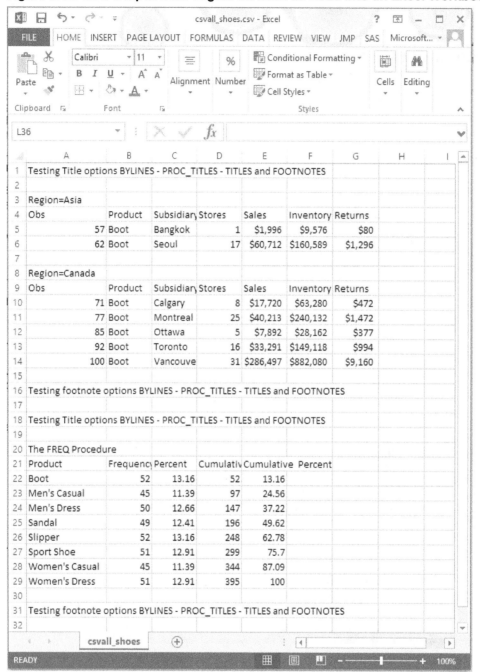

Figure 8.7 and Figure 8.9 show the output of the ODS CSV processing, while Figure 8.8 and Figure 8.10 show the output from the CSVALL processing.

## Example 8.8.3 Write Currency Values as Unformatted Numbers

In the following example, the tagset template CSVALL is used to convert formatted dollar values directly to numeric values. The source input file had only whole dollar amounts, so the output CSV file will have only whole dollar amounts. Figure 8.11 shows the code to use the option CURRENCY_AS_NUMBER to convert currency values to numbers. The output appears in Figures 8.12 and 8.13. This option strips the currency symbol and displays only the numbers. The numbers displayed here are whole dollar amounts because the input file deals with whole dollar amounts.

**Figure 8.11: SAS Code to Create a CSV Test File Using the currency_as_number Option.**

```
/* convert currency to numbers         */
/* No titles or footnotes are assigned */

ODS TAGSETS.CSVALL BODY='c:\temp\csvall\csvall_shoes.csv'
        OPTIONS(CURRENCY_AS_NUMBER='Yes');

PROC PRINT DATA=sashelp.shoes
       (WHERE=(product = 'Boot' and region in('Asia', 'Canada')));
        BY region;
RUN;
ODS TAGSETS.CSVALL CLOSE;
```

**Figure 8.12: ODS Output File Generated with CSVALL and "CURRENCY_AS_NUMBER" Active.**

```
csvall_shoes.csv - Notepad
File  Edit  Format  View  Help
The SAS System

"Region=Asia"
"Obs","Product","Subsidiary","Stores","Sales","Inventory","Returns"
"57","Boot","Bangkok",1,1996,9576,80
"62","Boot","Seoul",17,60712,160589,1296

"Region=Canada"
"Obs","Product","Subsidiary","Stores","Sales","Inventory","Returns"
"71","Boot","Calgary",8,17720,63280,472
"77","Boot","Montreal",25,40213,240132,1472
"85","Boot","Ottawa",5,7892,28162,377
"92","Boot","Toronto",16,33291,149118,994
"100","Boot","Vancouver",31,286497,882080,9160
```

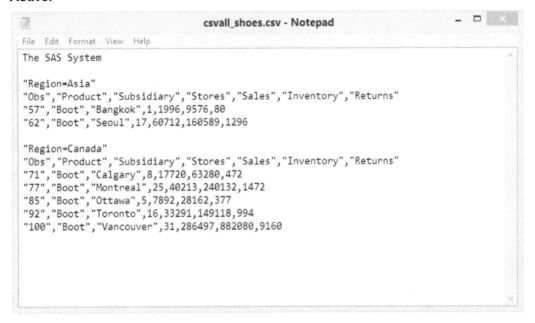

Compare Figure 8.12 with Figure 8.8 to see the difference in the way the dollar amounts are displayed.

### Figure 8.13: The File from Figure 8.12 Opened with Microsoft Excel.

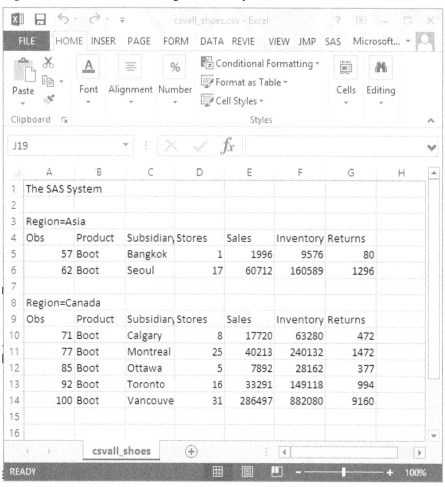

## Example 8.8.4 Change Delimiters When Outputting Data with CSV Tagset

In this example only the CSVALL tagset template will be demonstrated to show the enhanced headers. The following example deals with the delimiters used to separate the data values. In this case the "Tab" character ('09' hexadecimal) was used.

### Figure 8.14: SAS Code to Change the Delimiter for a CSV Text File to a Tab.

```
/* use a tab as delimiter                                */
/* a 'tab' character is hard to reproduce in code        */
/* so the standard convention is to use a hexadecimal    */
/* representation of a 'tab' character as '09'x          */
/* Both of these files are "Tab" delimited but the       */
/* CSV file will not open correctly when double clicking */
/* because Excel is expecting commas for delimiters      */
/* and commas exist in the test lines of the file.       */


ODS TAGSETS.CSVALL BODY='c:\temp\csv\csvall_shoes.csv'
        OPTIONS(DELIMITER='09'x);

PROC PRINT DATA=sashelp.shoes
        (where=(product = 'Boot' and region in('Asia', 'Canada')));
        BY region;
RUN;

ODS TAGSETS.CSVALL CLOSE;
```

```
ODS TAGSETS.CSVALL BODY='c:\temp\csv\csvall_shoes.tab'
        OPTIONS(DELIMITER='09'x);


PROC PRINT DATA=sashelp.shoes
        (where=(product = 'Boot' and region in('Asia', 'Canada')));
        BY region;
RUN;

ODS TAGSETS.CSVALL CLOSE;
```

The output CSV file 'c:\temp\csv\csvall_shoes.csv' shown in Figure 8.15 has a tab character as a delimiter. This may not seem very useful, but some older systems I have used have reacted in funny ways when commas appear in the numeric fields, even when they are enclosed in quotes. So, sometimes I use a tab as my delimiter.

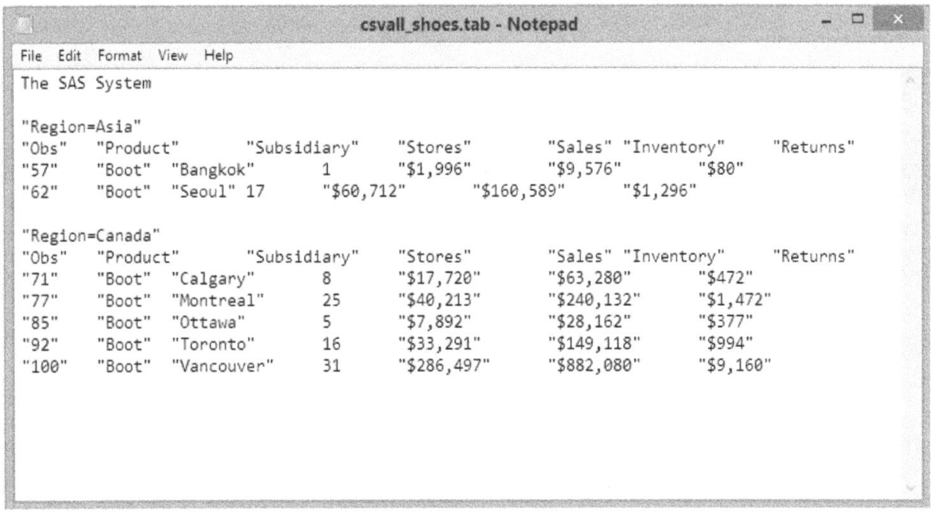

**Figure 8.15: The CSVALL_Shoes.csv File Opened with Notepad.**

```
csvall_shoes.csv - Notepad
File  Edit  Format  View  Help
The SAS System

"Region=Asia"
"Obs"   "Product"       "Subsidiary"    "Stores"        "Sales" "Inventory"     "Returns"
"57"    "Boot"  "Bangkok"       1       "$1,996"        "$9,576"        "$80"
"62"    "Boot"  "Seoul" 17      "$60,712"       "$160,589"      "$1,296"

"Region=Canada"
"Obs"   "Product"       "Subsidiary"    "Stores"        "Sales" "Inventory"     "Returns"
"71"    "Boot"  "Calgary"       8       "$17,720"       "$63,280"       "$472"
"77"    "Boot"  "Montreal"      25      "$40,213"       "$240,132"      "$1,472"
"85"    "Boot"  "Ottawa"        5       "$7,892"        "$28,162"       "$377"
"92"    "Boot"  "Toronto"       16      "$33,291"       "$149,118"      "$994"
"100"   "Boot"  "Vancouver"     31      "$286,497"      "$882,080"      "$9,160"
```

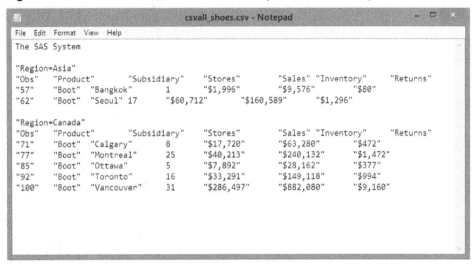

**Figure 8.16: The CSVALL_Shoes.tab File Opened with Notepad.**

```
csvall_shoes.tab - Notepad
File  Edit  Format  View  Help
The SAS System

"Region=Asia"
"Obs"   "Product"       "Subsidiary"    "Stores"        "Sales" "Inventory"     "Returns"
"57"    "Boot"  "Bangkok"       1       "$1,996"        "$9,576"        "$80"
"62"    "Boot"  "Seoul" 17      "$60,712"       "$160,589"      "$1,296"

"Region=Canada"
"Obs"   "Product"       "Subsidiary"    "Stores"        "Sales" "Inventory"     "Returns"
"71"    "Boot"  "Calgary"       8       "$17,720"       "$63,280"       "$472"
"77"    "Boot"  "Montreal"      25      "$40,213"       "$240,132"      "$1,472"
"85"    "Boot"  "Ottawa"        5       "$7,892"        "$28,162"       "$377"
"92"    "Boot"  "Toronto"       16      "$33,291"       "$149,118"      "$994"
"100"   "Boot"  "Vancouver"     31      "$286,497"      "$882,080"      "$9,160"
```

Figure 8.15 and Figure 8.16 appear exactly the same when opened with Notepad. However, when opened with Excel the files appear different. This is shown in Figure 8.17a and 8.17b below.

**Figure 8.17a: The CSVALL_Shoes.csv File Opened with Excel.**

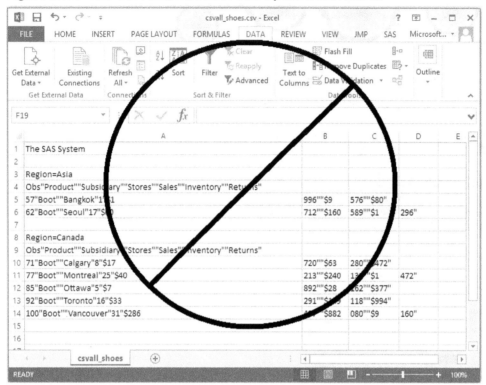

When a text file with the name that ends in CSV is written with tabs as delimiters and opened with Excel, any commas embedded in the file can cause problems. This is because the Excel workbook expects commas to separate fields when the file name ends in ".CSV".

**Figure 8.17b: The CSVALL_Shoes.tab File Opened with Excel.**

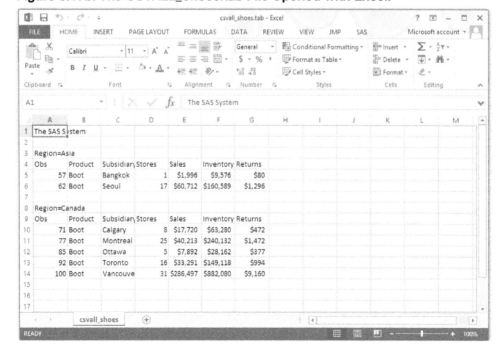

When a text file with the name that ends in TAB is written with tabs as delimiters and opened with Excel, any commas embedded in the file are treated as text and placed into the cells. This is because the Excel workbook expects tabs to separate fields when the file name ends in ".TAB".

## Example 8.8.5 Save Leading Zeroes in Character Fields Sent to Excel

Here is a clever little trick I found that has helped me to preserve leading zeros and force an Excel cell to hold a character value even if it "looks" numeric. These options are used together to convert numeric values in a SAS dataset to a test field in an Excel spreadsheet. The simple code below creates the dataset "by_type", which will do the trick. The option "PREPEND_EQUALS" will place an equals sign in front of quoted numeric values, but only when the "QUOTE_BY_TYPE" option is set to "YES" to place numeric values in quotes. I have found this to be a useful pair of options because Excel treats the value as a text field. So, it is left justified, and leading zeros are retained. Then, the numeric value is left in quotes in the CSV file, but is displayed as a number in Excel.

**Figure 8.18: Code to Retain Leading Zeros When Sending Character Values to Excel.**

```
ODS TAGSETS.CSVALL BODY='c:\temp\csv\by_type.csv'
        OPTIONS(QUOTE_BY_TYPE='Yes' PREPEND_EQUALS='Yes');

DATA by_type;
        a = "010";
        b = a * 2;
RUN;

PROC PRINT DATA=by_type;
RUN;

ODS TAGSETS.CSVALL CLOSE;
```

The code in Figure 8.18 produces a *.csv file, and it can be opened by Excel as a worksheet called "by_type". It has two character variables and one observation. The PROC PRINT output is sent to the CSV file using the CSVALL tagset template.

**Figure 8.19: CSV Output File Produced by the Code in Figure 8.18.**

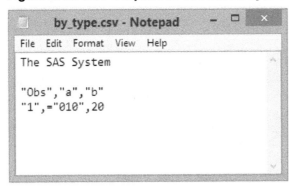

**Figure 8.20: Excel Output File Produced by the Code in Figure 8.18.**

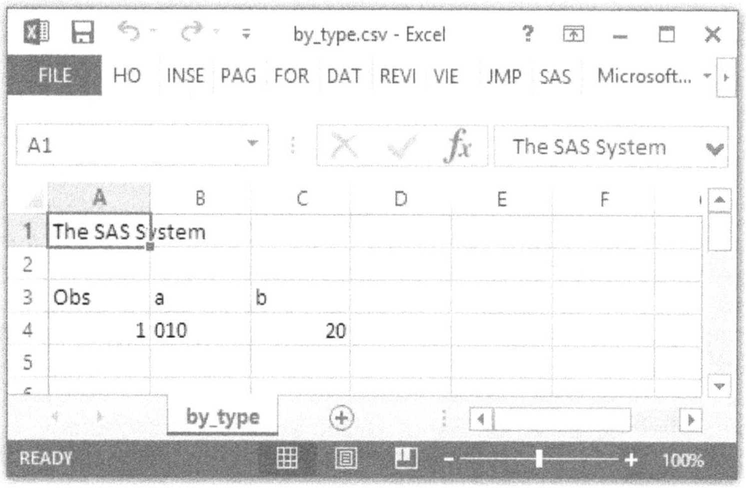

## 8.9 Syntax of ODS MSOFFICE2K Output Processes to Write HTML Files

The tagset template "MSOFFICE2K" is a Base SAS tagset template specifically designed to produce HTML output files that are intended to be loaded into Microsoft Office products. The parent tagset template for MSOFFICE2K is a version of the HTML tagset template. These tagset templates share the same options. The options for the MSOFFICE2K tagset template are described below, and in general they fall into several categories based upon what part of the output the option modifies.

### MSOFFICE2K Tagset Template Options Grouped by Features

I like to group the options by the parts of the output they affect, by the features if you will:

- Modify or display SAS outputs for titles, footnotes, bylines, summary, and table headers.
- Display of SAS log messages and tagset template Help messages.
- Page formatting and style sheet selection.
- Summary table formatting.

### ODS Syntax for MSOFFICE2K Tagset Template Use

The examples shown below will include ODS commands to execute the features for each example, but may not include all of the ODS options available for each command. Three examples are shown in this section that deal with the MSOFFICE2K tagset template. To get started let's use different ODS command sequences in Figure 8.21 that show three ways to execute the MSOFFICE2K processing. All three of these methods are part of Base SAS, and they all produce the same output.

**Figure 8.21: ODS MSOFFICE2K Calls Generate an Output HTML File.**

```
ODS MARKUP BODY='c:\temp\msoffice2k\shoes1.htm' TAGSET=MSOFFICE2K;
PROC PRINT DATA=sashelp.shoes;
RUN;
ODS MARKUP CLOSE;


*************************************************************;

ODS MSOFFICE2K BODY ='c:\temp\msoffice2k\shoes2.htm';
PROC PRINT DATA =sashelp.shoes;
RUN;
ODS MSOFFICE2K CLOSE;
```

```
*************************************************;
ODS TAGSETS.MSOFFICE2K BODY ='c:\temp\msoffice2k\shoes3.htm';
PROC PRINT DATA =sashelp.shoes;
RUN;
ODS TAGSETS.MSOFFICE2K CLOSE;
```

## 8.10 MSOFFICE2K Tagset Template Options

The following examples feature the MSOFFICE2K tagset templates, showing simple files being created and loaded into Excel using many of the options listed above. I have compiled a list of the options available for the MSOFFICE2K tagset template. The ODS Tagset MSOFFICE2K produces HTML formatted files.

**Table 8.4: MSOFFICE2K Options.**

| MSOFFICE2K Option | Description |
| --- | --- |
| **Options related to titles and footnotes** | |
| HEADER_DATA_ASSOCIATIONS | Associates data cells with header cells. (Used only in PROC Report.) |
| HEADER_DOTS | Add hidden dots to text in table headers. |
| PERCENTAGE_FONT_SIZE | Adjust titles and footnotes to be a percentage of font size. |
| **Options related to page formatting and style sheet selection** | |
| CSS_TABLE | Select type of style attributes to use. |
| PAGE_BREAK | Selects the type of page break to apply. |
| **Options related to summary tables** | |
| SUMMARY | Select quoted value to use as text on end of the table summary. |
| SUMMARY_AS_CAPTION | Select "YES" to use the table summary to create a table caption. |
| SUMMARY_BYVALS | Add the values of the BY variables to the table summary. |
| SUMMARY_BYVARS | Selecting "YES" will add the BY variable list to the table summary. |
| SUMMARY_PREFIX | Quoted value to be added to the end of the table summary. |
| **Options related to tagset template documentation** | |
| DOC | Write user Help information about the tagset template options to the system log. |

## 8.11 Overview of MSOFFICE2K Examples

The list of examples shown below generate HTML-formatted output text files. The recommended tagset for generating HTML files is the MSOFFICE2K tagset. This tagset template is a predecessor to the ExcelXP tagset. It also has a little more flexibility when it comes to what can be displayed. Some graphs are permitted in the MSOFFICE2K outputs. But the focus here is moving data.

**Table 8.5: MSOFFICE2K Examples.**

| Example Number | General Description |
| --- | --- |
| 8.12.1 | **Generating an HTML Output File with No Options.** This example is a simple output of the SASHELP Shoes dataset with the same output file being displayed in both Microsoft Excel and Internet Explorer. |
| 8.12.2 | **Generating HTML Output File Using the Summary_Vars Option.** This example shows how summary totals can be added to the output along with default page formatting. |

## 8.12 MSOFFICE2K Examples to Write HTML Files

### Example 8.12.1 Generating an HTML Output File with No Options

The following code produces an HTML-formatted file that is acceptable to Microsoft Excel, Microsoft Word, and many Internet browsers. While Microsoft Power Point cannot directly open one of these HTML files, after the output is opened by Microsoft Excel or Microsoft Word, the output can be copied and pasted into Microsoft PowerPoint.

**Figure 8.22: ODS Code to Use the MSOFFICE2K Tagset.**

```
ODS TAGSETS.MSOFFICE2K BODY ='c:\temp\msoffice2k\shoes3.htm';
PROC PRINT DATA =sashelp.shoes;
RUN;
ODS TAGSETS.MSOFFICE2K CLOSE;
```

The following figure shows the result of opening the file shoes3.htm using Microsoft Excel. Note that no special commands or options were used to produce this output.

**Figure 8.23: MSOFFICE2K Tagset Output with No Special Options Selected, Shown in Excel.**

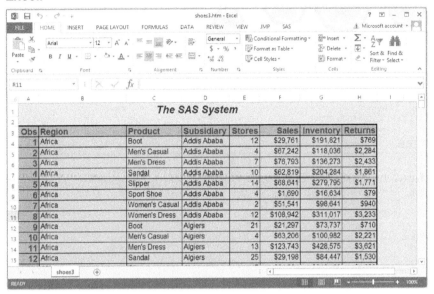

**Figure 8.24: MSOFFICE2K Tagset Output with No Special Options Selected, Shown in Internet Explorer.**

## Example 8.12.2 Generating an HTML File Using the Summary_Vars Option

The following code produces an HTML-formatted file that is acceptable to Microsoft Excel, Microsoft Word, and many Internet browsers. After the output is opened by Microsoft Excel or Microsoft Word, the output can be copied and pasted into Microsoft PowerPoint; however, if the amount of data exceeds one Power Point slide it may become unreadable, as shown in the code sample in Figure 8.25. The code in

Figure 8.25 is the output to the region "Asia" and allows the display of the summary information in one screen display. This example uses the default page formatting established on my system; yours may be different.

### Figure 8.25: Using Tagset MSOFFICE2K to Show and Summarize Data.

```
ODS TAGSETS.MSOFFICE2K BODY ='c:\temp\MSOFFICE2K\shoes4.htm'
     OPTIONS (summary_byvars='yes');
     PROC PRINT DATA =sashelp.shoes(where=(region='Asia'));
     BY region;
     SUM sales inventory returns;
     RUN;
ODS TAGSETS.MSOFFICE2K CLOSE;
```

### Figure 8.26: Output from the MSOFFICE2K Tagset Using Summary_byvars = 'YES'.

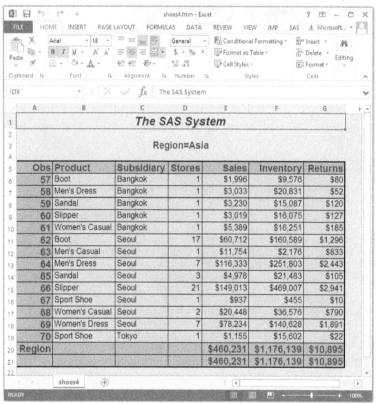

## 8.13 Syntax of the ODS EXCELXP Tagset Template Output Processes

The sole intent of the ExcelXP tagset template, as stated in the Help messages that are displayed, is to facilitate moving data from the SAS system to Microsoft Excel files. Not all versions of the ExcelXP tagset have the same features. However, each tagset version has only minor variations from one version to the next. Because the August 2013 version has the richest option set, it is explained in Table 8.7. Also, the ExcelXP tagset template must be accessed as a tagset. It cannot be referenced directly as an ODS destination. The output files are spreadsheets in XML format and cannot contain graphics.

The XML source documentation references that are listed below can be found in the Help output listing. The following SAS command will send the Help messages to the SAS Log window; valid options also include 'DOC' and 'ALL':

ODS TAGSETS.EXCELXP file='test.xml' options(doc='Help');

The SAS web site http://support.sas.com/rnd/base/topics/odsmarkup/ provides access to many of the ExcelXP versions listed in Table 8.6.

**Table 8.6: List of ExcelXP Tagset Releases.**

| Version | Comment |
|---|---|
| **V1.28** | Shipped with SAS V9.1.3M |
| **V1.37** | June 2006 version available on web site |
| **V1.62** | May 2007 version available on web site |
| **V1.72** | June 2007 version available on web site |
| **V1.86** | April 2008 version available on web site |
| **V1.94** | Shipped with SAS V9.2 * |
| **V1.116** | August 2010 version available on web site |
| **V1.122** | Shipped with SAS V9.3 * |
| **V1.130** | August 2013 version available on web site |

* Check your maintenance release to verify the ExcelXP tagset version shipped with your software.

The options for the ExcelXP tagset template are described below, and fall into several categories based upon what part of the output the option modifies.

### ExcelXP Tagset Template Feature Groups:
- Column features including width, repetition, and visibility of the column on the Excel spreadsheet.
- Data features ASCII dots, subtotals, Excel column filtering, numeric formatting, and formulas.
- Excel display features that include frozen row and column headers, alignment of missing values, page orientation, text wrapping in cells, and the Excel "ZOOM" level.
- Displaying Help text in the SAS Log window.
- Printing features controlling color printing, centering output, punctuation, print quality, scale, print order, page breaks, gridlines, and sizing the data to the pages.
- Row features including height and repetition.
- Title and footnote features that allow control of titles, footnotes including content, merging, and frequency.
- Workbook and worksheet features allowing the generation of blank sheets, table of contents sheets, index sheets, sheet names, sheet labels, and style features.

The SAS code below shows two ways to invoke the ExcelXP tagset. The ExcelXP tagset template is not available as an ODS destination.

### Figure 8.27: SAS Code Using the Markup Option to Invoke the ODS Tagset ExcelXP.

```
ODS MARKUP BODY='c:\temp\ExcelXP\shoes1.xml' TAGSET=ExcelXP;
PROC PRINT DATA=sashelp.shoes;
RUN;
ODS MARKUP CLOSE;
```

**Figure 8.28: SAS Code Using the Tagsets Option to Invoke the ODS Tagset ExcelXP.**

```
ODS TAGSETS.EXCELXP BODY='c:\temp\ExcelXP\shoes2.xml';
PROC PRINT DATA=sashelp.shoes;
RUN;
ODS TAGSETS.EXCELXP CLOSE;
```

## 8.14 ODS EXCELXP Tagset Options

The following examples feature the ExcelXP tagset template, showing simple files being created and loaded into Excel using many of the options listed above. The tagset template sub-options for the ExcelXP tagset template fall into the following general categories. The asterisk "*" indicates an option will be shown in an example below.

**Table 8.7: ExcelXP Options.**

| ExcelXP Option | Description |
| --- | --- |
| **Options related to titles and footnotes** | |
| EMBED_TITLES_ONCE * | If specified, embedded titles will appear only at the top of each worksheet. |
| EMBEDDED_FOOTNOTES | Put footnotes in the worksheet. |
| EMBEDDED_TITLES | Put titles in the worksheet. |
| MERGE_TITLES_FOOTNOTES | Merge left-justified titles and footnotes. |
| PRINT_FOOTER | If no footers or embedded footnotes are on, this is used as print footer. |
| PRINT_FOOTER_MARGIN | This is the footer margin as set in the page set-up dialog box. |
| PRINT_HEADER | If no titles or embedded titles are on, this is used as print header. |
| PRINT_HEADER_MARGIN | This is the header margin as set in the page set-up dialog box. |
| TITLE_FOOTNOTE_WIDTH | The number of columns titles and footnotes that are allowed to span. |
| **Options related to Excel display features** | |
| FROZEN_HEADERS* | Freeze rows from scrolling with the scrollbar. |
| FROZEN_ROWHEADERS | Freeze columns from scrolling with the scrollbar. |
| MISSING_ALIGN | Sets the alignment for missing values. |
| ORIENTATION* | Print orientation for the worksheet: portrait or landscape. |
| WRAPTEXT* | This value turns "WRAPTEXT" on and off for all style definitions. |
| ZOOM* | This value determines the zoom level on the worksheet. |
| **Options related to summary tables** | |
| SUMMARY | Select quoted value to use as text on end of the table summary. |
| SUMMARY_AS_CAPTION | Select "YES" to use the table summary to create a table caption. |
| SUMMARY_BYVALS | Add the values of the BY variables to the table summary. |
| SUMMARY_BYVARS | Selecting "YES" will add the BY variable list to the table summary. |
| SUMMARY_PREFIX | Quoted value to be added to the end of the table summary. |
| **Options related to Column features:** | |
| ABSOLUTE_COLUMN_WIDTH* | List of widths to use for each column. |
| COLUMN_REPEAT | Repeat columns across pages when printing. |
| DEFAULT_COLUMN_WIDTH | List of widths to use for each column in a table, if there are no widths. |

| ExcelXP Option | Description |
| --- | --- |
| HIDDEN_COLUMNS* | Range or list of column numbers to hide. |
| WIDTH_FUDGE | This value is used to calculate an approximate table column width. |
| WIDTH_POINTS | Override value for width calculations. |
| **Options related to Row features** | |
| AUTOFIT_HEIGHT | If selected, no row heights will be specified. |
| ROW_HEIGHT_FUDGE | A fudge value to add to the row height for each row. |
| ROW_HEIGHTS | Positional list of point sizes to use for row heights. |
| ROW_REPEAT | Repeat rows across pages when printing. |
| ROWCOLHEADINGS | This value turns on row and column headings for printing. |
| **Options related to Data features** | |
| ASCII_DOTS | Turn off and on leading dots in textual 'batch' output. |
| AUTO_SUBTOTALS | Add a subtotal function to the summary line of PROC PRINT. |
| AUTOFILTER* | Turn on auto filter for all columns or a range of columns. |
| AUTOFILTER_TABLE | Specify which table on the worksheet should get the filters. |
| CONVERT_PERCENTAGES | Remove percent sign, apply Excel percent format, and multiply by 100. |
| FORMULAS | Data values that start with an '=' will become formulas. |
| NUMERIC_TEST_FORMAT | Used for determining if a value is numeric or not. |
| THOUSANDS_SEPARATOR | The character used for indicating thousands in numeric values. |
| **Options related to Printing features** | |
| BLACKANDWHITE | This value turns on black and white for printing. |
| CENTER_HORIZONTAL | This value controls horizontal centering for printing. |
| CENTER_VERTICAL | This value controls vertical centering for printing. |
| CURRENCY_FORMAT | The currency format specified for Excel to use. |
| CURRENCY_SYMBOL | Detection of currency formats and symbols so that Excel outputs numbers. |
| DECIMAL_SEPARATOR | The character used for the decimal point. |
| DPI* | This value determines the dots per inch for printing. |
| DRAFTQUALITY | This value turns on draft quality for printing. |
| FITTOPAGE | Fit to page when printing. |
| GRIDLINES | This value turns on gridlines for printing. |
| PAGE_ORDER_ACROSS | If specified, the worksheet page order will be set to print across, and then down. |
| PAGEBREAKS | Insert page break lines in the worksheet. |
| PAGES_FITHEIGHT | Determine number of pages down to fit the worksheet when printing. |
| PAGES_FITWIDTH | Determine number of pages across to fit the worksheet when printing. |
| SCALE | This value determines the scale level for printing. |
| SHEET_INTERVAL* | Interval to divide the output between worksheets. |

| ExcelXP Option | Description |
| --- | --- |
| **Options related to Workbook features** | |
| BLANK_SHEET | Create a blank worksheet with the name given. |
| CONTENTS* | Create a worksheet that will contain a table of contents. |
| CONTENTS_WORKBOOK | Create a workbook table of contents and/or an index of workbooks and/or an index of worksheets. |
| INDEX* | Create a worksheet that will contain an index of worksheets. |
| MINIMIZE_STYLE | Minimize styles written to stylesheets. (Caution: Can cause unloadable XML files.) |
| SKIP_SPACE | Multiplier for the space that follows the different types of output. |
| SUPPRESS_BYLINES | Suppresses bylines in the worksheet. |
| SHEET_LABEL* | Replace the prefix of the worksheet name with this value. |
| SHEET_NAME* | Worksheet name to use for the next worksheet. |
| BLANK_SHEET | Create a blank worksheet with the name given. |
| **Options related to Tagset template documentation** | |
| DOC | Write user Help information about the tagset template options to the system log. |

## 8.15 Overview of EXCELXP Examples

Each of these examples writes an XML file that can be processed by Excel into a workbook.

### Table 8.8: Brief Description of ExcelXP Examples.

| Example Number | General Description |
| --- | --- |
| 8.16.1 | **Generating an XML Output File with No Options.** This example shows a simple ODS command and a PROC PRINT to output data into an XML file that can be read by Microsoft Excel. |
| 8.16.2 | **Adjusting Column Width Using Tagset Template Options.** The difference between this example and Example 8.16.1 is that the width of the columns has been adjusted by the ABSOLUTE_COLUMN_WIDTH option. This can aid in preparing a workbook for delivery. It is often time-consuming to adjust more than a few columns. |
| 8.16.3 | **Tagset Option to Hide Output Columns While Writing the File.** The added part of this example is the hidden columns. This is another task that can become repetitive. I like to let the program do this work for me. |
| 8.16.4 | **Apply an Excel "AUTOFILTER" to Selected Output Columns.** This example applies a filter to three Excel columns in the spreadsheet that is generated. I have selected the filter in the Excel image to show the contents of the filter. |
| 8.16.5 | **Using Multiple Options to Produce a "Ready-to-Print" Spreadsheet.** By applying multiple options, you can enhance your printable output and make your spreadsheets look better. This can help you get a spreadsheet ready to print before you open the spreadsheet. |
| 8.16.6 | **Creating a Table of Contents in an Excel Workbook.** This example produces a table of contents on a worksheet. You can use this table of contents to jump to the worksheet page that contains the data described. |

| Example Number | General Description |
|---|---|
| 8.16.7 | **Methods of Naming Excel Worksheets.** This example allows you to assign your own name to a worksheet. It moves away from letting ExcelXP assign it for you. |
| 8.16.8 | **Splitting One Report onto Multiple Excel Worksheets.** This example uses the "Sheet_Names" option with a BY statement in the PROC PRINT routine to apply a prefix to the names of worksheets generated. |
| 8.16.9 | **Methods of Placing Labels in Excel Worksheet Names.** By using the SHEET_LABEL option with the BY statement in a procedure, you can create worksheet names that include both the prefix and the BY variable value as part of the worksheet name. |
| 8.16.10 | **Use SHEET_INTERVAL= BYGROUP to Create Worksheets.** This example uses the BYGROUP setting of the SHEET_INTERVAL option to create a worksheet for each BY group of the output. Individual procedures may create separate output worksheet groups using this option. |
| 8.16.11 | **Use SHEET_INTERVAL= PROC to Create Worksheets.** This example uses the PROC setting of the SHEET_INTERVAL option to create a worksheet for each procedure group of the output. This will produce worksheets with multiple tables on a given worksheet based upon the number of procedures that are executed. |
| 8.16.12 | **Build Separate Worksheets with Titles on Each Sheet.** This example will place print-ready titles and footnotes into the Excel workbook. Each sheet could have unique titles and footnotes. |

## 8.16 EXCELXP Examples to Write XML Files

### Example 8.16.1 Generating an XML Output File with No Options

The code in Figure 8.29 produces an XML-formatted file that is acceptable to Microsoft Excel, and many Internet browsers. The output file can be opened with Excel, and this file has no special options set to enhance the output. Figure 8.30 shows the output in an Excel worksheet.

**Figure 8.29: ExcelXP Code to Write an Excel Workbook File with No Options Set.**

```
ODS TAGSETS.EXCELXP BODY='c:\temp\ExcelXP\shoes1.xml';
PROC PRINT DATA=sashelp.shoes;
RUN;
ODS TAGSETS.EXCELXP CLOSE;
```

**Figure 8.30: Output for Example Generating an ExcelXP File with No Options Set.**

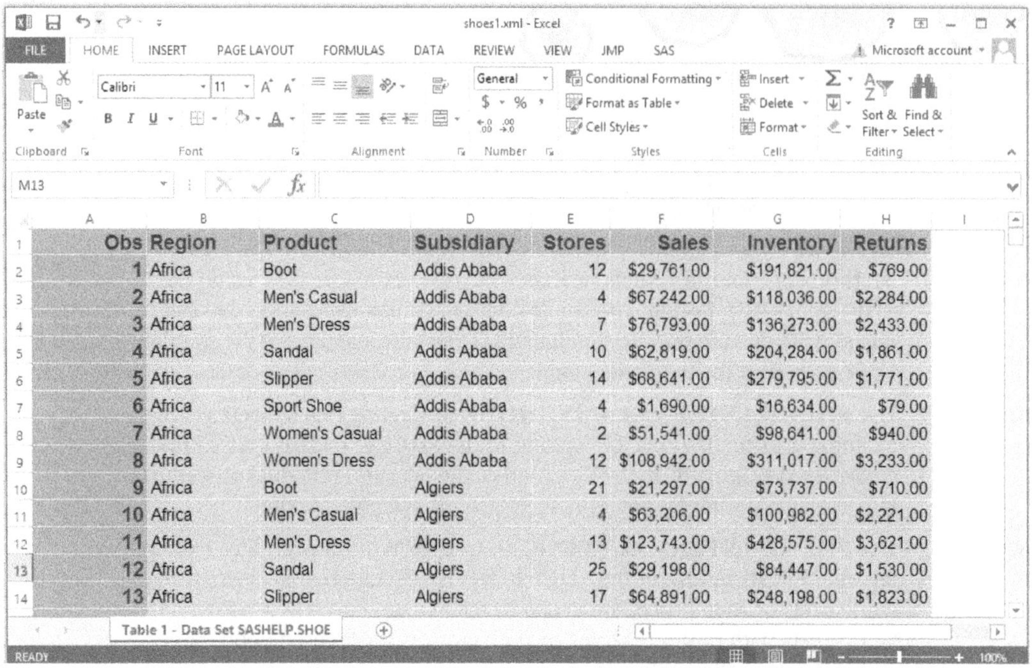

| Obs | Region | Product | Subsidiary | Stores | Sales | Inventory | Returns |
|---|---|---|---|---|---|---|---|
| 1 | Africa | Boot | Addis Ababa | 12 | $29,761.00 | $191,821.00 | $769.00 |
| 2 | Africa | Men's Casual | Addis Ababa | 4 | $67,242.00 | $118,036.00 | $2,284.00 |
| 3 | Africa | Men's Dress | Addis Ababa | 7 | $76,793.00 | $136,273.00 | $2,433.00 |
| 4 | Africa | Sandal | Addis Ababa | 10 | $62,819.00 | $204,284.00 | $1,861.00 |
| 5 | Africa | Slipper | Addis Ababa | 14 | $68,641.00 | $279,795.00 | $1,771.00 |
| 6 | Africa | Sport Shoe | Addis Ababa | 4 | $1,690.00 | $16,634.00 | $79.00 |
| 7 | Africa | Women's Casual | Addis Ababa | 2 | $51,541.00 | $98,641.00 | $940.00 |
| 8 | Africa | Women's Dress | Addis Ababa | 12 | $108,942.00 | $311,017.00 | $3,233.00 |
| 9 | Africa | Boot | Algiers | 21 | $21,297.00 | $73,737.00 | $710.00 |
| 10 | Africa | Men's Casual | Algiers | 4 | $63,206.00 | $100,982.00 | $2,221.00 |
| 11 | Africa | Men's Dress | Algiers | 13 | $123,743.00 | $428,575.00 | $3,621.00 |
| 12 | Africa | Sandal | Algiers | 25 | $29,198.00 | $84,447.00 | $1,530.00 |
| 13 | Africa | Slipper | Algiers | 17 | $64,891.00 | $248,198.00 | $1,823.00 |

## Example 8.16.2 Adjusting Column Width Using Tagset Template Options

This example shows how you can adjust the column width of each column in a worksheet by listing the size of each column as part of an option called ABSOLUTE_COLUMN_WIDTH. Each column is listed as a number describing the width of each column.

**Table 8.9: Options Used in Example 8.16.2.**

| ExcelXP Option | Setting in Example | Result |
|---|---|---|
| ABSOLUTE_COLUMN_WIDTH | '4,19,10.5,10,5,8,8,8' | Set column widths for each column. |

**Figure 8.31: ODS Code to Define Column Sizes While Writing the Output File.**

```
ODS TAGSETS.EXCELXP BODY ='c:\temp\ExcelXP\shoes2.xml'
    OPTIONS(ABSOLUTE_COLUMN_WIDTH='4,19,10.5,10,5,8,8,8');
    PROC PRINT DATA=sashelp.shoes;
    RUN;
ODS TAGSETS.EXCELXP CLOSE;
```

The code in Figure 8.31 will set the column widths to pre-selected values. Each number represents the width of a column. The values are assigned in the order listed. If more columns exist than numbers in the list, then the list is reused until all columns are assigned a width. Look at the differences in the column widths in Figure 8.30 and Figure 8.32.

### Figure 8.32: The Output Excel Worksheet Creates with Varying Sizes.

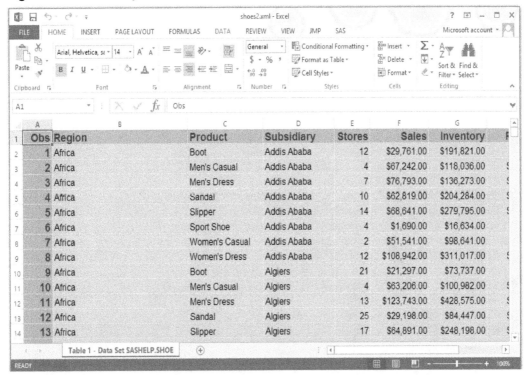

## Example 8.16.3 Tagset Option to Hide Columns While Writing the File

This next example uses the code from Example 8.16.2 and adds an option to hide some of the columns. The HIDDEN_COLUMNS option allows you to create a list of columns that you want to hide and does it without opening the Excel file. This little bit of automation can save a lot of time when preparing the Excel workbook for delivery to a user.

### Table 8.10: Options Used in Example 8.16.3.

| ExcelXP Option | Setting in Example | Result |
| --- | --- | --- |
| ABSOLUTE_COLUMN_WIDTH | '4,19,10.5,10,5,8,8,8' | Set column widths for each column. |
| HIDDEN_COLUMNS | '6-8' | Hide the output columns 6, 7, and 8 in the output Excel workbook. |

### Figure 8.33: Example That Builds on Example 8.16.2 and Also Hides Some of the Columns.

```
ODS TAGSETS.EXCELXP BODY='c:\temp\ExcelXP\shoes3.xml'
    OPTIONS(ABSOLUTE_COLUMN_WIDTH='4,19,10.5,10,5,8,8,8'
            HIDDEN_COLUMNS='6-8' );
    PROC PRINT DATA=sashelp.shoes;
RUN;
ODS TAGSETS.EXCELXP CLOSE;
```

**Figure 8.34: Output with Column Widths and Hidden Columns 6, 7, and 8.**

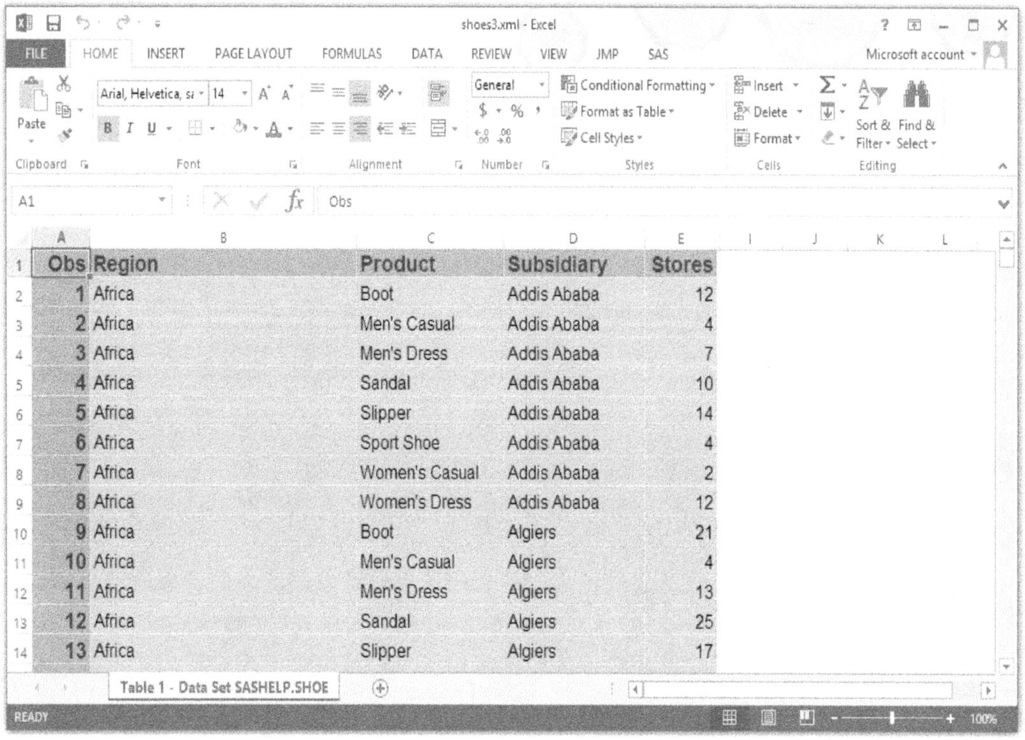
shoes3.xml - Excel
FILE HOME INSERT PAGE LAYOUT FORMULAS DATA REVIEW VIEW JMP SAS
Microsoft account
A1 Obs
Obs Region Product Subsidiary Stores
1 Africa Boot Addis Ababa 12
2 Africa Men's Casual Addis Ababa 4
3 Africa Men's Dress Addis Ababa 7
4 Africa Sandal Addis Ababa 10
5 Africa Slipper Addis Ababa 14
6 Africa Sport Shoe Addis Ababa 4
7 Africa Women's Casual Addis Ababa 2
8 Africa Women's Dress Addis Ababa 12
9 Africa Boot Algiers 21
10 Africa Men's Casual Algiers 4
11 Africa Men's Dress Algiers 13
12 Africa Sandal Algiers 25
13 Africa Slipper Algiers 17
Table 1 - Data Set SASHELP.SHOE
READY

I want you to notice here that columns F through H are hidden.

## Example 8.16.4 Apply an Excel "AUTOFILTER" to Selected Output Columns

The following code will set the column widths and turn "AUTOFILTER" on for columns 6, 7, and 8. (See columns "F", "G", and "H" below.) AUTOFILTER is an option in Excel that indicates special processing for an Excel spreadsheet column. The processing features in Excel include showing all values, selecting rows to show when only some of the values are visible, and hiding unwanted values. Columns set up in Excel worksheets for the AUTOFILTER processing have an arrow icon in the title row.

**Table 8.11: Options Used in Example 8.16.4.**

| ExcelXP Option | Setting in Example | Result |
|---|---|---|
| ABSOLUTE_COLUMN_WIDTH | '4,11,10.5,10,5,8,8,8' | Set column widths for each column. |
| AUTOFILTER | '6-8' | Turn on the filtering for columns 6, 7, and 8 in the output Excel workbook. |

**Figure 8.35: Using the ExcelXP AUTOFILTER Option to Enhance Several Columns.**

```
ODS TAGSETS.EXCELXP BODY='c:\temp\ExcelXP\shoes4.xml'
      OPTIONS(ABSOLUTE_COLUMN_WIDTH='4,19,10.5,10,5,8,8,8'
            AUTOFILTER='6-8'
            );
      PROC PRINT DATA=sashelp.shoes;
RUN;
ODS TAGSETS.EXCELXP CLOSE;
```

Figure 8.36 has one of the columns with the filter applied highlighted by opening the Filter tab to show some of the options available for selection in that column.

**Figure 8.36: Excel Output Showing a Filtered Column Created by ExcelXP Directly.**

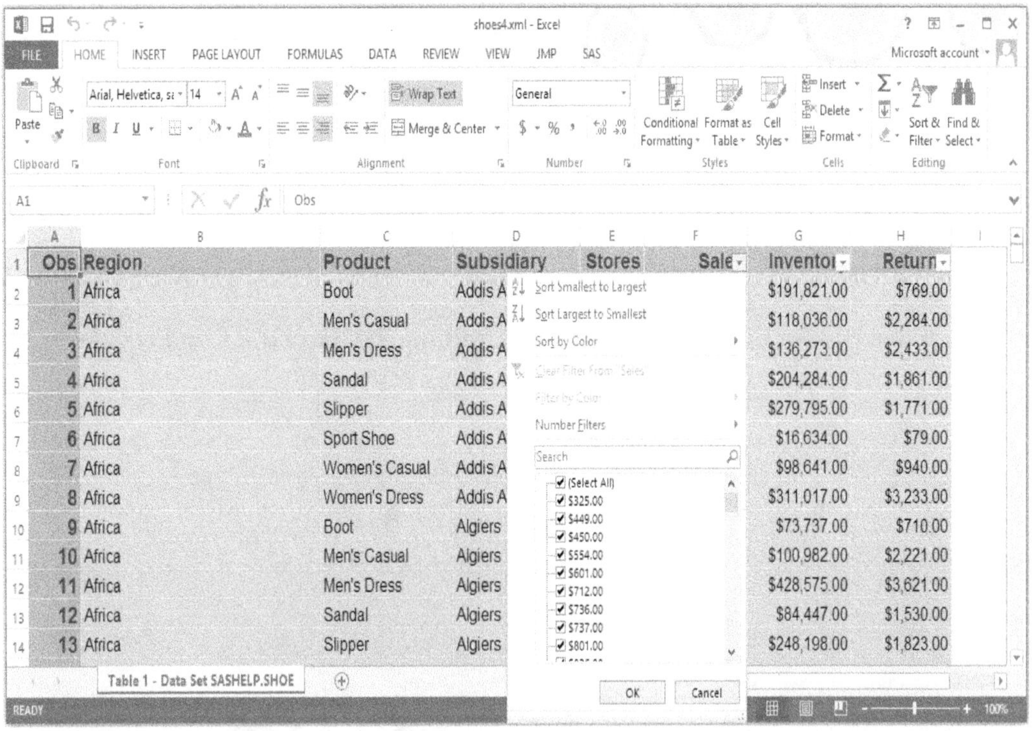

### Example 8.16.5 Using Multiple Options to Produce a "Ready-to-Print" Spreadsheet

This example uses several features of the ExcelXP tagset to produce a customized output worksheet in the Excel output workbook.

**Table 8.12: Options Used in Example 8.16.5.**

| ExcelXP Option | Setting in Example | Result |
| --- | --- | --- |
| ABSOLUTE_COLUMN_WIDTH | '4,11,10.5,10,5,8,9,8' | Set column widths for each column. |
| FROZEN_HEADERS | '1' | Freeze the first row of the spreadsheet when scrolling. |
| ORIENTATION | 'Landscape' | Pre-set the page output to landscape for printing. |
| WRAPTEXT | 'Yes' | Allow text in an Excel cell to wrap onto the next line. |
| ZOOM | '50' | Shrink the size of the displayed spreadsheet to 50% of the original size. |

### Figure 8.37: Code to Prepare a Worksheet for Printing Using ExcelXP Automatically.

```
ODS TAGSETS.EXCELXP BODY='c:\temp\ExcelXP\shoes5.xml'
      OPTIONS(ABSOLUTE_COLUMN_WIDTH='4,9,9,10,5,8,9,8'
              FROZEN_HEADERS='1'
              ORIENTATION='Landscape'
              WRAPTEXT='Yes'
              ZOOM='50');
       PROC PRINT DATA=sashelp.shoes;
       RUN;
ODS TAGSETS.EXCELXP CLOSE;
```

The Excel 2010 screen print shown below contains the spreadsheet image and the "Page Setup" popup menu image that can be found in the toolbar "File" and "Page Layout" options. The same "Page Setup" screen can be found on the "Page Setup" and "Page Layout" tabs when using Excel 2003 or Excel 2007.

### Figure 8.38 Excel Output Showing Print Ready Options That ExcelXP Can Provide.

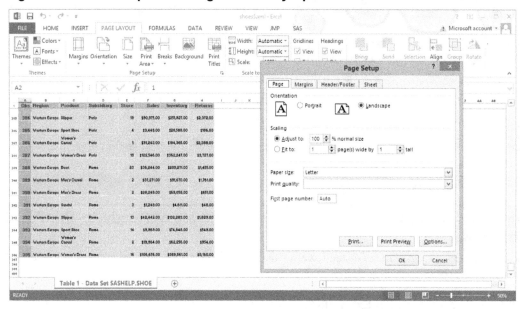

In Figure 8.38 I want to point out a few of the changes that show up here.

- Column "C" was defined a little smaller for this example. See the WRAPTEXT option detail.
- We are looking at the bottom of the file, but the header row is still visible because the ODS FROZEN_HEADERS option says to freeze the first line.
- The "PAGE SETUP" page indicates that the output when printed would print in Landscape mode.
- The WRAPTEXT option forces long lines to wrap onto the next line, and column "C" was shortened to get it to wrap. All rows are also output at the same height.
- The ZOOM option allows the image of the spreadsheet to fit on one half of the page, and the slider bar in the right-hand corner is scaled to 50%.

## Example 8.16.6 Creating a Table of Contents in an Excel Workbook

The following code will perform functions that have not been shown yet. First, the PROC PRINT step of the SAS code will create several worksheets in the workbook. One sheet for each value of the "BY variable" will be output by SAS. Second, the ODS commands will set the CONTENTS option in the ExcelXP tagset template, and this will cause the creation of a sheet in the output workbook with hyperlinks to each page in the workbook. Clicking on any of the hyperlinks (for example, "Region=Africa") will open the worksheet for that data. The ExcelXP option "BY" causes multiple worksheets to be created in this workbook.

**Table 8.13: Options Used in Example 8.16.6.**

| ExcelXP Option | Setting in Example | Result |
|---|---|---|
| CONTENTS | 'Yes' | Turn on the option to produce a worksheet with a "CONTENTS" page that lists all of the pages in the workbook, and provides a hyperlink to access each page. |

**Figure 8.39: EXCELXP Options to Create a Table of Contents in an XML Document.**

```
ODS TAGSETS.EXCELXP BODY='c:\temp\ExcelXP\shoes6.xml'
     OPTIONS(CONTENTS='Yes');
     PROC PRINT DATA=sashelp.shoes;
     BY region;
     RUN;
ODS TAGSETS.EXCELXP CLOSE;
```

**Figure 8.40: Result of the Execution of SAS Code in Figure 8.39.**

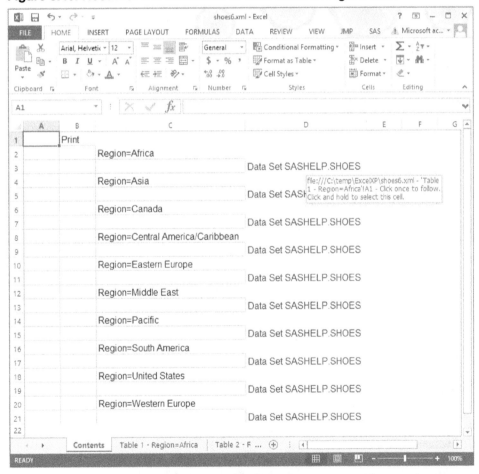

I have highlighted the hyperlink that the table of contents generated when writing this Excel workbook. When you are looking at the table of contents page, holding the mouse over the cell that contains either a "Region" name or the dataset name will expose the hyperlink instruction fly-over. Simply click on one of the cells to select the data for viewing.

## Example 8.16.7 Methods of Naming Excel Worksheets

Up until Example 8.16.5, the examples for ExcelXP had output only a single worksheet in the output Excel workbook, and the selection of the sheet name was left up to ExcelXP. Example 8.16.6 created more than one output worksheet, but ExcelXP still assigned the names of the worksheets. This example will use the SHEET_NAME option to give SAS control over creating the worksheet names.

**Table 8.14: Options Used in Example 8.16.7.**

| ExcelXP Option | Setting in Example | Result |
| --- | --- | --- |
| SHEET_NAME | 'All Regions' | Create a worksheet with the specified name. When only one worksheet is output, this is the name generated, up to 32 characters. |

**Figure 8.41: SAS Code to Assign a Worksheet Name.**

```
ODS TAGSETS.EXCELXP BODY='c:\temp\ExcelXP\shoes7.xml'
     OPTIONS(SHEET_NAME='All Regions' );
     PROC PRINT DATA=sashelp.shoes;
     RUN;
ODS TAGSETS.EXCELXP CLOSE;
```

By using this SAS ExcelXP option, you can rename the spreadsheets directly when they are being built, without opening Excel to change the sheet name. See Figure 8.42.

**Figure 8.42: Excel Workbook with a Sheet Name Assigned as the File Was Written.**

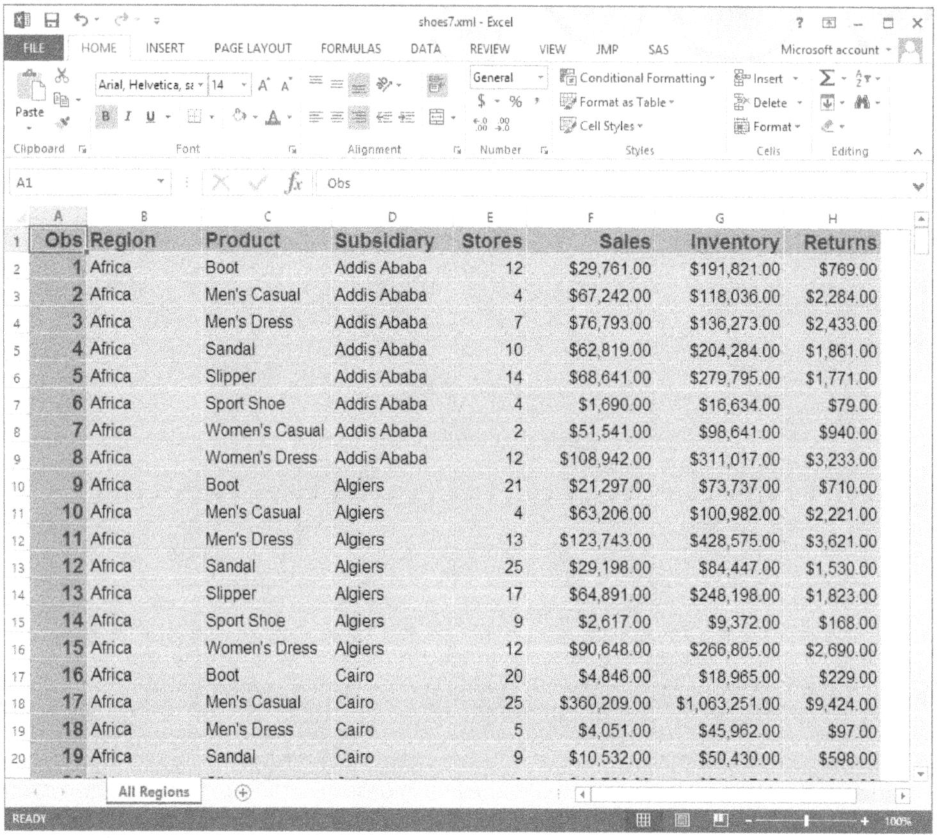

| Obs | Region | Product | Subsidiary | Stores | Sales | Inventory | Returns |
| --- | --- | --- | --- | --- | --- | --- | --- |
| 1 | Africa | Boot | Addis Ababa | 12 | $29,761.00 | $191,821.00 | $769.00 |
| 2 | Africa | Men's Casual | Addis Ababa | 4 | $67,242.00 | $118,036.00 | $2,284.00 |
| 3 | Africa | Men's Dress | Addis Ababa | 7 | $76,793.00 | $136,273.00 | $2,433.00 |
| 4 | Africa | Sandal | Addis Ababa | 10 | $62,819.00 | $204,284.00 | $1,861.00 |
| 5 | Africa | Slipper | Addis Ababa | 14 | $68,641.00 | $279,795.00 | $1,771.00 |
| 6 | Africa | Sport Shoe | Addis Ababa | 4 | $1,690.00 | $16,634.00 | $79.00 |
| 7 | Africa | Women's Casual | Addis Ababa | 2 | $51,541.00 | $98,641.00 | $940.00 |
| 8 | Africa | Women's Dress | Addis Ababa | 12 | $108,942.00 | $311,017.00 | $3,233.00 |
| 9 | Africa | Boot | Algiers | 21 | $21,297.00 | $73,737.00 | $710.00 |
| 10 | Africa | Men's Casual | Algiers | 4 | $63,206.00 | $100,982.00 | $2,221.00 |
| 11 | Africa | Men's Dress | Algiers | 13 | $123,743.00 | $428,575.00 | $3,621.00 |
| 12 | Africa | Sandal | Algiers | 25 | $29,198.00 | $84,447.00 | $1,530.00 |
| 13 | Africa | Slipper | Algiers | 17 | $64,891.00 | $248,198.00 | $1,823.00 |
| 14 | Africa | Sport Shoe | Algiers | 9 | $2,617.00 | $9,372.00 | $168.00 |
| 15 | Africa | Women's Dress | Algiers | 12 | $90,648.00 | $266,805.00 | $2,690.00 |
| 16 | Africa | Boot | Cairo | 20 | $4,846.00 | $18,965.00 | $229.00 |
| 17 | Africa | Men's Casual | Cairo | 25 | $360,209.00 | $1,063,251.00 | $9,424.00 |
| 18 | Africa | Men's Dress | Cairo | 5 | $4,051.00 | $45,962.00 | $97.00 |
| 19 | Africa | Sandal | Cairo | 9 | $10,532.00 | $50,430.00 | $598.00 |

I did compress the width of the columns to display all of the columns but the name was assigned when the file was generated.

## Example 8.16.8 Splitting One Report onto Multiple Excel Worksheets

This example uses the SHEET_NAME option with a BY statement. It causes the string specified to be used as a prefix for the sheet names. A numeric value is added to the prefix to generate the sheet names by ExcelXP, as shown below. This output report splits one PROC PRINT report into several worksheets.

**Table 8.15: Options Used in Example 8.16.8.**

| ExcelXP Option | Setting in Example | Result |
| --- | --- | --- |
| SHEET_NAME | 'One Regions' | Create a worksheet with the specified name. When more than one worksheet is output, unique worksheet names are generated by adding a number on the end of the sheet name, up to 32 characters. |

**Figure 8.43: Code to Create an Excel Workbook with Multiple Worksheets and User-Supplied Prefixes.**

```
ODS TAGSETS.EXCELXP BODY='c:\temp\ExcelXP\shoes8.xml'
        OPTIONS(SHEET_NAME='One Region');
    PROC PRINT DATA=sashelp.shoes;
        BY region;
    RUN;
ODS TAGSETS.EXCELXP CLOSE;
```

**Figure 8.44: Excel Output Using the "Sheet_Name" Prefix.**

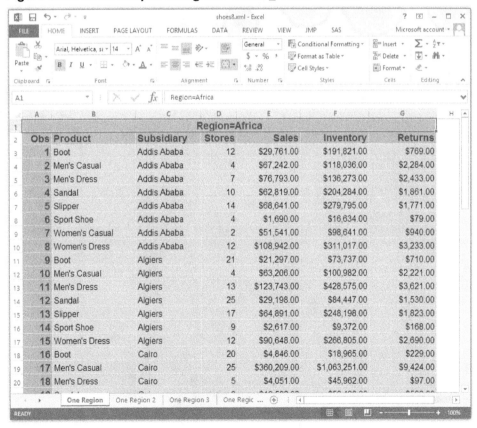

Notice that the worksheet has a "BY" variable title on the top of the worksheet identifying the data on the worksheet.

### Example 8.16.9 Methods of Placing Labels in Excel Worksheet Names

Using the "SHEET_LABEL" option with a BY statement causes the string specified by the option to be used as a prefix that is combined with a numeric value and the BY statement variable name and value to generate the sheet names by ExcelXP, as shown below as shown in Figure 8.46. The first sheet name is "One Region 1 – Region=Africa".

**Table 8.16: Options Used in Example 8.16.9.**

| ExcelXP Option | Setting in Example | Result |
| --- | --- | --- |
| SHEET_LABEL | 'One Region' | Create a worksheet with the specified name. When more than one worksheet is output, the SHEET_LABEL, a number, and the BY variable and value are used for the sheet name, up to 32 characters. |

**Figure 8.45: Code That Uses EXCELXP to Add a Label to All Sheet Names Generated.**

```
ODS TAGSETS.EXCELXP BODY='c:\temp\ExcelXP\shoes9.xml'
        OPTIONS(SHEET_LABEL='One Region');
    PROC PRINT DATA=sashelp.shoes;
    BY region;
    RUN;
ODS TAGSETS.EXCELXP CLOSE;
```

**Figure 8.46: Code That Uses the "Sheet_Label" Option to Add a Label to All Sheet Names Generated.**

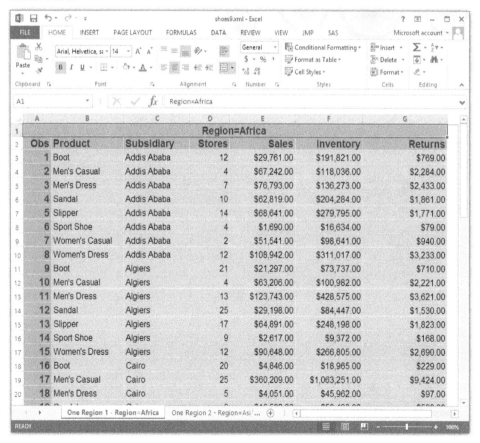

Notice that the worksheet names and the top of the worksheet both have the "BY" value that describes the variable name and its value.

## Example 8.16.10 Use SHEET_INTERVAL= BYGROUP to Create Worksheets

This example uses the SHEET_INTERVAL option to build a separate worksheet at different intervals. This ExcelXP option has several values that produce different results. See Table 8.18 for details. Here in this example we will use the "BYGROPUP" setting and in the next example, 8.16.11, we will use the "PROC" setting.

**Table 8.17: Options Used in Example 8.16.10.**

| ExcelXP Option | Setting in Example | Result |
| --- | --- | --- |
| SHEET_INTERVAL | 'BYGROUP' | Create a worksheet with the specified name. When more than one worksheet is output, the BY variable and value are used for the sheet name, up to 32 characters. |

**Table 8.18: Descriptions of Other Options Available for the SHEET_INTERVAL Option.**

| SHEET_INTERVAL | Result |
| --- | --- |
| **TABLE** | Split the outputs into new pages when a new "TABLE" starts. This is the default. |
| **PAGE** | Split the outputs into new pages when a new "PAGE" starts. |
| **BYGROUP** | Split the outputs into new pages when a new "BYGROUP" starts. |
| **PROC** | Split the outputs into new pages when a new "PROC" starts. |
| **NONE** | Does not apply the option. |

When you are using the SHEET_INTERVAL='BYGROUP' option in conjunction with the PROC PRINT "BY" statement, each BY group is printed on a separate worksheet with the value of the BY group variable as part of the name of the worksheet.

**Figure 8.47: Code to Use "SHEET_INTERVAL" for Labeling Worksheet Names.**

```
ODS TAGSETS.EXCELXP BODY='c:\temp\ExcelXP\shoes10.xml'
     OPTIONS(SHEET_INTERVAL='BYGROUP');
     PROC PRINT DATA=sashelp.shoes;
     BY region;
     RUN;
ODS TAGSETS.EXCELXP CLOSE;
```

Figure 8.48 shows the result of running the code in Figure 8.47 and causes the data for each region to be on a separate Excel worksheet.

## Figure 8.48: Result of Using "SHEET_INTERVAL" to Add a BYGROUP Name to a Sheet Name.

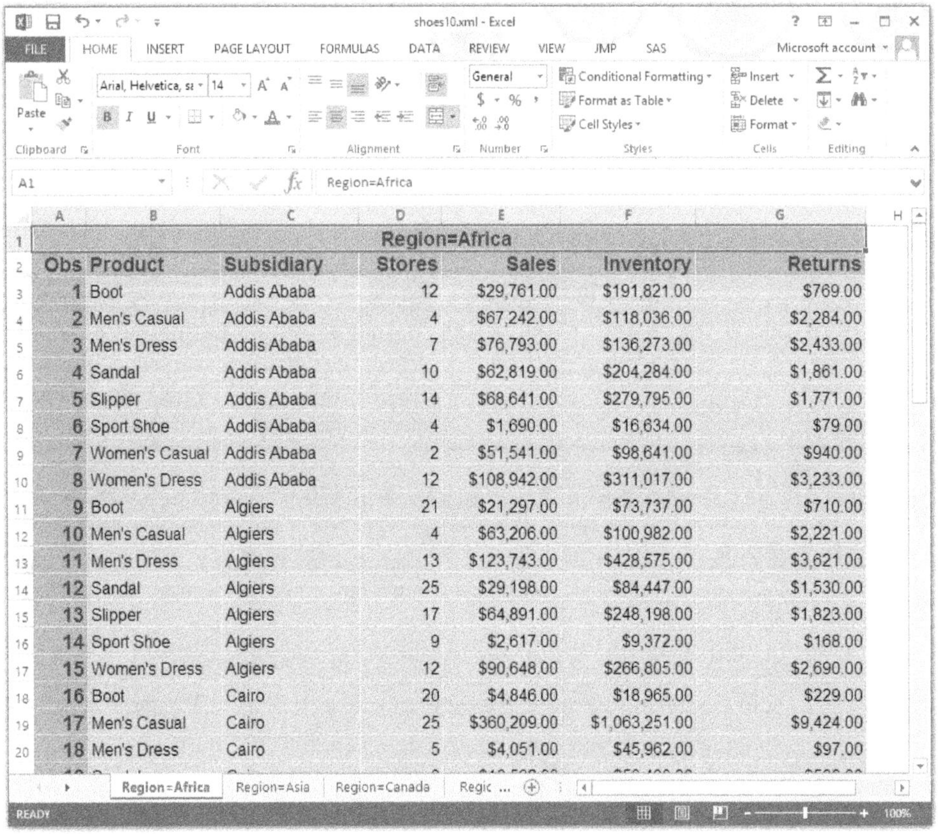

| Obs | Product | Subsidiary | Stores | Sales | Inventory | Returns |
|---|---|---|---|---|---|---|
| | | | Region=Africa | | | |
| 1 | Boot | Addis Ababa | 12 | $29,761.00 | $191,821.00 | $769.00 |
| 2 | Men's Casual | Addis Ababa | 4 | $67,242.00 | $118,036.00 | $2,284.00 |
| 3 | Men's Dress | Addis Ababa | 7 | $76,793.00 | $136,273.00 | $2,433.00 |
| 4 | Sandal | Addis Ababa | 10 | $62,819.00 | $204,284.00 | $1,861.00 |
| 5 | Slipper | Addis Ababa | 14 | $68,641.00 | $279,795.00 | $1,771.00 |
| 6 | Sport Shoe | Addis Ababa | 4 | $1,690.00 | $16,634.00 | $79.00 |
| 7 | Women's Casual | Addis Ababa | 2 | $51,541.00 | $98,641.00 | $940.00 |
| 8 | Women's Dress | Addis Ababa | 12 | $108,942.00 | $311,017.00 | $3,233.00 |
| 9 | Boot | Algiers | 21 | $21,297.00 | $73,737.00 | $710.00 |
| 10 | Men's Casual | Algiers | 4 | $63,206.00 | $100,982.00 | $2,221.00 |
| 11 | Men's Dress | Algiers | 13 | $123,743.00 | $428,575.00 | $3,621.00 |
| 12 | Sandal | Algiers | 25 | $29,198.00 | $84,447.00 | $1,530.00 |
| 13 | Slipper | Algiers | 17 | $64,891.00 | $248,198.00 | $1,823.00 |
| 14 | Sport Shoe | Algiers | 9 | $2,617.00 | $9,372.00 | $168.00 |
| 15 | Women's Dress | Algiers | 12 | $90,648.00 | $266,805.00 | $2,690.00 |
| 16 | Boot | Cairo | 20 | $4,846.00 | $18,965.00 | $229.00 |
| 17 | Men's Casual | Cairo | 25 | $360,209.00 | $1,063,251.00 | $9,424.00 |
| 18 | Men's Dress | Cairo | 5 | $4,051.00 | $45,962.00 | $97.00 |

Sheet tabs: Region=Africa | Region=Asia | Region=Canada | Regio...

## Example 8.16.11 Use SHEET_INTERVAL= PROC to Create Worksheets

This example uses the SHEET_INTERVAL option to build a separate worksheet at different intervals. This ExcelXP option has several values that produce different results. See Table 8.18 for details. Here in this example we will use the "PROC" setting, and in the last example, 8.16.10, we used the "BYGROUP" setting. This example writes two worksheets, one for each procedure executed.

### Table 8.19: Options Used in Example 8.16.11.

| ExcelXP Option | Setting in Example | Result |
|---|---|---|
| SHEET_INTERVAL | 'PROC' | Create a worksheet with the specified name. When more than one worksheet is output, the procedure name and a number are used for the sheet name, up to 32 characters. |

### Figure 8.49: Code to Create a Multi-Sheet Workbook with Data for Each Procedure on a Worksheet.

```
ODS TAGSETS.EXCELXP BODY='c:\temp\ExcelXP\shoes11.xml'
     OPTIONS(SHEET_INTERVAL='PROC');
     PROC PRINT DATA=sashelp.shoes;
     BY region;
     RUN;

*****************************************************;
```

```
        PROC FREQ DATA=sashelp.shoes;
        WEIGHT sales;
        TABLE region / LIST;
        RUN;
ODS TAGSETS.EXCELXP CLOSE;
```

**Figure 8.50: Page 1 of a Multi-Page Excel Worksheet Generated with SHEET_INTERVAL = "PROC".**

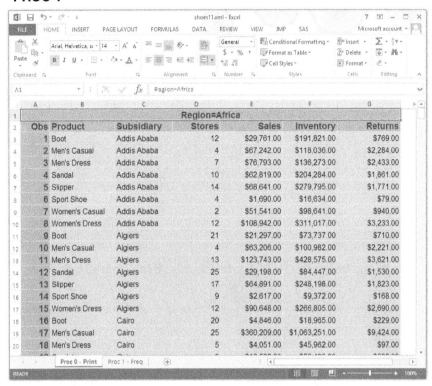

**Figure 8.51. Page 2 of a Multi-Page Excel Worksheet Generated with SHEET_INTERVAL = "PROC".**

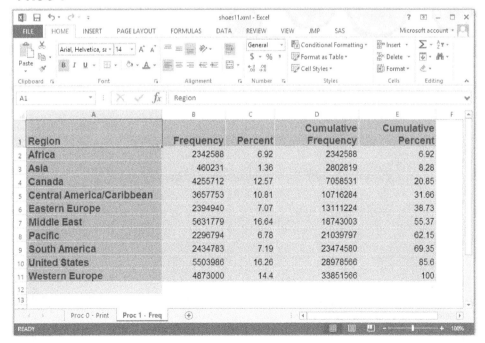

## Example 8.16.12 Build Separate Worksheets with Titles on Each Sheet

The next example uses the EMBED_TITLES_ONCE option. This example builds a workbook with separate sheets for each BY group, and the option places the title and footnote from the SAS code into the print header and print footer for the Excel pages.

### Table 8.20: Options Used in Example 8.16.12.

| ExcelXP Option | Setting in Example | Result |
| --- | --- | --- |
| EMBED_TITLES_ONCE | 'yes' | This option causes the SAS Title and Footnote values to be used as the Excel page setup Header and Footnote values. |

### Figure 8.52: Code to Put Heading and Footnote Information into Excel Worksheets.

```
ODS TAGSETS.EXCELXP BODY='c:\temp\ExcelXP\shoes12.xml'
   options(EMBED_TITLES_ONCE='yes');

   TITLE1 'This is my title line';
   FOOTNOTE1 'This is my footnote line';

   PROC PRINT DATA=sashelp.shoes;
   BY region;
   RUN;
ODS TAGSETS.EXCELXP CLOSE;
```

### Figure 8.53: EXCELXP Code to Preset the Print Features of an Excel Workbook.

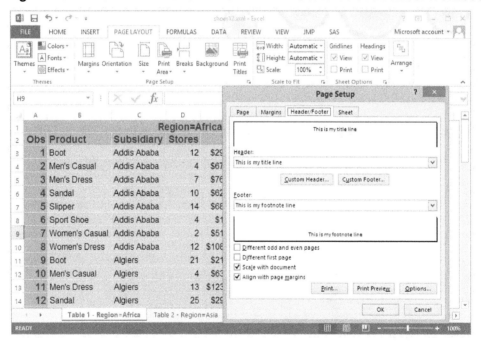

Notice here that I opened the Excel workbook and called up the "Page Setup" window manually to show that Excel has preloaded values from the SAS TITLE and FOOTNOTE statements from the SAS code in Figure 8.52.

## 8.17 The New ODS Destination EXCEL for Writing Workbooks

The last topic I want to discuss in this chapter is a new entry into the ODS arena of data transfer to Excel. ExcelXP is a tagset that has many features and can be changed by you. It is delivered as source code within the SASHELP.Tmplmst tagset files. This can be modified or a new version can be downloaded at any time. But, there is something new on the horizon. It is called the EXCEL ODS Destination. This will be a process included within SAS ODS features that is not available for updating by users. It was first shipped with SAS 9.4 TS Level 1M1. The outputs of this ODS destination are native format Excel workbooks for the version 2007 format and later.

As with all things SAS sends out in an "Experimental" mode, this ODS destination is likely to change before it is in a production mode. I will show you it exists, but will show no options because they may change.

**Figure 8.54: ODS EXCEL Destination Code to Produce an Excel Workbook.**

```
ODS EXCEL FILE='c:\temp\Excel\shoes_native.xlsx';

  PROC PRINT DATA=sashelp.shoes;
  BY region;
  RUN;

ODS EXCEL CLOSE;
```

**Figure 8.55: Output Native Excel File from the Code in Figure 8.54.**

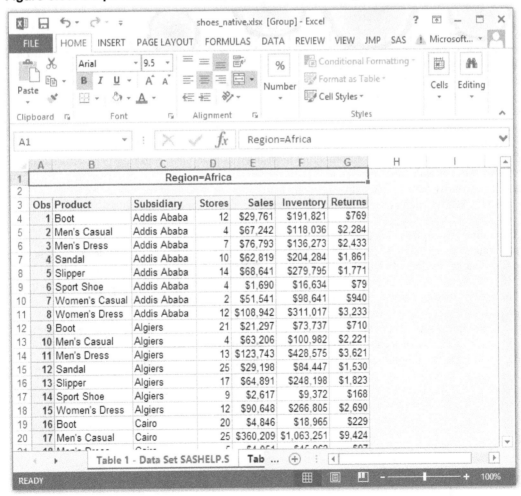

### Figure 8.56: Demonstration That the XLSX File Is Truly a Native Excel XLSX File.

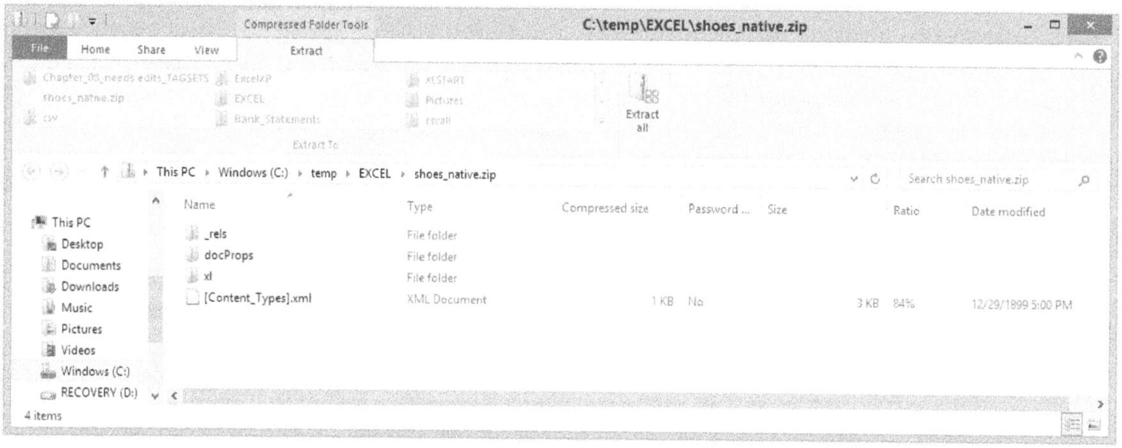

By changing the name to *.zip and opening the "zip" file you can view the XML code segments.

## 8.18 Conclusion

This chapter covers a lot of ground, ranging from the first ODS tools to write files that Excel could read to new tools that can build native format output files simply by declaring an output filename and the data to send.

While this chapter did not show examples of all of the CSV and CSVALL tagset template options, the featured examples may be useful in transporting data using text-based files. Most applications that I have seen that use CSV files use the files as a simple dump of the data from a SAS dataset. Perhaps some of the features shown here will add some punch to the CSV file you create in the future. These CSV files can actually be used to format a simple report for an Excel workbook.

I also did not show examples of all of the MSOFFICE2K tagset template options. The featured examples should be useful in transporting data using text-based files. Most applications that I have seen that use HTML files use the files to display information on an Internet or intranet web page. These files can be read by Excel workbooks and, since they are in HTML format, any version of SAS can write these files. Some of the features shown here should help you create better looking reports in the future. As with the CSV files, these HTML files can actually be used to format a report for an Excel workbook, too.

I did show many examples using the EXCELXP tagset template options. The featured examples can transport data to Excel using XML-based files. Most applications that I have written using these features create files that are loaded into Excel and rewritten as an XLS or XLSX Excel file. Excel will reformat these files in internal format from an XML file to an Excel formatted file. These files can be read by Excel workbooks and since they are in XML format any version of SAS can write these files.

Finally, I have introduced you to the future of ODS to Excel processing by showing you the new "EXCEL" destination for ODS. I have seen some of the new features but dare not go further because they may change. I will let it be enough the make you aware that the changes are coming.

# Chapter 9: Accessing Excel with OLE DB or ODBC Application Program Interfaces (API Methods)

| | |
|---|---|
| 9.1 Introduction | 149 |
| 9.2 Purpose | 149 |
| 9.3 Concept of the OLE DB or ODBC API Processes | 149 |
| 9.4 Guidelines for Setting Up OLE DB or ODBC Connections | 150 |
| 9.5 List of Examples | 150 |
| 9.6 Examples | 151 |
|     Example 9.1 Assign a Libref to an Excel Worksheet with the OLE-DB Dialog Box | 151 |
|     Example 9.2 Using LIBNAME Prompt Mode to Build an OLE-DB Connection | 152 |
|     Example 9.3 Using an OLE-DB init_string to Open an Excel Workbook | 154 |
|     Example 9.4 Using PROC CONTENTS to Verify Excel to OLE DB Connection | 154 |
| 9.7 Conclusion | 156 |

## 9.1 Introduction

The ability of SAS to reference external data sources has been enhanced by having access to application program interface (API) standards and implementations. SAS uses the Microsoft API interface routines (along with others) to communicate between SAS and other database systems. We are interested in the ability to communicate with the Microsoft application Excel. OLE DB is an API interface that can interface with many applications. The really great thing is that, as SAS programmers, we do not need to know how to access these API routines, or anything about them. API routines are the building blocks of an operation system. They allow an application to be programmed by using a standard interface--hence the name Application Program Interface. By opening an API through the OLE DB interface, a programmer could have direct SQL access to an OLE DB object, like an Excel workbook.

## 9.2 Purpose

This chapter shows you how to access Excel files from SAS using the OLE DB or ODBC API interfaces. While newer methods do exist, these are being included here because they are still a viable method of accessing Excel data from SAS.

## 9.3 Concept of the OLE DB or ODBC API Processes

One of the first places, and perhaps the main place, that the OLE DB interface surfaces is in the LIBNAME statement. There are specific options that address the OLE DB interface to Excel. Options that apply to OLE DB access to files other than Excel files are not shown here, but are easy to find in the SAS Help menus. Select "OLEDB" or "OLE DB" and the name of the database you are using. Options that apply to Microsoft JET or ACE providers are likely to be able to access Excel files. You can connect to OLE DB services or directly to the data provider, like the Microsoft JET or ACE providers. The syntax of the OLE DB options for the LIBNAME statement is shown below.

## 9.4 Guidelines for Setting Up OLE DB or ODBC Connections

OLE DB LIBNAME syntax is shown below in Table 9.1. The engine name OLEDB is required to access the OLE DB methods.

```
LIBNAME libref OLEDB <connection-options> <LIBNAME-options>;
```

### Table 9.1: OLE DB LIBNAME Syntax Options.

| Object | Description |
| --- | --- |
| Libref | This is a one- to eight-character SAS name that is used to reference the connection within the SAS code. This may point to a database, schema, server, tables, or a view supported by another system. |
| OLEDB | This is a constant that indicates that the OLE DB tools will be used to reference the external data files. The connection options and LIBNAME options are not required to be submitted to activate the LIBNAME statement.<br><br>A dialogue box will open and allow you to enter the required values when only the following is submitted:<br><br>LIBNAME *libref* OLEDB; |
| <connection-options> | These are options that specify how and where to connect to the external data source. |
| <LIBNAME-*options*> | These options are used to decide how to tune performance, access external records, and determine how to deal with variable names and other tasks. |

The examples in this chapter show how to connect to an OLE DB database management system. Space does not permit showing an example of each option. Generally speaking, there are two methods to connect with OLE DB. One is interactively by using the PROMPT= connection option (or getting it to appear because you omitted required information). Additional packages exist that allow interaction with other DBMS processing, but they are not discussed here. The following SAS software is needed to use the features explained in this chapter that access Excel files from SAS:

- SAS/ACCESS Interface to PC Files
- SAS/ACCESS Interface to ODBC
- SAS/ACCESS Interface to OLE DB

## 9.5 List of Examples

### Figure 9.5.1: List of Examples for PROC EXPORT.

| Example Number | General Description |
| --- | --- |
| 9.1 | **Assign a Libref to an Excel Worksheet with the OLE-DB Dialog Box.** This example shows the steps to assign an OLE DB connection to a libref. |
| 9.2 | **Using LIBNAME Prompt Mode to Build an OLE-DB Connection.** This example shows how to capture the init_string into the SAS log. The example also assigns the libref and opens the Excel workbook. |
| 9.3 | **Using an OLE-DB init_string to Open an Excel Workbook.** This example shows how to use the init_string to assign a file to the libref and open the Excel workbook. |
| 9.4 | **Using PROC CONTENTS to Verify Excel to OLE DB Connection.** This example prints the output from PROC CONTENTS. |

## 9.6 Examples

### Example 9.1 Assign a Libref to an Excel Worksheet with the OLE-DB Dialog Box

The SAS code to request that a dialog box be displayed is simple. The following code produces a dialog box:

```
libname my_lib OLEDB;
```

There is no real wizard for this processing, but the dialog boxes do a good job of allowing you to work your way through the process to connect to an OLE DB connection.

**Figure 9.1: Starting the Process of Assigning an OLE DB Connection.**

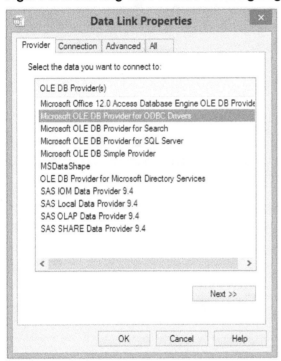

The options available will depend upon the software installed on your computer. For these examples we will use SAS 9.4 software on a 64-bit O/S. Other versions of SAS provide a similar window when running on a Windows operating system. The same window appeared on SAS 9.3 on a 32-bit O/S. You may have to try accessing more than one option on your computer to get the connection to work.

Pressing "Next" or "Connection" will display a screen similar to Figure 9.2. We will enter the required values onto this screen to point to the data files we want to process. Of course, you will enter values that are relevant to your data, your computer, or your network configuration. The four screen shots on the next page step through the process of setting up the connection.

**Figure 9.2: Data Link Properties Connection Tab.**

In Figure 9.2, you can pick a data source that relates to Excel files to proceed, and then enter the full path and file name of an Excel file in box 3 as shown in Figure 9.2.

**Figure 9.3: Data Link Properties Connection.**

After completing the connection manually, you can save the information that is needed for the SAS software. That way you can repeat the process. As long as your underlying configuration does not change, the commands needed to access the data will not change. Of course, if the file name changes, then that will need to change in your access string.

## Example 9.2 Using LIBNAME Prompt Mode to Build an OLE-DB Connection

The procedure required to capture the "init_string" is to capture the value of the SAS system macro variable "sysdbmsg". The "%put %superq(SYSDBMSG);" instruction works well. See the log output below.

Chapter 9: Accessing Excel with OLE DB or ODBC Application Program Interfaces (API Methods) 153

**Figure 9.4: SAS Code to Ask for a Prompt to Assign an OLE DB Connection to a Libref.**

```
LIBNAME my_lib OLEDB;
%put %superq(SYSDBMSG);
```

**Figure 9.5: SAS Log Showing Results of Assigning an OLE DB Connection and Saving the init_string.**

```
NOTE: This session is executing on the X64_8PRO  platform.
NOTE: Additional host information:X64_8PRO WIN 6.2.9200  Workstation
NOTE: SAS initialization used:
      real time           2.00 seconds
      cpu time            1.28 seconds

1     LIBNAME my_lib OLEDB;
NOTE: Libref MY_LIB was successfully assigned as follows:
      Engine:         OLEDB
      Physical Name:
2     %put %superq(SYSDBMSG);
OLEDB: Provider=MSDASQL.1;Persist Security Info=True;Data Source=Excel
Files;Initial Catalog=C:\My_Excel_Files\Excel_Shoes.xlsb
```

Figure 9.5 shows the init_string needed to process an OLE DB connection. Your output may vary. On the log listing above, the %PUT command wrote out the "OLEDB" string generated by the LIBNAME statement. The actual string is prefixed by "OLEDB:" and one or more spaces. This information can be captured (all except for the OLEDB: and any spaces) to be used as the INIT_STRING= value. Each of the parameters provided is separated by a semicolon. Spaces are not shown between the parameters, but they are permitted. Also, notice that the active libraries on the Explorer window now show a libref open for an Excel file.

**Figure 9.6: Data Read from the OLE DB Connection into a SAS Dataset.**

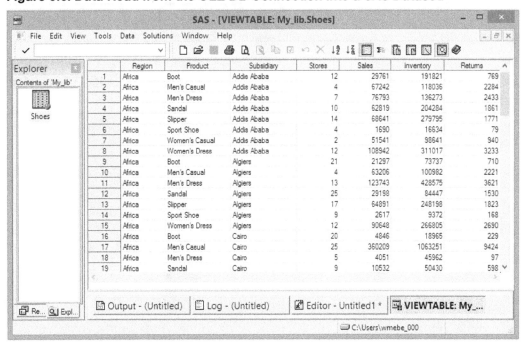

## Example 9.3 Using an OLE-DB init_string to Open an Excel Workbook

By using the output from the "%put %superq(SYSDBMSG);" to build a new LIBNAME statement that looks something like this (with one parameter per line), we can reassign the libref again using SAS code in our programs.

### Figure 9.7: SAS Code to Open an OLE-DB Connection to an Excel File.

```
libname my_lib oledb preserve_tab_names=yes
init_string="
Provider=MSDASQL.1;
Persist Security Info=True;
Data Source=Excel Files;
Initial Catalog=C:\My_Excel_Files\Excel_Shoes.xlsb";
%put %superq(SYSDBMSG);
```

The output SAS log looks something like Figure 9.8 with the actual init_string crossed out so that no one can copy your information.

### Figure 9.8: LIBNAME Assignment Using the OLE DB init_string= Option.

```
1       libname my_lib oledb preserve_tab_names=yes
2       init_string=X
3       XXXXXXXXXXXXXXXXXX
4       XXXXXXXXXXXXXXXXXXXXXXXXXX
5       XXXXXXXXXXXXXXXXXXXXXX
6       XXXXXXXXXXXXXXXXXXXXXXXXXXXXXXXXXXXXXXXXXXXX;
NOTE: Libref MY_LIB was successfully assigned as follows:
      Engine:         OLEDB
      Physical Name:
7       %put %superq(SYSDBMSG);
```

## Example 9.4 Using PROC CONTENTS to Verify Excel to OLE DB Connection

Using PROC CONTENTS you can determine if the data from the Excel workbook was able to be read by SAS. See Figure 9.9 for the code.

### Figure 9.9: PROC CONTENTS Code to Verify That the LIBNAME Connection Using OLE DB Is Working.

```
proc contents data=my_lib._all_;
run;
```

**Figure 9.10: Page One of the PROC CONTENTS Output.**

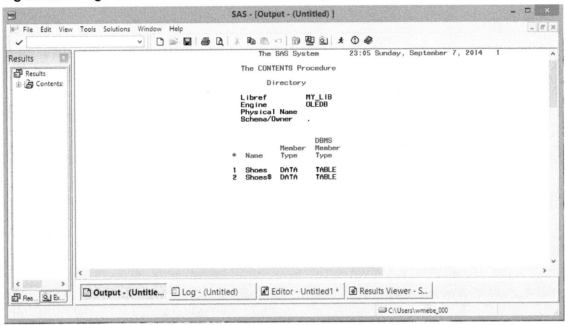

**Figure 9.11: Page Two of the PROC CONTENTS Output.**

### Figure 9.12: Page Three of the PROC CONTENTS Output.

```
                              The SAS System            23:05 Sunday, September 7, 2014    3
                              The CONTENTS Procedure

Data Set Name         MY_LIB.'Shoes$'n       Observations          .
Member Type           DATA                   Variables             7
Engine                OLEDB                  Indexes               0
Created               .                      Observation Length    0
Last Modified         .                      Deleted Observations  0
Protection                                   Compressed            NO
Data Set Type                                Sorted                NO
Label
Data Representation   Default
Encoding              Default

                     Alphabetic List of Variables and Attributes

        #   Variable     Type   Len   Format    Informat   Label
        6   Inventory    Num    8                          Inventory
        2   Product      Char   255   $255.     $255.      Product
        1   Region       Char   255   $255.     $255.      Region
        7   Returns      Num    8                          Returns
        5   Sales        Num    8                          Sales
        4   Stores       Num    8                          Stores
        3   Subsidiary   Char   255   $255.     $255.      Subsidiary
```

## 9.7 Conclusion

The examples in this chapter showed how to use the dialog boxes to make a connection and how to capture the information about that connection to create a LIBNAME statement. The LIBNAME connections open up all of the other uses and routines to use the data and procedures available to SAS users. The "preserve_tab_names=yes" on the LIBNAME statement allowed the SHOES$ table in the Excel file to be visible to PROC CONTENTS, and showed an example of how to submit LIBNAME options for processing. As mentioned before, full Excel pages are displayed with a "$" suffix, while Excel named ranges do not have a suffix.

# Chapter 10: Using PROC SQL to Access Excel Files

| | |
|---|---|
| 10.1 Introduction | 157 |
| 10.2 Purpose | 158 |
| 10.3 Basic Syntax of the SQL Procedure | 158 |
| 10.4 A Simple Explanation of SQL "PASS-THROUGH" Processing | 160 |
| 10.5 Overview of the Examples | 160 |
|     10.5.1 List of Examples | 160 |
| 10.6 Examples | 160 |
|     Example 10.1 LIBNAME Assignments to Access Excel Using PROC SQL | 160 |
|     Example 10.2 Create an Excel File, Read It with SQL, and Then Compare the Files | 161 |
|     Example 10.3 Use PROC SQL to Read a Subset of Records from an Excel Workbook | 162 |
|     Example 10.4 Use PROC SQL Pass-Through Facilities to Process an Excel File | 162 |
|     Example 10.5 Read a Pre-defined Range of Cells from an Excel Workbook | 163 |
|     Example 10.6 Calculate a New Variable within the SQL Code and Sort the Output | 165 |
|     Example 10.7 Examine the Contents and Structure of an Excel Workbook with a "PCFILES::" Special Query | 165 |
| 10.7 Conclusion | 166 |

## 10.1 Introduction

Many books have been written about SQL in general and the SAS SQL procedure in particular, and within those books many pages are devoted to Microsoft Access and SQL. However, the number of pages dedicated to using PROC SQL to access Excel files is far more limited. SQL and PROC SQL are considered tools for accessing database file systems where the files are interconnected by primary keys, secondary keys, and foreign keys, and the results are output as right-joined, left-joined and full-joined data sets. However, Excel is often overlooked by SQL users. Unlike entering data into an Excel spreadsheet, getting data into a database system is generally not as easy as double clicking on the file and starting to type. When you want to add data to a Microsoft Access database, you usually need a tool to enter the data; it is hard to open an Access database and just start typing. Because of that, the "rows" and "columns" of files in a Microsoft Access database (they are really called tables) are predictably uniform and considered fair game for tools like PROC SQL.

Excel is a more open set of files and data storage units (workbooks and worksheets). These are frequently edited and updated just by typing data into the files. This produces data files that are rarely uniform. They also often have columns with mixed data types. But, as this document has shown, there are ways to generate uniform Excel files and read data from these files. This chapter is about using PROC SQL to read and write data to Excel files. The ins and outs of learning PROC SQL are covered in many other documents. Therefore, this book will concentrate on showing you how to open Excel workbooks for moving data by reading and writing between SAS datasets and Excel workbooks.

## 10.2 Purpose

Within this chapter I intend to show how PROC SQL can locate and access the data within and about a Microsoft Excel workbook. The Structured Query Language standard that defines SQL and PROC SQL has been developed over a long period of time. The SQL standard is an independent query language intended to run on multiple operating systems and computer hardware. Some implementations of SQL extend the features and options to meet the needs and features of their system. For example, some date functions of Oracle SQL software do not produce the same output as similar date features of the SAS SQL implementation. While on the surface this seems like a problem, the SQL standard allows a feature called "Pass-Through". This allows the user to send instructions from one SQL platform to another. An example of this would be a SAS SQL user sending a command to retrieve a date value as it was formatted in Oracle directly to a SAS SQL table that was created.

## 10.3 Basic Syntax of the SQL Procedure

```
PROC SQL<options-list>
     CONNECT TO data-source-name AS <alias>
     <(connect-statement-arguments)>,<(database-connection-arguments)>
        DISCONNECT FROM <data-source-name> <alias>
           EXECUTE (data-source-specific-SQL-statement)
              BY <data-source-name> <alias>
                 SELECT column-list
                    FROM CONNECTION TO data source-name AS <alias>
<database-connection-arguments;>
```

### Ordering of the SELECT Statement Commands

When you build an SQL SELECT statement, the SQL syntax for clauses that you use dictates that they must appear in this order. The SQL system will verify that the commands are in the proper order before proceeding to execute the submitted SQL commands. These clauses are not discussed here because they are a part of the normal SQL syntax and do not directly relate to using Excel specifically. Only the SELECT and FROM clauses are required in order to use an SQL SELECT statement.

SELECT

FROM

WHERE

GROUP BY

HAVING

ORDER BY

In the SQL syntax definition above, the clauses below are used to connect to an external database.

- CONNECT TO
- DISCONNECT FROM
- EXECUTE BY
- CONNECTION TO

We will use these clauses to connect to Excel files. They are also useful to connect to other database systems like Microsoft Access. We will examine some PCFILES query options and JET/ACE query options. For you to connect to an external database using the pass-through facility, the following must occur:

Start PROC SQL.
Use the PCFILES engine with a CONNECT clause (and optionally any alias).
Provide any arguments and/or attributes required.
Issue the CONNECTION TO statement.

The following SAS PROC SQL syntax is used to connect to another data source, usually a different data base system. The CONNECTION TO statement below allows you to connect to another database system on either your current computer or another remote computer that could be thousands of miles away.

**Figure 10.1: SQL Code Syntax.**

```
SELECT column-list
FROM CONNECTION TO data-source-name AS <alias>
<database-connection-arguments;>
```

The "CONNECTION TO" option in SQL usually means you want to connect to something other than a SAS dataset. Example 10.4 below will show how to connect to an Excel workbook using this method. Knowing how to do this kind of connection is really useful, since this is a book about moving data between SAS and Excel.

The SQL command in Figure 10.2 is a little different from the others. Notice the "SELECT FROM" keywords that precede the CONNECTION TO statement. The power of this statement comes from the fact that the "SELECT FROM" pair can have nested commands. The code in Figure 10.2 has two "SELECT" and two "FROM" commands.

**Figure 10.2: SQL Code Syntax.**

```
SELECT column-list (SELECT
       FROM . . .
           WHERE . . .
           GROUP BY . . .
HAVING . . .
ORDER BY)
FROM CONNECTION TO ....<other SQL commands>
```

## The PCFILES Special Queries

These queries have a special format that includes the case-sensitive constant "PCFILES::" (yes, both colons are required) in front of the query name and arguments. Also note that the arguments themselves are case-sensitive and must be entered that way. An argument "OneTwoThree" is not the same as "onetwothree". When submitted to the query, the parameters in Table 10.6.1 below are enclosed in quotation marks like ("parameter"). Example 10.7 deals with some of these queries.

## The JET or ACE Special Queries

These queries have a special format that includes one of the case-sensitive constants "JET::" or "ACE::" in front of the query name and arguments. Also note that the arguments themselves are case-sensitive and must be entered that way. An argument "OneTwoThree" is not the same as "onetwothree". When submitted to the query, the parameters are enclosed in quotation marks ("parameter").

## 10.4 A Simple Explanation of SQL "PASS-THROUGH" Processing

SQL "PASS-THROUGH" processing usually involves sending commands from one version of SQL to another for execution. Just because there is a Structured Query Language Standard that each implementation of "SQL" follows, that does not mean that all SQL versions are the same. The SAS version of SQL has some features that do not adhere strictly to the standard. This is the same for other versions of SQL like Oracle, IBM, MySQL, and many others. Because SQL is designed to be a standardized language, it was necessary to implement a method of bridging the gap between what one language can understand and the language version of SQL that is used to access data stored on a specific computer. This method is called the "PASS-THROUGH" processing. It allows a SAS user to pass commands that are specific to Oracle to a computer storing data in an Oracle database, even if the commands would not have the correct syntax if executed as a SAS SQL command.

## 10.5 Overview of the Examples

### 10.5.1 List of Examples

Table 10.5.1: List of Examples for Using PROC SQL.

| Example Number | General Description |
| --- | --- |
| 10.1 | **LIBNAME Assignments to Access Excel Using PROC SQL.** This example shows how to code a LIBNAME statement when using PROC SQL with SAS 9.4. |
| 10.2 | **Create an Excel File, Read It with SQL, and Then Compare the Files.** This example creates an Excel file, reads the file using SAS PROC SQL, and compares the files. |
| 10.3 | **Use PROC SQL to Read a Subset of Records from an Excel Workbook.** This example uses a SAS WHERE clause to read a subset of the records from an Excel worksheet. |
| 10.4 | **Use PROC SQL Pass-Through Facilities to Process an Excel File.** This example uses a pass-through connection to send commands to a different SQL version. Here, PROC SQL is sending commands to Microsoft SQL to get data from the Excel worksheet. |
| 10.5 | **Read a Pre-defined Range of Cells from an Excel Workbook.** |
| 10.6 | **Calculate a New Variable within the SQL Code and Sort the Output.** |
| 10.7 | **Examine the Contents and Structure of an Excel Workbook with a "PCFILES::" Special Query.** |

## 10.6 Examples

### Example 10.1 LIBNAME Assignments to Access Excel Using PROC SQL

The code in Figure 10.3 shows the LIBNAME construct available when using PROC SQL for the three Excel formats that can be used as input and then output with PROC SQL in SAS 9.4. The macro version of Excel files (*.xlsm) is not supported, and older versions of SAS may support only some of these configurations.

### Figure 10.3: SAS LIBNAME Statements Valid Using 64-bit SAS on a 64-bit O/S with SAS 9.4.

```
LIBNAME out_data PCFILES PATH="C:\My_Excel_Files\Excel_Shoes.xlsb";
LIBNAME out_data PCFILES PATH="C:\My_Excel_Files\Excel_Shoes.xlsx";
LIBNAME out_data PCFILES PATH="C:\My_Excel_Files\Excel_Shoes.xls";
```

The *.xlsm version of the Excel files does not work in this configuration for a LIBNAME statement.

## Example 10.2 Create an Excel File, Read It with SQL, and Then Compare the Files

This example creates an Excel file with a LIBNAME statement, and a SAS DATA step then reads it into a SAS dataset with PROC SQL and compares the files. The first file to be created will be an Excel 2007 formatted file. Notice the *.xlsb extension on the output file. This extension is required. There are no macros allowed, and the Excel workbook is in binary format.

### Figure 10.4: SAS Code and SQL to Create an Excel Workbook, Read the Worksheet, and Compare.

```
LIBNAME out_data PCFILES PATH="C:\My_Excel_Files\Excel_Shoes.xlsb";
   DATA out_data."Shoes"n;
      SET SASHELP.Shoes;
   RUN;
   PROC SQL;
      CREATE TABLE New_shoes AS
      SELECT * FROM out_data."Shoes"n;
   QUIT;
   PROC COMPARE BASE=SASHELP.shoes COMP=New_Shoes;
   RUN;
```

The output report from PROC COMPARE shows some minor differences in the files. The labels are not output and minor differences appear in the FORMAT and INFORMAT values, but the data values are all equal. The next two figures (10.5 and 10.6) are partial output listings and show the differences.

The following is true of the output shown in Figure 10.5:

- It relates to the SAS dataset SASHELP.SHOES.
- The variables were created in this file without formats or informats for the character variables.
- Some of the numeric variables were assigned the DOLLAR12 format and informat to aid in the display of the data.
- Some of the variables were also given descriptive labels when the SAS dataset was created.

### Figure 10.5: SAS Variable Characteristics of SASHELP.SHOES Dataset.

```
Variable     Dataset         Type   Length  Format     Informat    Label
----------   --------------  -----  ------  ---------  ----------  ---------------
Region       SASHELP.SHOES   Char       25
Product      SASHELP.SHOES   Char       14
Subsidiary   SASHELP.SHOES   Char       12
Stores       SASHELP.SHOES   Num         8                         Number of Stores
Sales        SASHELP.SHOES   Num         8  DOLLAR12.  DOLLAR12.   Total Sales
Inventory    SASHELP.SHOES   Num         8  DOLLAR12.  DOLLAR12.   Total Inventory
Returns      SASHELP.SHOES   Num         8  DOLLAR12.  DOLLAR12.   Total Returns
```

The following is true of the output shown in Figure 10.6:

- It is the data sent to the Excel file by the SAS DATA step and reread from Excel by PROC SQL.
- The variable names were assigned as labels.
- The character variables now have formats and informats.
- The numeric variables do not have formats or informats.

**Figure 10.6: SAS Variable Characteristics of SAS Dataset Read by PROC SQL.**

```
Variable     Dataset           Type   Length  Format     Informat    Label
----------   ---------------   -----  ------  ---------  ----------  -----------
Region       WORK.NEW_SHOES    Char      25   $25.       $25.        Region
Product      WORK.NEW_SHOES    Char      14   $14.       $14.        Product
Subsidiary   WORK.NEW_SHOES    Char      12   $12.       $12.        Subsidiary
Stores       WORK.NEW_SHOES    Num        8                          Stores
Sales        WORK.NEW_SHOES    Num        8                          Sales
Inventory    WORK.NEW_SHOES    Num        8                          Inventory
Returns      WORK.NEW_SHOES    Num        8                          Returns
```

## Example 10.3 Use PROC SQL to Read a Subset of Records from an Excel Workbook

The code in Figure 10.7 uses the Excel workbook that was created in Example 10.2 by opening it with a LIBNAME statement. Then PROC SQL read a subset of the records and created a SAS dataset with those Excel rows. The comparison output is not shown because it looks very similar to Figure 10.6, but these SAS datasets contain only the data where Region equals Africa.

**Figure 10.7: SAS Variable Characteristics.**

```
LIBNAME out_data  pcfiles path="C:\My_Excel_Files\Excel_Shoes.xlsb";

PROC SQL;
   CREATE TABLE New_shoes AS
   SELECT *
       FROM out_data."SHOES$"n
       WHERE region='Africa';
QUIT;

PROC COMPARE
BASE=SASHELP.shoes(where=(region='Africa'))
COMP=New_Shoes;
RUN;
```

## Example 10.4 Use PROC SQL Pass-Through Facilities to Process an Excel File

This example uses the PCFILES engine for the connection to Excel. The code shown in Figure 10.8 is reading a named range of cells (called my_range) assigned by Excel to the cells B1 to G17 of an Excel workbook that contains the shoes data.

In the SAS code the PATH= option takes the place of the LIBNAME statement for defining the Excel file. The range named "**my_range**" is shown in Figure 10.8 below. This example takes advantage of the fact that the first row of data values from the Excel file are used as the variable names. Other ranges that do not include the first data row may produce unpredictable SAS variable names while opening the file because of the fact that the SAS normally uses first row of the input range to define the variable names. See Example 10.5 for a workaround the converts a blank row in the names range to special SAS variable names.

### Figure 10.8: SAS SQL Reading a Named Range of Excel Cells.

```
PROC SQL;
   CONNECT TO pcfiles AS db (PATH="C:\My_Excel_Files\Shoes_1.xls");
   SELECT * FROM CONNECTION TO db
        (
         SELECT * FROM "my_range"n
              );
   DISCONNECT FROM db;
   QUIT;
```

### Figure 10.9: The Input Excel File.

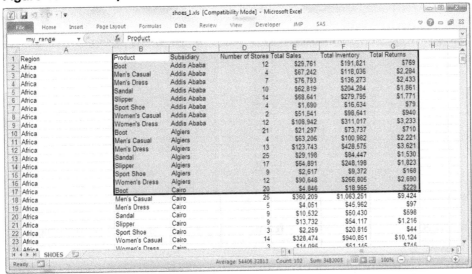

The SQL code in Figure 10.8 does not create a file. Instead, it just writes the data to the output listing window (or with the default settings for SAS 9.3 and above, it will write the data to an HTML display). Notice that the range "my_range" is case-sensitive in the code and here includes the top row of the Excel worksheet, but not the first column of the worksheet. The workbook in Figure 10.9 was the input file for this task.

## Example 10.5 Read a Pre-defined Range of Cells from an Excel Workbook

This example uses a PCFILES engine for the connection to Excel. The code reads a named range of cells (called New_range2) assigned by Excel. This range includes cells copied from [B10 to E24] to [I10 to L24] of an Excel workbook that contains the shoes data. This named range does not include a title row with variable names, but it does include a blank row of cells at the top of the range (starting at I9). This simulates a case where you may have to read a small group of cells from a spreadsheet, or even several groups of cells. SAS changes the way that SQL reads the spreadsheet and creates the SAS variable names. When the first row is blank, the names and labels are assigned the values F1, F2, F3 etc.

### Figure 10.10: Excel File with a New Named Range.

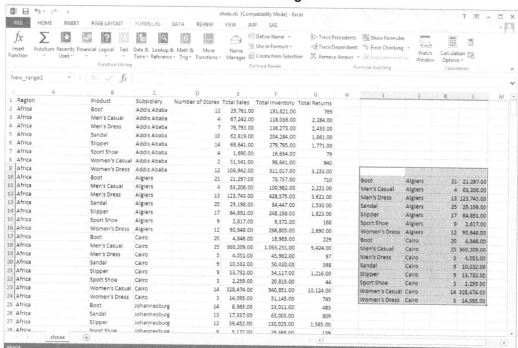

### Figure 10.11a: PROC SQL Code to Read a Named Range from an Excel Workbook.

```
PROC SQL;
   CONNECT TO pcfiles AS db (PATH="C:\My_Excel_Files\Shoes_1.xlsx");
   CREATE TABLE New_shoes1 as
   SELECT * FROM CONNECTION TO db
         (
           SELECT * FROM "New_range2"n
         ) ;
   DISCONNECT FROM db;
QUIT;
```

The SQL code in Figure 10.11a reads the Excel named range and converts the blank row in the named range to variable names F1, F2, F3, etc., the SAS variable names are generated, as shown in Figure 10.11b.

**Figure 10.11b: SAS Output Dataset from Reading the Named Range from Excel.**

| | F1 | F2 | F3 | F4 |
|---|---|---|---|---|
| 1 | Boot | Algiers | 21 | 21297 |
| 2 | Men's Casual | Algiers | 4 | 63206 |
| 3 | Men's Dress | Algiers | 13 | 123743 |
| 4 | Sandal | Algiers | 25 | 29198 |
| 5 | Slipper | Algiers | 17 | 64891 |
| 6 | Sport Shoe | Algiers | 9 | 2617 |
| 7 | Women's Dress | Algiers | 12 | 90648 |
| 8 | Boot | Cairo | 20 | 4846 |
| 9 | Men's Casual | Cairo | 25 | 360209 |
| 10 | Men's Dress | Cairo | 5 | 4051 |
| 11 | Sandal | Cairo | 9 | 10532 |
| 12 | Slipper | Cairo | 9 | 13732 |
| 13 | Sport Shoe | Cairo | 3 | 2259 |
| 14 | Women's Casual | Cairo | 14 | 328474 |
| 15 | Women's Dress | Cairo | 3 | 14095 |



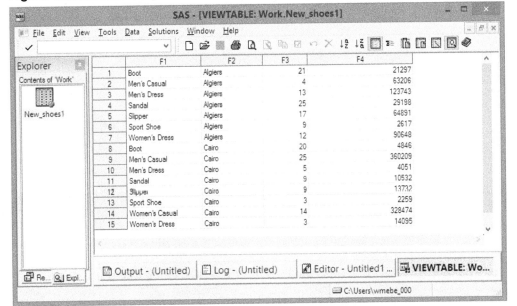

## Example 10.6 Calculate a New Variable within the SQL Code and Sort the Output

This example shows how to calculate a new variable within the SQL code. Here the calculation is done after the data is returned from the connection to Excel. There is no table created here; the data is just printed.

**Figure 10.12: SQL Code to Calculate a New Variable from an Excel Workbook.**

```
Proc SQL ;
   CONNECT TO PCFILES AS db (PATH="C:\My_Excel_Files\Shoes_2.xls");
   SELECT * ,Total_Sales - Total_Returns as net_sales
FROM CONNECTION TO DB
(SELECT * FROM "Shoes$"n
     WHERE region='Asia'
     ORDER BY product);
   DISCONNECT FROM DB;
   QUIT;
```

## Example 10.7 Examine the Contents and Structure of an Excel Workbook with a "PCFILES::" Special Query

Figure 10.13 shows a set of three "PCFILES::" special queries in one PROC SQL call. The first query shows the columns in the worksheet "Shoes$", the second shows the columns in all of the Worksheets and ranges in the workbook, and the third query shows all of the tables in the workbook.

**Figure 10.13: SQL Code to Examine the Structure of an Excel Workbook.**

```
PROC SQL ;
   CONNECT TO pcfiles AS db (PATH="C:\My_Excel_Files\Excel_Shoes.xlsb");
   CREATE TABLE work1 AS
         SELECT * FROM CONNECTION TO DB
                (PCFILES::SQLColumns "","","Shoes$","");
   CREATE TABLE work2 AS
         SELECT * FROM CONNECTION TO DB
                (PCFILES::SQLColumns "","","","");
   CREATE TABLE work3 AS
         SELECT * FROM CONNECTION TO DB
                (PCFILES::SQLTables "","","","");
   QUIT;
```

The results of the SQL code in Figure 10.13 is a set of three output tables, as shown in Figure 10.14. These tables contain information that is very similar to the output of PROC CONTENTS. The CREATE TABLE command could be eliminated and the result would print on the output listing destination. The outputs shown below are the SAS "Form View" to show variable names, but not all values.

**Figure 10.14: SAS Form View Showing the Result of the SQL Code in Figure 10.13.**

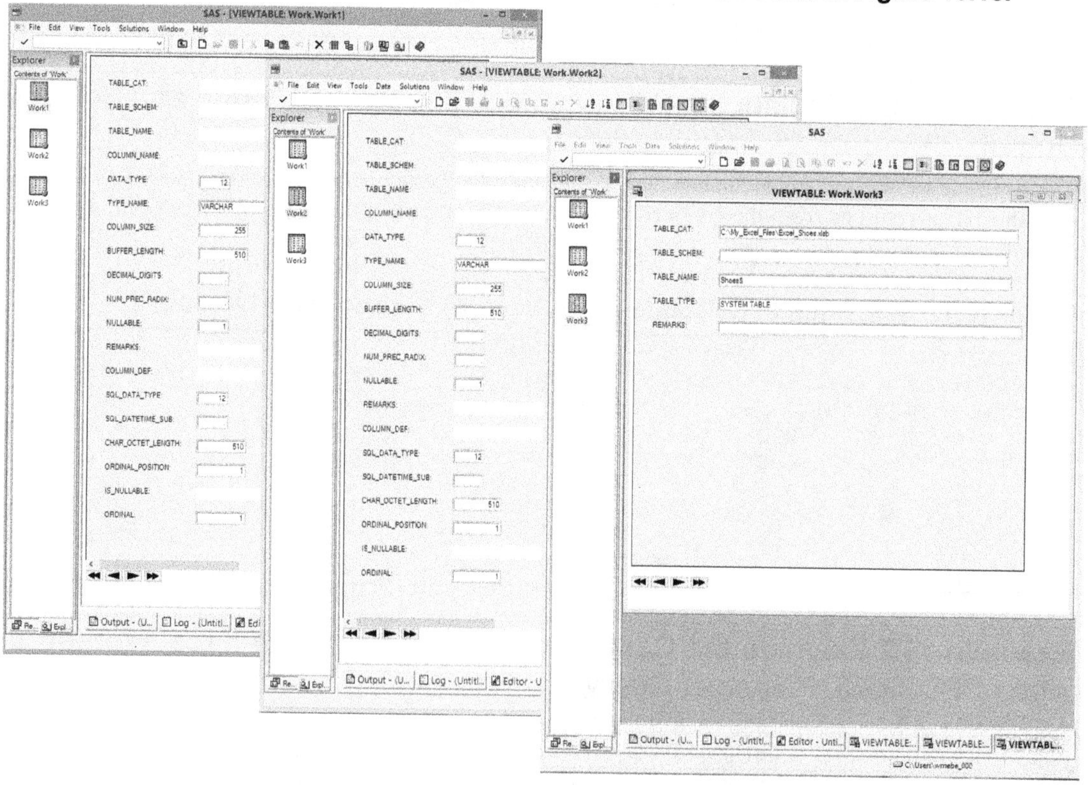

## 10.7 Conclusion

Many people use PROC SQL with their SAS programming applications; here, I have shown examples of how to use PROC SQL when reading and writing to an Excel workbook. When you write your programs, you should take into consideration both your hardware and the software that is available. This chapter presented alternatives to using PROC CONTENTS, and showed how to read data from Excel in several different ways. It is my hope that you will find these examples to be efficient and maintainable by you and the person that you might hand the program off to. Many books have been written about using different versions of SQL. It is becoming more popular as a programming tool. The SAS version of SQL called PROC SQL is also a popular topic for SAS authors. The few pages in this chapter looked at the usage of SQL as it relates directly to usage with Excel workbooks.

# Chapter 11: Using DDE to Read and Write to Excel Workbooks

**11.1 Introduction .................................................................................................... 167**
**11.2 Purpose ........................................................................................................... 167**
**11.3 Basic Concept of the DDE Client-Server Environment ............................... 168**
    11.3.1 How the DDE Client-Server Relationship Works ....................................... 168
    11.3.2 General DDE Syntax and Options .............................................................. 168
**11.4 List of User-Written SAS Macros That Can Enhance DDE Processing ....... 171**
    11.4.1 SAS Macro to Start Excel ........................................................................... 171
    11.4.2 SAS Macro to SAS to Issue Commands to Excel ...................................... 172
    11.4.3 SAS Macro to Define a Range of Excel Cells for Processing ................... 172
    11.4.4 SAS Macro to Save the Contents of an Excel Workbook .......................... 174
    11.4.5 SAS Macro to Close Excel Workbook ....................................................... 174
    11.4.6 SAS Macro to Write All or Selected Variables to an Excel Output Workbook . 175
**11.5 List of Examples ............................................................................................ 177**
**11.6 Examples ........................................................................................................ 177**
    Example 11.6.1 The Hello World Project ............................................................. 177
    Example 11.6.2 The Hello World Project When the Excel Workbook Is Closed .......... 179
    Example 11.6.3 The Hello World Project Using NOTAB and LRECL= Options ........... 180
    Example 11.6.4 Writing "Hello World" to an Excel File Using DDE Macros ................ 182
    Example 11.6.5 Writing a SAS Dataset to an Excel File Using the SAS_2_EXCEL DDE Macro ......................................................................................................................... 184
**11.7 Conclusion ..................................................................................................... 187**

## 11.1 Introduction

Dynamic Data Exchange (DDE) has been around for a long time. DDE is a feature of the Microsoft Office operating system and is available only if your computer is running Microsoft products. This is a method of trading data between clients and servers. In this case, SAS is the client and Excel is the server. The magic of DDE is that it boils down to using SAS as a way to enter data directly into an open Excel spreadsheet. You can also write to Microsoft Word and many other products. Note that the Excel spreadsheet must be open, and SAS must be running, because the data transfer is a "real time" "do it right now" type of a data transfer. But, never fear, the SAS application can open the Excel program and close it when you issue simple "X" commands.

## 11.2 Purpose

DDE is the oldest of the automated data transfer methods. The roots of DDE date back to Excel 4. Many programmers and companies still use this method today, but it seems to work best only if you know exactly where the input and output Excel data cells are within the worksheets. You as a programmer need to make your programs keep track of exactly which Excel worksheet cells are to be read from or written to.

Microsoft Excel 2007 and 2010 have a feature that allows them to ignore DDE commands. So, if DDE does not work, check to see if you have the proper settings to allow Microsoft products to process DDE commands. This option is found by opening the "Excel Options" and looking under the "Advanced" features. Also, DDE is a "real-time" application, so sometimes your SAS program has to wait. This can be accomplished through the SAS SLEEP function. SAS Enterprise Guide can execute DDE commands. However, SAS Enterprise Guide does not have access to the SAS "X" command; therefore, you must open the Excel files manually before running DDE commands when using SAS Enterprise Guide.

## 11.3 Basic Concept of the DDE Client-Server Environment

### 11.3.1 How the DDE Client-Server Relationship Works

A client-server relationship as it relates to computers means that one computer or program issues commands, or requests data, from another computer or program. That action in turn fills the request and returns the data to the original computer or program. When SAS uses DDE to communicate with Excel, the two programs are working together to transfer data or commands between SAS and Excel when they are running programs. Here are some of the actions available as DDE commands:

- Use the SAS "X" command to open an Excel workbook. (This is not really a DDE command, but the workbook needs to be open, and SAS needs to know how to reference the Excel file.)
- Read data from specific cells in the Excel worksheet.
- Write data to specific cells in the Excel worksheet.
- Open a DDE session.
- Close a DDE session.
- Save the output Excel file.

### 11.3.2 General DDE Syntax and Options

The DDE filename has two main forms: the first form points to Excel cells; the second form issues commands. The general syntax form is shown below.

Format 1:     **FILENAME fileref DDE 'DDE-triplet' <DDE-options>;**

Format 2:     **FILENAME fileref DDE "excel|system";**

**Figure 11.1: DDE Filename Options for Reading and Writing Data.**

| Format 1 - FILENAME fileref DDE 'DDE-triplet' <DDE-options>; | | |
|---|---|---|
| Fileref | This is the standard 8-character pointer used to identify any file that is not a SAS file. | |
| DDE | Is the engine name for the fileref to use for the output commands. | |
| 'DDE-TRIPLET' | Has the general form of 'application-name|topic!item'. An example would be 'excel|[book1.xls]Sheet1!r1c1:r1c2'. This "Triplet" served as the pointer to a location within the Excel file. The SAS "INPUT" and "PUT" commands start the process of transferring the actual data. | |
| | The Application Name is separated by "|" from the topic value. | The example shows 'excel' as the Application Name. |
| | The Topic Name is separated from the "item" by a "!". | The example shows '[book1.xls]Sheet1' as the Topic Name. |

| Format 1 - FILENAME fileref DDE 'DDE-triplet' <DDE-options>; | | |
|---|---|---|
| | The Item value is the last part of the triplet. | Here the item value is "r1c1:r1c2", which points to two cells in worksheet "SHEET1" (cells a1-a2). |
| | <DDE-options> | DDE Options describe additional features and functions that apply to the DDE Triplet. These features are described here. |
| | COMMAND | Some systems (other DDE server applications) may not accept the "system" command, and this is an alternate form of issuing commands to the system. This is not required for use with Excel. |
| | HOTLINK | This option opens an active link between SAS and Excel. Every time a cell in the Excel range is modified, SAS can detect the change by repeatedly checking the Excel libref with a SAS INPUT statement. When there are no changes, a zero length record is returned. |
| | LRECL= | This option specifies the length of the DDE character string records. The default on Windows is 256 bytes, but can be changed to be anywhere in the range of 1 byte to 1 gigabyte-1 (1,073,741,823 bytes – (2**30)-1), |
| | NOTAB | This keeps SAS from placing tab characters (ascii value hex '09') in output streams of data, and causes SAS to ignore tabs in input streams of data. This allows character strings with spaces included to be stored in a single Excel cell. When this option is used, SAS can accept delimiters other than a tab character in the input data. |

| Format 1 - FILENAME fileref DDE 'DDE-triplet' <DDE-options>; | | |
|---|---|---|
| | RECFM= | This option identifies the record format of the data. One of the following options can be specified on Windows. |
| | | V or D – This is the default and indicates a variable length record. |
| | | F – This is for fixed length records. |
| | | N – This is for binary input data. The data is read as a byte stream, and, without specifying LRECL, the data will be read in groups of 256 bytes. |
| | | P – This indicates a print format. |

**Figure 11.2: DDE Filename Options for Issuing Commands to Excel.**

| Format 2 - FILENAME fileref DDE "excel\|system"; | | |
|---|---|---|
| fileref | This is the standard 8-character pointer used to identify any file that is not a SAS file. | |
| DDE | Is the engine name for the fileref to use for the output commands. | |
| `'excel\|system'` | The Application Name is separated by "\|" from the topic value. | **"excel"** is the name of the server type. |
| | The Item value is the last part of the command. | **"system"** – A special topic name. The use of this syntax "excel\|system" allows SAS to act as a client application and use SAS INPUT and PUT commands to issue commands to Excel, the server application pointed to by the SAS FILENAME statement. |

## 11.4 List of User-Written SAS Macros That Can Enhance DDE Processing

This section describes the SAS macros that were written as a companion to the book. These macros are available on the SAS Author page. What follows here are written descriptions of the macros, their parameters, and their functions. These macros execute a DATA step, and therefore must be executed between DATA steps and procedure calls within your code, rather than within them. The macros were not tested with file or sheet names that had embedded blanks; this was left as an enhancement that you could work on to make the macros your own. The file type (*.xls or *.xlsx) must be the same for the input and output filenames because the file output formats are different and no checks are made to convert the Excel file formats. Because of the complexity of these examples they will be stored on the SAS Author page as Appendix C.

**Figure 11.4.1: Summary of SAS Macros to Process DDE Commands.**

| Macro Name | General Description |
| --- | --- |
| Start_Excel | Use the SAS X command to open an Excel workbook. |
| Open_cmd | Open an Excel DDE-triplet to issue Excel commands from SAS. |
| Out_range | Open an Excel DDE-triplet to transfer data between SAS and Excel. |
| Save_excel | This macro will save an Excel workbook, overwriting an existing one. |
| Close_excel | This macro will close an Excel workbook, preventing further updates to this session. |
| SAS_2_Excel | This macro uses all of the macros defined here to write out data from a SAS dataset to an Excel workbook. |

### 11.4.1 SAS Macro to Start Excel

**Macro Name:**

Start_Excel

**FUNCTION:**

Open an Excel workbook

**PARAMETERS:**

This macro accepts the following parameters:

- Full Excel Path and File name
- Length of time to wait for the Excel file to open

This macro accepts the full path and file name of an existing Excel file to open for use by SAS to interface with Excel. The macro verifies that a path and file name were provided, but does not verify that the path or file exists. The SAS code shows two possible locations of where the Excel executable file exists to aid the user in finding the Excel application location.

When the "X" command is executed by SAS to start Excel, the SAS program pauses to allow Excel to become fully operational before returning control back to the SAS runtime environment. If the default waiting period of 8 seconds is not the right amount of time, then the "wait" parameter can be adjusted to fine-tune the wait time needed.

### Result:

This macro opens the Excel file, but does not establish any filename connections to the Excel file. In other words, no DDE-triplets are opened by this macro. No data ranges or command paths are established by this macro.

### RESTRICTIONS:

If this macro fails to start the Excel workbook, it may be because the time for the operating system to respond exceeded the wait time parameter. Increase the amount of time for the wait time parameter and resubmit.

## 11.4.2 SAS Macro to SAS to Issue Commands to Excel

### Macro Name:

Open_cmd

### FUNCTION:

Open an Excel DDE-triplet to issue Excel commands from SAS.

### PARAMETERS:

This macro accepts the following parameters:

- Full Excel Path and File name
- Length of time to wait for the FILENAME command

This macro accepts a fileref name used to build an Excel DDE-triplet to issue commands to Excel from SAS. The macro verifies that a FILEREF name was provided, but does not verify that the FILEREF is valid. Any valid SAS FILEREF can be provided. If the FILEREF is the same as one already used, the new FILEREF will override the previous FILEREF and DDE-triplet if the FILEREF pointed to a DDE-triplet.

After the SAS FILENAME command is executed, the SAS program waits to allow Excel to become fully operational before returning control back to the SAS runtime environment. If the default waiting period of eight seconds is not the right amount of time, then the "WAIT" parameter can be adjusted to fine-tune the wait time needed.

### Result:

This macro establishes a FILENAME connection to the Excel file using the DDE "excel|system" triplet to open a command path to Excel.

### RESTRICTIONS:

If this macro fails to start the Excel workbook, it may be because the time for the operating system to respond exceeded the wait time parameter. Increase the amount of time for the wait time parameter and resubmit.

## 11.4.3 SAS Macro to Define a Range of Excel Cells for Processing

### Macro Name

Out_range

### FUNCTION:

Open an Excel DDE-triplet to transfer data between SAS and Excel.

## PARAMETERS:

This macro accepts the following parameters:

- file_ref — FILEREF name up to 8-character SAS reference name
- path_name — full path of Excel file
- file_name — full name of Excel file
- Sheet_name — Workbook sheet name (default name = Sheet1)
- Start_row_Col — Starting input/output data row and column (top-left of Excel range)
- End_row_col — Ending input/output data row and column (bottom-right of Excel range)
- no_tab — YES/NO answer, where YES = use notab option and NO = do not use notab
- L_REC_L — Logical Record Length value in the range of 1 byte to 1 gigabyte-1

This macro accepts a fileref name used to build an Excel DDE-triplet to point to a data range of Excel worksheet cells from SAS. The macro verifies that a FILEREF name was provided, but does not verify that the FILEREF is valid. Any valid SAS fileref can be provided. If the FILEREF is the same as one already used, it will override the previous FILEREF and DDE triplet if the FILEREF pointed to a DDE triplet.

The second parameter is the full path and file name of an existing Excel file to open for use by SAS to interface with Excel. The macro verifies that a path and file name are provided, but does not verify that the path or file exists.

The third parameter is the full file name of an existing Excel file to open for use by SAS to interface with Excel. The macro verifies that a path and file name are provided, but does not verify that the path or file exists.

The Sheet_name parameter is optional and points to the sheet name "Sheet1" as the default sheet name in the Excel file. The macros were not tested with file or sheet names that had embedded blanks; this was left as an enhancement that you can work on to make the macros your own.

The "Start_row_col" parameter points to the upper left Excel cell in the DDE range to be defined. The format of these values must be the R1C1 Excel formula format. The value used to point to cell "A1" as the upper left cell is "R1C1" and the cell "D10" is referenced as "R10C4" for the lower right cell of the range. (The notation stands for "Row" number "Column" number.)

The "End_row_col" parameter points to the lower-right Excel cell in the DDE range to be defined. The format of these values must be the R1C1 Excel formula format. The value used to point to cell "A1" as the upper left cell is "R1C1" and the cell "D10" is referenced as "R10C4" for the lower right cell of the range. (The notation stands for "Row" number "Column" number.)

The NOTAB parameter accepts a value of "NO" to apply the NOTAB feature to the SAS FILENAME command. Any other value for the parameter will suppress the "NOTAB" option on the FILENAME statement. This parameter is optional.

The L_REC_L parameter allows you to change the size of the output character fields from 256 bytes to the amount you specify in this parameter. (L_REC_L stands for Logical Record Length). The Excel default length is 256 bytes.

### Result:

This macro resets the FILENAME to point to a specific set of cells in the Excel workbook. SAS INPUT and PUTcommands control whether data is read from Excel or written to Excel.

## RESTRICTIONS:

The format of the row and column values must be the R1C1 excel formula format.

## 11.4.4 SAS Macro to Save the Contents of an Excel Workbook

### Macro Name:
Save_Excel

### FUNCTION:
This macro issues an Excel "SAVE.AS" command after turning off the Excel error messages. This will overwrite an existing Excel file without notifying you that the file was overwritten.

### PARAMETERS:
This macro accepts the following parameters:

- file_ref      - FILEREF name up to 8-character SAS reference name
- path_name      - full path of Excel file
- file_name      - full name of Excel file

This macro accepts an open fileref name, which is used to issue commands to Excel via DDE. The FILEREF must point to an "excel|system" triplet that is currently open. The macro verifies that a FILEREF name was provided but does not verify that the FILEREF is valid.

The second parameter is the full path of an existing Excel file that is opened for use by SAS to interface with Excel. The macro verifies that a path is provided, but does not verify that the path exists.

The third parameter is the file name of an existing Excel file that is opened for use by SAS to interface with Excel. The macro verifies that a file name is provided, but does not verify that the file exists.

### Result:
The contents of the Excel workbook are saved.

### RESTRICTIONS:
Directory paths and file names are assumed to have spaces within the values and are quoted within the macro. Enclosing parameters in quotes may produce unpredictable results.

## 11.4.5 SAS Macro to Close Excel Workbook

### Macro name:
Close_excel

### FUNCTION:
This macro issues an Excel "QUIT" command after turning off the Excel error messages. This will exit the Excel file without notifying you that the Excel is closing.

### PARAMETERS:
This macro accepts an open fileref name, which is used to issue commands to Excel via DDE. The FILEREF must point to an "excel|system" triplet that is currently active. The macro verifies that a FILEREF name is provided but does not verify that the FILEREF is valid.

### Results:
This macro will close the Excel workbook to further activity.

## RESTRICTIONS:
The FILEREF provided to this macro must be open, or unpredictable results may occur.

---

## 11.4.6 SAS Macro to Write All or Selected Variables to an Excel Output Workbook

### Macro Name:
SAS_2_Excel

### FUNCTION:
This macro is designed only to transfer data from SAS to Excel. The macro inspects the contents of the input SAS file and builds the DDE commands required to write the selected variables from the input SAS dataset to the output Excel workbook and worksheet. All records are written to the output Excel file, but not all variables need to be selected from the input SAS file. The order of the variables selected is the same as the order of the variables in the source SAS dataset. Therefore, if a specific order is required for the output variables, it must be established within the source SAS file prior to using this macro. A RETAIN statement can usually work to create the proper order if it needs to be changed.

### PARAMETERS:
- SAS_input_file      - SAS input file
- Excel_template_path      - Excel format file path
- Excel_template_name      - Excel format file name
- Excel_template_sheet      - Excel format sheet name
- Excel_output_path      - Excel output file path
- Excel_output_name      - Excel output file name
- Excel_starting_row      - First Excel output row
- Excel_starting_column      - First Excel output col
- notab=notab      - switch tab/notab output

The first parameter, "SAS_input_file", is the name of a SAS file. This is a fully qualified SAS name. It can be a file from the Work library, or another predefined libref, or from the USER libref or from a user-defined libref. This macro allows the inclusion of SAS dataset options that deal with dataset variable names like DROP, KEEP, or RENAME. But, data set options that deal with the data values are not acceptable to this macro, like the WHERE= clauses. You could add a feature to this macro that would make those acceptable.

The "Excel_template_path" parameter refers to a directory path where an input Excel file exists that can have preexisting formatting to be used as a template for the output Excel file layout of the output Excel dataset.

The "Excel_template_name" parameter refers to an input Excel file that can have preexisting formatting to be used as a template for the output Excel file layout of the output Excel dataset.

The "Excel_template_sheet" parameter is optional and points to the sheet name. The name "Sheet1" is used as the default sheet name in the Excel file.

The "Excel_output_path" parameter is the directory path of the output file. This can be the same as the input parameter "Excel_template_path" because there is no restriction on the usage. This macro was not tested with file or sheet names that had embedded blanks.

The "Excel_output_name" parameter is the file name of the output file. This can be the same as the input parameter "Excel_template_name" because there is no restriction on the usage. This macro was not tested with file or sheet names that had embedded blanks.

The "Excel_starting_row" parameter is an integer that specifies the row number of the "Top Left Cell" of the Excel range. Only the numeric portion of the R1C1 format of the cell formula is needed for the parameter. These values are substituted into the R1C1 excel formula format. The value used to point to cell "A1" as the upper left cell is "R1C1" and the lower right hand cell location is calculated from the PROC CONTENTS listing of the input SAS file. This also implies that the macro will not work with SAS files that are stored on sequential storage devices, like magnetic tape.

The "Excel_starting_column" parameter is an integer that specifies the Column number of the "Top Left Cell" of the Excel range. Only the numeric portion of the R1C1 format of the cell formula is needed for the parameter. These values are substituted into the R1C1 excel formula format. The value used to point to cell "A1" as the upper left cell is "R1C1" and the lower right hand cell location is calculated from the PROC CONTENTS listing of the input SAS file. This also implies that the macro will not work with SAS files that are stored on sequential storage devices, like magnetic tape.

The NOTAB parameter for this macro provides a switch to turn on or off the placement of a tab character between output data records when writing to the DDE destination.

## FUNCTION:

This macro inspects the contents of the input SAS file and builds the DDE commands required to write the selected variables from the input SAS dataset to the output Excel workbook and worksheet. All records are written to the output Excel file but not all variables need to be selected from the input SAS file. The order of the variables selected is the same as the order of the variables in the source SAS dataset. Therefore, if a specific order is required for the output variables, it must be established within the source SAS file prior to using this macro. A RETAIN statement can usually work to create the proper order if it needs to be changed.

The NOTAB parameter for this macro provides a default value of "NOTAB=NOTAB" and can be changed by providing the parameter in the macro call as "NOTAB=". The FILENAME statement in the SAS_2_Excel macro uses the NOTAB parameter macro value in the FILENAME statement proper. Providing a null value for the NOTAB parameter just placed a space into the FILENAME statement. The default is to place the characters "NOTAB" into the FILENAME statement. This applies the NOTAB feature to the SAS FILENAME command. Any other value for the parameter will cause a syntax error in the FILENAME statement. This parameter is optional.

### Results:

Data from a SAS dataset will be written to an Excel workbook in the specified rows and columns of the output Excel worksheet.

## RESTRICTIONS:

This macro makes use of SAS macro variable names that begin or end with an underscore to avoid conflicts with user-selected variable names. If a specific order is required for the output variables, it must be established within the source SAS file prior to using this macro. All of the parameters except "Excel_template_sheet" and "NOTAB" are checked to see if they are provided, but are not checked for validity.

## 11.5 List of Examples

**Figure 11.5.1: List of Examples for PROC EXPORT.**

| Example Number | General Description |
|---|---|
| 11.6.1 | **The Hello World Project.** Write the simple phrase "Hello world" to an Excel workbook using DDE when the workbook is open. |
| 11.6.2 | **The Hello World Project When the Excel Workbook Is Closed.** Write the simple phrase "Hello World" to an Excel workbook using DDE when the workbook is closed. |
| 11.6.3 | **The Hello World Project Using NOTAB and LRECL= Options.** |
| 11.6.4 | **Writing "Hello World" to an Excel File Using DDE Macros.** |
| 11.6.5 | **Writing a SAS Dataset to an Excel File Using the SAS_2_EXCEL DDE Macro.** |

Many of the examples below use the path to Excel to open the Microsoft Office EXCEL.EXE program. These paths are the typical default paths for the EXCEL.EXE program. These may be needed to execute Excel from within SAS by using the SAS "X" commands to access an operating system command.

**Figure 11.5.2: Common Default Locations for Excel Software.**

| | Common Default Locations for Microsoft Excel Software |
|---|---|
| Excel 2013 | C:\Program Files\Microsoft Office 15\root\office15\EXCEL.EXE |
| Excel 2010 | C:\Program Files\Microsoft Office\Office14\EXCEL.EXE |
| Excel 2007 | C:\Program Files\Microsoft Office\Office12\EXCEL.EXE |
| Excel 2010 (32-bit) | C:\Program Files (x86)\Microsoft Office\Office14\EXCEL.EXE |
| Excel 2007 (32-bit) | C:\Program Files (x86)\Microsoft Office\Office12\EXCEL.EXE |

**NOTE:** When using Excel 2013, I found that I needed to suppress the "Splash Screen" by using an Excel command line switch when starting Excel. The switch I used was "/x" to suppress the splash screen. In addition, I was not able to get the DDE process to work with any spaces in the file path or name. My testing was not exhaustive, so you may find other combinations that work. This "/x" switch and/or unquoted file path and name may not be required for other versions of Excel.

## 11.6 Examples

### Example 11.6.1 The Hello World Project

Figure 11.6.1 shows how to write the words "Hello" and "World" to an open Excel spreadsheet. Note that the code does not have a defined path to find the Excel file. This will work only if the open Excel workbook file is located in the same directory that SAS is using. The SAS display window has a small icon and directory path shown in the bottom tray that indicates the current folder (see Figure 11.6.2). I found that by double-clicking on the folder name I could get a dialogue box to allow me to change the current folder. This is in lower left corner for SAS 9.3 and in the lower right corner for 9.2 and 9.4. The code in Figure 11.6.1 does not open an Excel file; the Excel file Book1 must have been opened in the current SAS directory prior to running this SAS code.

### Figure 11.6.1: SAS Code for the DDE Hello World Test.

```
* this worked on 32 and 64 bit o/s with Excel 97-2010 (xls, and xlsx);
* this code looks for the Excel file in the current folder;
* The Excel file must be open when the SAS code is executed;
FILENAME ddewrite DDE 'excel|[book1.xls]Sheet1!r1c1:r1c2';
DATA OUT;
   FILE ddewrite;
   x='Hello World';
   PUT x;
RUN;
```

### Figure 11.6.2: SAS Code Window with Current Directory Location Highlighted in Red Circle.

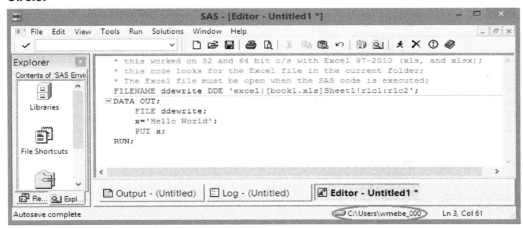

A double-click on the folder name in the SAS window causes a window similar to the following to appear. These screens are from SAS 9.4.

### Figure 11.6.3: SAS Window to Allow You to Change the Current SAS Folder.

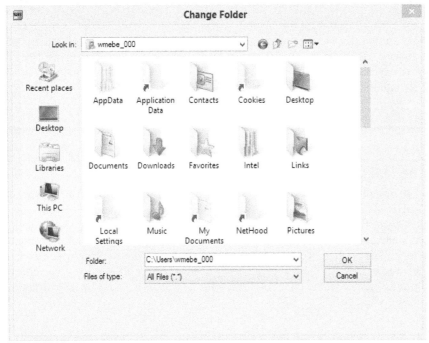

**Figure 11.6.4: Excel File with "Hello" "World" in Cells R1C1 and R1C2 (A1:B1).**

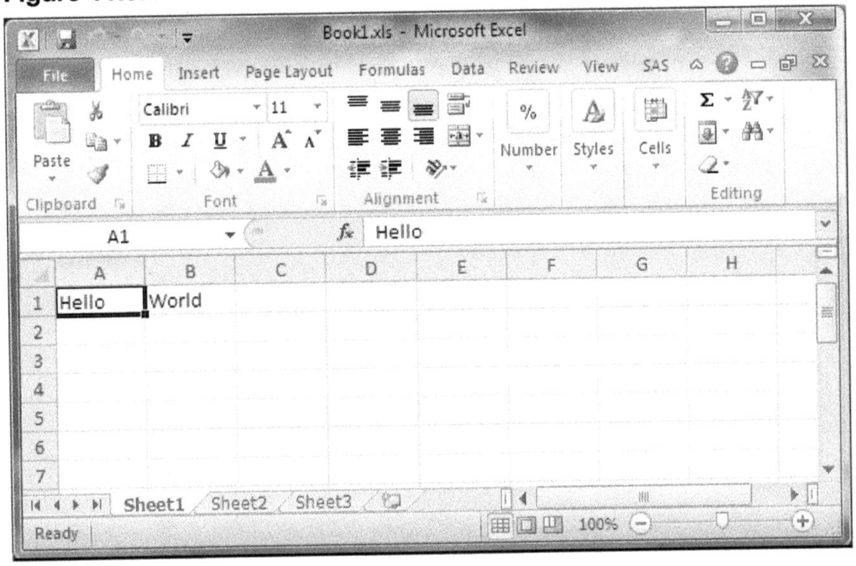

Being able to find the words "Hello" and "World" in Excel completes this project. In fact, the words appear while the SAS program executes or shortly after it finishes running. The space between the words caused the data to be placed into two separate cells in the Excel worksheet. This simple example made two assumptions:

- The Excel file was in the same directory that SAS was using.
- The Excel file was open.

The SAS FILENAME syntax that will point to an open Excel file in another directory follows:

**Figure 11.6.5: SAS FILENAME Statement to Assign ddewrite to a File.**

```
FILENAME ddewrite DDE 'excel|C:\Excel_Files\[Book1.xls]Sheet1!r1c1:r1c2';
```

## Example 11.6.2 The Hello World Project When the Excel Workbook Is Closed

This example is very similar to the first example except that it starts with the Excel file closed. The SAS "X" command is used to issue an operating system command to start the Excel application before running the DDE commands from SAS. Some of the newer versions of Microsoft Excel take several seconds to open and activate the add-ins that are installed. The number and type of add-ins may vary for each computer installed in your area. Since SAS and Excel must both be running at the same time, it is important to wait long enough for Excel to be fully active before issuing a DDE command.

Figure 11.5.2 has a list of common default locations for where Microsoft Excel is installed. Your location may be different. Ask your Information Technologies Department if your location is different.

The SAS code in Figure 11.6.6 shows how to open an Excel workbook using the SAS X command, and then how to write "Hello World" to the Excel worksheet. This example is using Excel 2013 on a 64-bit system. One of the new features of Excel 2013 is "Splash Screen" that allows you to choose the type of workbook or template you want to open. Previous versions of Excel also had "Splash Screens", but none of them waited until you entered a response.

### Figure 11.6.6: SAS Code to Process the "Hello World" Project with a Closed Excel File as Input.

```
OPTIONS NOXWAIT NOXSYNC;
/* Use to X command to start Excel 2013*/

%let Excel_path =
'C:\Program Files\Microsoft Office 15\root\office15\EXCEL.EXE';

* no spaces in unquoted file path or name;
%let Excel_file = C:\Excel_Files\Book2.xlsx;

X "&Excel_Path. /x  &excel_file.";

DATA _NULL_;
   x = sleep(10); * Allow enough seconds for Excel to open;
RUN;

FILENAME ddewrite DDE 'excel| C:\Excel_Files\[Book2.xlsx]Sheet1!r1c1:r1c2';

DATA OUT;
   FILE ddewrite;
   x='Hello World';
   PUT x;
RUN;
```

Once again, the simple act of finding the words "Hello" and "World" complete this exercise. Note here that each word is in a separate cell of the Excel worksheet, just like they were in Figure 11.6.4. Also, this example opened Excel before writing the output text, and did not close Excel when finished. Figure 11.6.7 shows the results.

### Figure 11.6.7: Excel File with "Hello" "World" in Cells R1C1 and R1C2 (A1:B1).

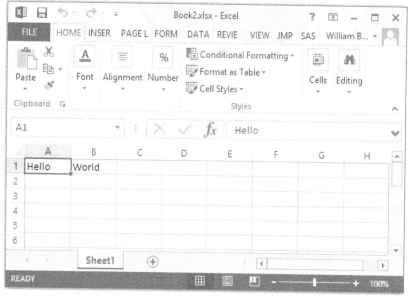

## Example 11.6.3 The Hello World Project Using NOTAB and LRECL= Options

This example is also very similar to the first example except that it starts with the Excel file closed. The SAS "X" command is used to issue an operating system command to start the Excel application before running the DDE commands from SAS. Some of the newer versions of Microsoft Excel take several seconds to open and activate the add-ins that are installed. The number and type of add-ins may vary for each computer installed in your area. Since SAS and Excel must both be running at the same time, it is important to wait long enough for Excel to be fully active before issuing a DDE command.

Figure 11.5.2 has a list of common default locations for where Microsoft Excel is installed; your location may be different. Ask your Information Technologies Department if your location is different.

This example also introduces two new features of the DDE commands. The options NOTAB and LRECL are introduced in this example. The NOTAB option limits the output of ASCII tab characters ('09'x). Typically, the tab characters cause the data following a tab to be placed into the next Excel cell of the worksheet. The LRECL (Logical RECord Length) option limits the number of characters output into an individual cell in the Excel worksheet. The word "World" is truncated to "Worl" because the constant "Hello World" is 11 bytes long and the LRECL=10 allowed only 10 bytes to be output.

**Figure 11.6.8a: SAS DDE Code to Write 10 Characters "Hello Worl" to the Output Excel File in One Cell.**

```
OPTIONS NOXWAIT NOXSYNC;
/* Use to X command to start Excel 2013*/

%let Excel_path =
'C:\Program Files\Microsoft Office 15\root\office15\EXCEL.EXE';

* no spaces in file path or name;
%let Excel_file = C:\Excel_Files\Book3.xlsx;

X "&Excel_Path. /x  &excel_file.";

DATA _NULL_;
   x = sleep(10); * Allow enough seconds for Excel to open;
RUN;

FILENAME ddewrite DDE 'excel|C:\Excel_Files\[Book3.xlsx]Sheet1!r1c1:r1c2'
                  NOTAB LRECL=10;

/* using NOTAB will 'Hello World' text to be placed on one cell   */
/* using LRECL=10 to write 11 characters will truncate the string */
DATA OUT;
   FILE ddewrite;
   x='Hello World';
   PUT x;
RUN;
```

**Figure 11.6.8b: SAS DDE Code to Write 10 Characters "Hello Worl" to the Output Excel File in One Cell.**

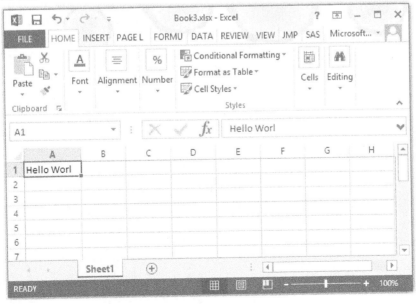

## Example 11.6.4 Writing "Hello World" to an Excel File Using DDE Macros

In general, computer programs have the following steps:

1. Start running.
   a. Open input files.
   b. Open output files.
      i. Read input values.
      ii. Calculate values.
      iii. Output values.
      iv. Repeat until there is no more input.
   c. Close output files.
   d. Close input files.
2. End.

A SAS DATA step usually takes care of most of these operations for you. Therefore, many SAS programmers rarely do more than start a DATA step (which names an output file), declare a SET statement (which names an input file), and calculate the values needed. At the end of the DATA step, a RUN or a PROC statement outputs the data records. When SAS runs out of input data, it closes the input and output files for you. When you use DDE, you have to open and close the files yourself. You also have to find a way to determine where the input and output data exists in your Excel files.

Examples 11.6.4 and 11.6.5 will use the macros described in Section 11.4 and explain how to automate some of the input/output processed. This will give you a framework upon which to build your own DDE interface tools. Before we examine the next example, you might want to review Section 11.4 to better understand the macros presented in this chapter and stored on my SAS Author's page. They are used to move data from SAS to Excel.

In Example 11.6.4, the five macros (start_Excel, Open_cmd, out_range, save_Excel, and Close_Excel) defined in Section 11.4 above will be used to write two sets of data to an Excel workbook. The commands used to do the writing are exactly the same, but the DDE-triplet defined by the macro is different. The NOTAB option is set differently for each of the DDE-triplets. The data will be written to the same Excel workbook, but in different locations on the same worksheet. We will start with a blank Excel workbook. The following SAS macro calls invoke the macros described in Section 11.4 above. The implementation of the code is stored on the SAS website for my Author's page. More elaborate and feature-rich macros exist to do the similar DDE functions, but these are good enough to get you started.

http://support.sas.com/publishing/authors/benjamin.html

**Figure 11.6.9: SAS DDE Code to Write "Hello World" to the Output Excel File Twice.**

```
*******************************************************************;
**     Example 4 - Simple Execution Code Example                 **;
*******************************************************************;

*===================================================================;
*  Define the file name                                            *;
*===================================================================;
   %let my_file = "C:\Excel_Files\Book1.xls";

*===================================================================;
*  Start Excel and open the file defined above as My_file          *;
*===================================================================;
   %start_excel (file_name=&my_file,wait=5); /* Excel output file name */

*===================================================================;
*  Set up a filename to use to issue commands to Excel via DDEs    *;
*===================================================================;
   %Open_cmd(file_ref=commands,wait=5); /* command stream excel fileref*/
```

```
*===================================================================;
*  Set up a filename command to point to a range of Excel cells    *;
*===================================================================;
   %out_range
      (file_ref      = Hello_W,           /* fileref to use for output */
       path_name     = C:\Excel_Files\,   /* Excel file path           */
       file_name     = Book1.xls,         /* Excel file name           */
       Start_row_Col = R1C1,              /* must use the R1C1 format  */
       End_row_col   = R1C2,              /* must use the R1C1 format  */
       no_tab        = YES,               /* change tab setting option */
       L_REC_L       = 20          );     /* change output Logical len */

*===================================================================;
*  Set up a filename command to point to a range of Excel cells    *;
*===================================================================;
   %out_range
      (file_ref      = Hello_X,           /* fileref to use for output */
       path_name     = C:\Excel_Files\,   /* Excel file path           */
       file_name     = Book1.xls,         /* Excel file name           */
       Start_row_Col = R4C1,              /* must use the R1C1 format  */
       End_row_col   = R4C2,              /* must use the R1C1 format  */
       no_tab        = NO,                /* change tab setting option */
       L_REC_L       = 30          );     /* change output Logical len */

*===================================================================;
* This data step writes data to the defined Excel range using the file*;
* command and the put command. Filerefs "Hello_w" and "Hello_x" are   *;
* defined by the out_range macro                                      *;
*===================================================================;
   data _null_;

      Hello = 'Hello';
      World = 'World';

      file Hello_w;
      put hello $char6.  World $char6.;

      file Hello_x;
      put hello $char6.  World $char6.;
   run;
*===================================================================;
*  This macro saves the output file                                 *;
*===================================================================;
   %save_Excel
      (file_ref  = commands,           /* fileref to use for output*/
       path_name = C:\Excel_Files\,    /* Excel file path          */
       file_name = Book1.xls);         /* Excel file name          */
*===================================================================;
*  This macro closes the Excel file.                                *;
*===================================================================;
   %Close_Excel (file_ref  = commands);   /* fileref to close      */
run;
```

**Figure 11.6.10: The Excel Output of the DDE Macros.**

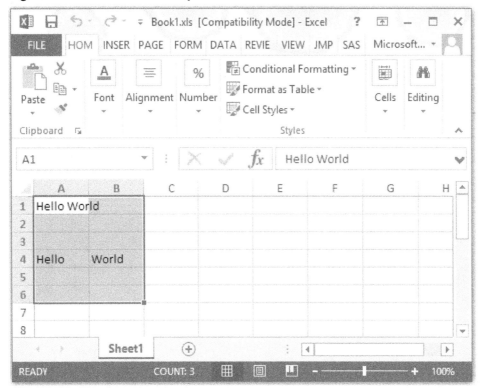

Yes, this looks like it is more trouble than it is worth! But the five macros used here and described in Section 11.4 are just building blocks. They can be used independently, or they can be combined to form a more comprehensive tool that allows the work to be abstracted to the level of "start here and do the work". The next example will demonstrate what I mean by reducing all of this work into a single macro call.

## Example 11.6.5 Writing a SAS Dataset to an Excel File Using the SAS_2_EXCEL DDE Macro

This example will once again use the Sashelp.shoes dataset; however, it will be processed by using PROC MEANS to create an output file with a summary of some of the variables found in the original file. Since for this example we are also assuming that you need a formatted report for periodic delivery to other users, we will also create an output format template that the data will be inserted into for the report. Below is a sample report layout. (Column E is for showing the row number on the right side of the report.) The final macro (SAS_2_EXCEL) in the DDE macro set described in Section 11.4 does not hard code any values. All values used by the macro are either derived by the macro or sent to it by you. The output data spaces start at Row 6 and Column 1. Figure 11.6.11 is the Excel template file. It is simply an Excel workbook with a predefined format and space for the new data to be placed. This Excel workbook template is provided as Shoe_Summary_Template.xlsx.

**Figure 11.6.11: The Excel Template for the Example 11.5 Report.**

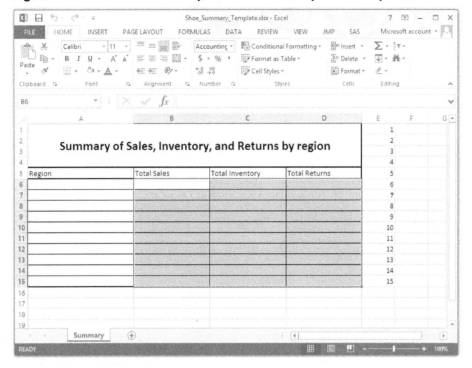

Figure 11.6.13 includes SAS code to execute PROC MEANS, a list of the SAS_2_EXCEL macro default values, and the coded parameters to execute the example.

**Figure 11.6.12: Parameters Descriptions for the SAS_2_EXCEL Macro.**

| Parameter Name / Default | Description |
| --- | --- |
| SAS_input_file | SAS input data to send to Excel |
| Excel_template_path | Directory path to the input Excel file |
| Excel_template_name | File name of the Input (template) Excel file |
| Excel_template_sheet =Sheet1 | Sheet name of the Excel input (Template) worksheet |
| Excel_output_path | Directory path to the output Excel file |
| Excel_output_name | File name of the Output (final) Excel file |
| Execl_starting_row | Top left row of the output field |
| Excel_starting_column | Top left column of the output field |
| NOTAB = NOTAB | NOTAB suppresses the output of tabs until requested |

**Figure 11.6.13: The SAS Macro Call and Parameters for the DDE Macro SAS_2_EXCEL.**

```
*******************************************************************;
*   Example 5 - Writing a full SAS file using a template and DDE macros *;
*******************************************************************;
proc means data=sashelp.shoes(keep=region sales inventory returns)
           noprint sum nway;
by region;
output  sum=sales inventory returns out=means_output;
run;


*******************************************************************;
*******************************************************************;
*   Sample macro call with notes about parameters                  *;
*******************************************************************;
```

## 186 Exchanging Data between SAS and Microsoft Excel

```
*macro SAS_2_Excel                     /* SAS Macro Name                          */
 (SAS_input_file           =,          /* SAS input data to send to Excel         */
  Excel_template_path      =,          /* Excel file path                         */
  Excel_template_name      =,          /* Excel file name                         */
  Excel_template_sheet =Sheet1,        /* Excel file sheet name                   */
  Excel_output_path        =,          /* Output Excel file Path                  */
  Excel_output_name        =,          /* Name of the output Excel file           */
  Execl_starting_row       =,          /* Top left cell of Excel output range     */
  Excel_starting_column=,              /* Bottom right cell of Excel output       */
  notab=notab                          /* turn off the tab output option          */
 );

***************************************************************;
**   Actual macro call with parameters                       **;
***************************************************************;

%SAS_2_Excel               /* SAS_input_File - may include dataset options*/
 (SAS_input_file        = means_output(drop=_TYPE_ _FREQ_),
  Excel_template_path   = C:\Excel_Files\,
  Excel_template_name   = Shoe_Summary_Template.xlsx,
  Excel_template_sheet  = Summary,
  Excel_output_path     = C:\Excel_Files\,
  Excel_output_name     = Current_Shoe_Summary.xlsx,
  Execl_starting_row    = 6,
  Excel_starting_column = 1,
  notab=notab
 );
run;
```

Figure 11.6.14 is the output from executing the SAS_2_EXCEL macro. When I wrote the macro, I was using a computer based on Windows XP and using SAS 9.2 writing to Microsoft Excel 2007. When working on my final draft of this chapter I was using Microsoft Windows 8.1, SAS 9.4, and Microsoft Excel 2013. The only change I needed to make to the macro was to point the macro to the new directory path for Microsoft Excel 2013.

### Figure 11.6.14: Output of DDE Processing.

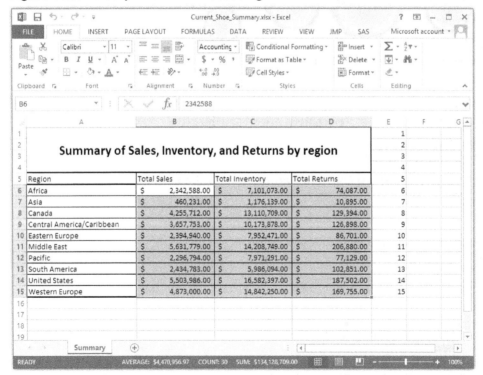

In Example 11.6.4, the five macros defined in Section 11.4 are building blocks that the SAS_2_EXCEL macro uses. Example 11.6.5 used the SAS_2_EXCEL macro, which in turn uses all five of those building block macros.

## 11.7 Conclusion

DDE is an old technology, but many people still find it useful. I have often found that generating code that can be reused is far more useful than rewriting the same code. I build libraries of reusable code; then I include and use the code. I set it up so that one piece of code can be used for many applications. SAS macros work well to provide reusable code. The SAS_2_EXCEL macro described in this chapter is one of those pieces of code. I can use any input Excel file and produce any output Excel file with minimal changes to the parameters. The SAS_2_EXCEL macro call identifies the input and output files, the locations to change, the movement of data, and the formatting of the data. The macro call does not say how it will be done. The pages of SAS code that define the DDE macro SAS_2_EXCEL fills in the "HOW". DDE is the kind of application that requires small changes to do just about everything, These macros help to make that job simpler.

# Chapter 12: Building a System of Excel Macros Executable by SAS

| | |
|---|---:|
| **12.1 Introduction** | **189** |
| **12.2 Purpose** | **190** |
| **12.3 General Design of a Tool to Control Excel Macros from SAS** | **190** |
|     12.3.1 Prepare a SAS File and Execute Excel to Process the Output | 191 |
|     12.3.2 Prepare Excel to Open the File Output by SAS | 192 |
|     12.3.3 Prepare Excel Macros to Reformat the Excel Workbooks | 194 |
| **12.4 Automate the Tool So That SAS Creates a Formatted Excel Output Workbook** | **197** |
|     12.4.1 Eliminate the Manual Steps from the Processing | 197 |
|     12.4.2 Create a SAS Output File with More Data and Control Information | 202 |
|     12.4.3 Create an Excel Macro to Process the Output SAS File | 203 |
|     12.4.4 Build an Excel Graph Using an Excel Macro | 207 |
| **12.5 Conclusion** | **209** |

## 12.1 Introduction

Chapters 12, 13, and 14 each define separate, but increasingly powerful tools that combine SAS and Excel features and that integrate your ability to transfer data between SAS and Excel. At the beginning of each of these chapters I have placed a short list of all of the tools--just in case you look at only one of the chapters today.

This chapter presents SAS code and Excel code that creates a fully integrated system to generate macro-free Excel workbooks, without downloading any external data or code from the Internet, using SAS 8 and Excel 97 and later. Software advances today have moved more and more to what I like to call "Push-Button" processes. I prefer to write software that allows me to turn over the processing to someone who only needs to know how to push the button. Not too long ago it would have taken a lot of effort to connect to a network; today you open the box for a new device, turn on the power, and often see more than one network.

When you use the ODS tagset template "EXCELXP", the output file is a *.xml text file that Excel can open by double-clicking on the file. This is great for many things, but SAS cannot include SAS graphs in *.xml files. This chapter will introduce a simple 'Hello World' project that will demonstrate how to use the SAS ODS tagset template EXCELXP to create a simple *.xml file, have SAS start Excel, use the *.xml file as input, and then modify the file and display the resulting graph for us to examine. The only action required on the part of the SAS user (after the initial setup of the code) is to run the SAS code. All of the Excel activity will be pre-programmed.

**Table 12.1.1: Tools Described in Chapters 12, 13, and 14.**

| Tool | Chapter | Description |
|---|---|---|
| **Personal Workbook Tool** SAS tool to run personal Excel macros, already included in Excel (under Macros → Record Macro → Store macro in). | 12 | This tool uses the SAS "X" command to execute Excel macros in your Personal Excel "Xstart" directory. It allows Excel workbooks to be delivered without embedded macros. |
| **Macro Library Tool** SAS tool to run externally stored Excel macros. You can use the "Macro Library Tool" within Excel to take the same macro you would have created in the Personal Workbook Tool and export it to a departmental library. | 13 | This tool uses the SAS "X" command and the Windows operating system scripting language to control processing of Excel macros. It allows Excel workbooks to be delivered without embedded macros. |
| **Excel Workbook Tool** An Excel tool to store parameters for SAS programs and either execute the SAS code or place the code into a directory for execution. (My_Excel_Tool, available in the example code and data folder on the author's SAS Press page) | 14 | This Excel workbook tool will save parameters for a SAS program and either execute or copy the code to a directory. It uses features of the other two tools and allows for storage of SAS code in a production-type area so the original code is not modified when the reports are processed. . |

## 12.2 Purpose

On a computer, a "Hello World" project is a *proof-of-concept* program demonstrating the ability to do something like write to a disk and read it back. What I will show is the ability to retain control of program execution starting within a SAS program and continuing after Excel begins processing. That will be followed by Excel processing the code I want to execute. This ability to establish a process in which you can maintain control is an advantage you can exploit. It gives programmers a path from which they can branch out and do more.

## 12.3 General Design of a Tool to Control Excel Macros from SAS

The code we will develop here is a push-button process. There are several steps that need to be programmed to make the code work. The object of the tool is to run a SAS program, have the SAS program load a file into Excel, and take control of processing the workbook and all of its worksheets before Excel allows the user to modify anything. This is the dream of every programmer who has ever lost a leading zero when Excel opened his or her workbook. But, enough of the dreaming. Let's define this concept and put it into action.

Here we will describe two processes. The first is the "Hello World" program, and the second is a real-world application of the steps. The task of having SAS create an Excel workbook is simple when you use ODS. The tagset "EXCELXP" is designed directly for this task. The next part is getting Excel to do what you want it to do. This is accomplished with Excel macros that you store on your computer, not in the generated workbooks. Table 12.3.1 shows a list of the tasks needed for this "Hello World" project.

**Table 12.3.1: Tasks to Create a "Hello World" Project.**

| Task | System | Comments |
|---|---|---|
| Create SAS program | SAS | Use ODS tagset EXCELXP to create a *.xml file with a label containing "Greetings World" in cell A1 and "Hello" in cell A2. |
| Create an Excel macro | Excel | Record a personal Excel macro to write "World" in cell B2. |
| Create a VBA code module | Excel | Write a personal VBA code module to run the Excel macro. |
| Run SAS program to create the *.xml file | SAS | Create the input *.xml file to be sent to Excel. |
| Run Excel using *.xml file as input | SAS | Use the SAS "X" command to start Excel. |

## 12.3.1 Prepare a SAS File and Execute Excel to Process the Output

The first thing we need to do is create a simple SAS program to run the ODS tagset EXCELXP. This step will generate a *.xml file that we can open with Excel by using the SAS "X command". All we need is one variable and one observation. We will use PROC PRINT to add a LABEL to the variable. The following SAS code will create the SAS dataset, open the *.xml file using ODS, print the data to the *.xml file, and open the *.xml file using Excel.

**Figure 12.3.1: SAS Code to Create and Open an Excel File.**

```
** Create a simple SAS File                                **;
DATA Hello_to_the_world;
   Greeting = 'Hello';
RUN;

** Default paths to program Excel.exe - your path may vary  **;
* 2003 = C:\Program Files\Microsoft Office\OFFICE11\EXCEL.EXE;
* 2007 = C:\Program Files(x86)\Microsoft Office\Office12\EXCEL.EXE;
* 2010 = C:\Program Files (x86)\Microsoft Office\Office14\EXCEL.EXE;
* 2013 = C:\Program Files\Microsoft Office 15\root\office15\EXCEL.EXE;

** Start ODS and name the output xml file                   **;
ODS TAGSETS.EXCELXP FILE="c:\my_excel_files\Hello_world_test.xml";

** Set up ods options to call the print procedure   **;
ODS TAGSETS.EXCELXP OPTIONS(Sheet_Name='Greeting') Style=minimal;

** Create an xml file using ods output and proc print   **;
PROC PRINT data = Hello_to_the_world
   label split = '*' noobs;

   Label greeting = 'Greetings*World';
   * increase row 1 font size;
   Var   greeting / style(head) = {Font_Size = 2};
RUN;

** Close the ExcelXP tagset to release the output file   **;
ODS TAGSETS.EXCELXP close;

** Choose a version of Excel to run - pick default path **;
** NOTE The file name needs to be quoted if it contains spaces, etc **;

* Excel 2013 64-Bit;
X ' "C:\Program Files\Microsoft Office 15\root\office15\EXCEL.EXE"
    "c:\my_excel_files\Hello_world_test.xml" ';

RUN;
```

The final result is an Excel workbook with a label that reads "Greetings World" in cell A1 and the word "Hello" in cell A2, as shown in Figure 12.3.2a. The background view in this Excel file is slightly different than usual because the ODS option "Style=minimal" was used when creating the *.xml file. Other style options are available and produce different results. The font size of the headers and the text in the cells is also different in each cell.

**Figure 12.3.2a: Excel Output of the Code from Figure 12.3.1.**

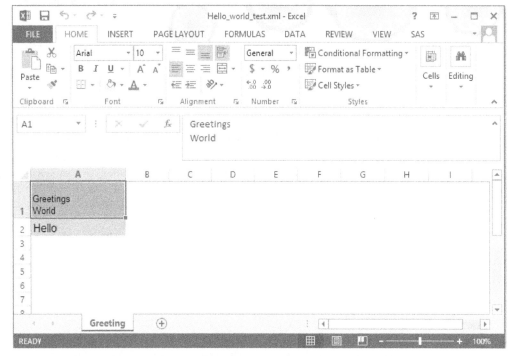

### 12.3.2 Prepare Excel to Open the File Output by SAS

The SAS code was the easy part; now on to the Excel setup. Most Excel users know that Excel will record a macro, and the macro can be associated with a key-stroke command that will run the macro. These macros are usually stored within the current workbook, or deleted before exiting Excel. What most users do not know is that Excel has a way to store macros so that they are available to be used on all workbooks on a given computer, and not stored with the workbook. Furthermore, there is a way to give those macros control of Excel before the spreadsheets are visible to the user. An Excel Personal Macro Workbook is stored in a location that Excel loads before accessing the spreadsheets. See Figure 12.3.2b. These commands (macros) are available before other workbooks are loaded into Excel. The Windows default file names and locations are listed in Table 12.3.2.

**Figure 12.3.2b: Excel Record Macro Window to Input Macro Name, Hot-Key, and Storage Location.**

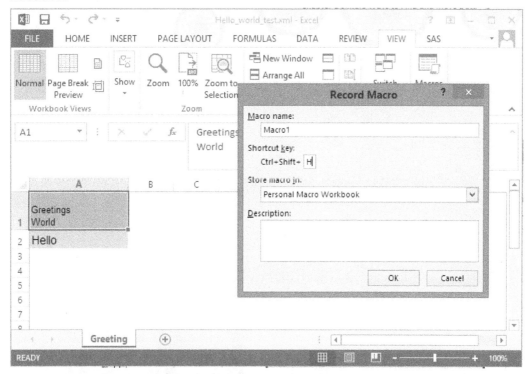

**Table 12.3.2: Default Locations of Excel Personal Macro Workbooks.**

| Software | Usual Default Location - Your location may vary by either Excel version or operating system version. Check with your IT Department to verify the location. |
|---|---|
| | **Common paths on older operating systems.** |
| Excel 2003 | C:\Program Files\Microsoft Office\Office11\XLSTART |
| Excel 2007 | C:\Program Files\Microsoft Office\Office12\XLSTART |
| | **The XLSTART directory is considered a trusted location. To locate it, check the following.** |
| For Excel = | To locate the XLSTART directory path: |
| 2007, 2010, and 2013 | With Excel open, select "Excel Options", and then select the "Trust Center" button. |
| | Select "Trust Center Settings", and then select the "Trusted Locations" button. |
| | Look for the path for the Excel XLSTART file. (See Figure 12.3.3a.) |

### Figure 12.3.3a: Excel Trusted Locations Window.

When you create or update your Personal Macro Workbook (it can hold data, too), you will be prompted in the following way if your Personal Macro Workbook has changes that need to be saved when you exit Excel. (See Figure 12.3.3b.) As mentioned above, this workbook can store many macros and make them available every time you open Excel. When you no longer need to access these macros, locate and delete the PERSONAL.XLS (or PERSONAL.XLSB) file in your XLSTART directory, as mentioned above. Any Excel file in the XLSTART directory will be opened every time you open Excel.

### Figure 12.3.3b: Excel Notice When Creating First Macro in Personal Macro Workbook.

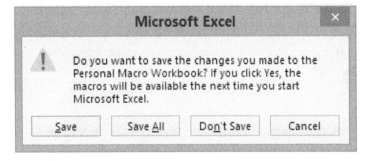

This notice appears when you are exiting Excel and reminds you that it will be available the next time you open Excel. Macros stored here will be available every time you open Excel until the macro is deleted or the PERSONAL.xlsb (or .xls) file is deleted. The VBA code shown will reference only the "PERSONAL.XLSB" Excel files.

## 12.3.3 Prepare Excel Macros to Reformat the Excel Workbooks

The next step is to record a macro. We started recording MACRO1 in Figure 12.3.2b. Now we will continue with that process. We started by assigning the name "Macro1" and the hot-key CNTL/SHIFT/H to the Excel macro we wanted to create. If we do this in a new workbook called "Book1," all that needs to be done is type "World" in cell "B2" and then stop recording. The results look like Figure 12.3.4.

### Figure 12.3.4: The Results of the Recorded Named Macro1.

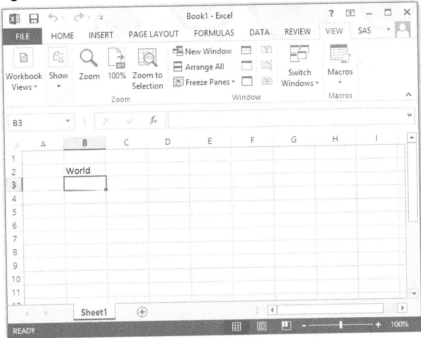

When we choose the "Personal Macro Workbook" as shown above in Figure 12.3.2b, the output macro is stored as a macro in a module. In this case it is called "Macro1" in "Module1." This macro will be stored in a hidden workbook called PERSONAL.XLS (or PERSONAL.XLSB) in your XLSTART directory.

### Figure 12.3.5: VBA Code for MACRO1.

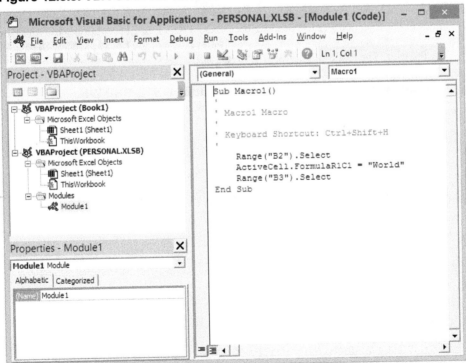

The Excel Project window (Figure 12.3.5) shows that the Macro1 code is stored in Module1 of the PERSONAL.XLSB workbook. Now we have the beginning SAS code and the Excel ending VBA code of the project. So, we can run the SAS program listed in Figure 12.3.1 the results are listed below in Figure 12.3.6a.

**Figure 12.3.6a: Results of Running the SAS Code in Figure 12.3.1.**

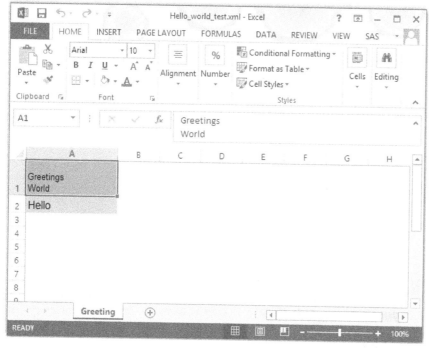

Figure 12.3.6a shows the way that Excel looks when the *.xml file opens. Note that the heading "Greetings World" is on two lines in the Excel output file because the PROC PRINT SPLIT= option wrote the label on two lines. We are almost there, but to get the word "World" into cell "B2" the hot key CNTL-SHIFT-H still needs to be typed by the user. The result of entering the "HOT-KEY" is shown in Figure 12.3.6b.

**Figure 12.3.6b: Results of Typing the Hot-Key Combination CNTL-SHIFT-H.**

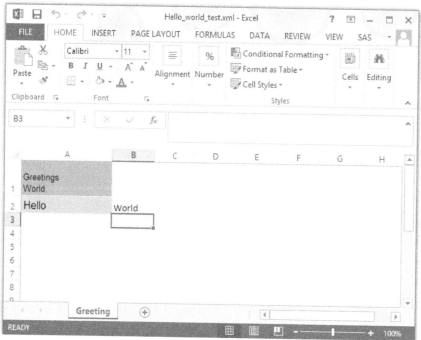

Now that we know the "Hello World" project can be done, we can move forward and take some of the manual steps out of the process.

## 12.4 Automate the Tool So That SAS Creates a Formatted Excel Output Workbook

The reason that computer programmers write programs is so that they do not have to do things manually. So, the next question becomes "How do we get this to run seamlessly?" Well, in order for us to take control so that we can make Excel do what we want, we need to find:

**"The First Place a Programmer Can Get Control of Excel"**

It turns out that once your PERSONAL.XLSB (or .xls) Macro Workbook is created, it is always the first workbook opened when Excel is started. The PERSONAL.XLSB (or .xls) workbook also contains our modules with our macros and an Excel structure called "ThisWorkbook" (see Figure 12.3.5). The VBA code module "ThisWorkbook" is given control of the Excel start-up activities. If there is a subroutine called "Workbook_open", is executed. This is where a programmer can capture control of Excel. The VBA code shown will reference only the "PERSONAL.XLSB" Excel files.

For this "Hello World" project, we will insert code into the "Workbook_open" subroutine to execute our macro to write "World" in cell B2 of the current sheet. See Figure 12.4.1 for the code.

### 12.4.1 Eliminate the Manual Steps from the Processing

The VBA code in Figure 12.4.1 defines a counter and two character (string) variables, and then checks the name of all open workbooks. If the name of the workbook is, in this case, "PERSONAL.XLSB", then no action is taken. If any other workbook is open, then Macro1 would execute and place the word "World" into cell B2 of the first worksheet. When opened from our SAS program, the generated workbook opens and the macro executes.

**Figure 12.4.1: VBA Code to Run the Workbook_open Subroutine.**

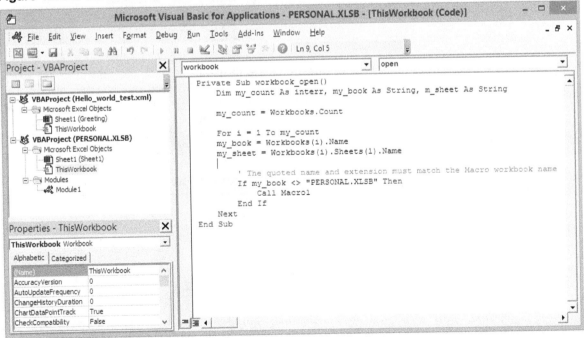

Figure 12.4.2a shows the results of running our SAS code and letting our macro (Macro1) execute. Like most other SAS programmers who are also Excel users, I determined that Excel likes to dance to its own tune, which means that as soon as someone tells you what Excel will do in a particular situation, it will seem to do something else. This is the case with getting the macros listed here to execute correctly (meaning how you want them to execute, not how Excel thinks they should run). To that end, it should be pointed out that if you double-click on an Excel file to open it, Excel may do its processing slightly differently than if the SAS System executes an "X" command to run Excel. Yes, that does not sound right, but try it for yourself.

**Figure 12.4.2a: Resulting Excel File When the SAS Program with the "X" Command Runs.**

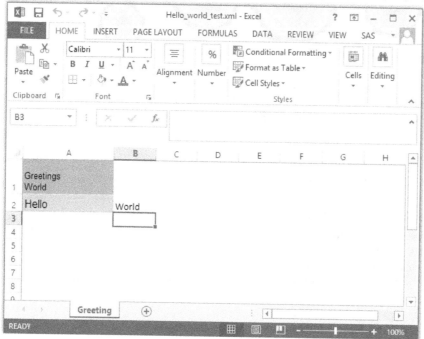

The code reproduced below includes a clever little trick. It is like the code that was shown in Figure 12.4.1 as the VBA subroutine called Workbook_Open. The upgraded version in Figure 12.4.2b below includes two message boxes to display information about what occurs when the code is running. This will display information contained in VBA code variables and pause the program. Since this is the first VBA code to execute, you do not have access to the Excel VBA debugger yet. However, you can print values in the message box so that you can examine them. This neat little debugging tool can be useful in figuring out what is going on with Excel at the start of the program (when you cannot start the debugger) to give you guidance about how to process the code. The VBA code shown will reference only the "PERSONAL.XLSB" Excel files.

**Figure 12.4.2b: Sample VBA Code to Trace the Path Through the VBA Code.**

```
Private Sub Workbook_open()
     Dim my_count As Integer, my_book As String, my_sheet As String
  My_count = Workbooks.Count

   'Define message
   Msg = "sub=workbook_open - workbook count= " + Str(my_count)

   'Define buttons
   Style = vbOKOnly

   'Define title
   Title = "Workbook name = " + ThisWorkbook.Name
```

```
      'wait for user
      Response = MsgBox(Msg, Style, Title)

   For I = 1 To my_count
      my_book = Workbooks(i).Name
      my_sheet = Workbooks(i).Sheets(1).Name

            Msg = "sub=workbook count loop - workbook counter= " + Str(i)
            Title = "Workbook name = " + my_book
            Response = MsgBox(Msg, Style, Title)

         if my_book <> "PERSONAL.XLSB" Then
             Call Macro1
         End If
   Next
End Sub
```

## An Explanation of the "Workbook_Open" Excel Macro in Figure 12.4.2b

- The code above (called Excel subroutine "Private Sub Workbook_open()") seems a little more complex than needed, and it is a little (but the message boxes can go away). If the macro had been a simple call to "Macro1", you might think it would have worked--but remember that now your computer has a new hidden workbook that Excel will always (yes, always) open.
- The Excel VBA desktop in Figure 12.4.1 above has TWO workbooks open, "PERSONAL.XLSB" and "Hello_World_Test.xml".
- One thing that has been relatively consistent is that when SAS uses the "X" command, both workbooks will be open ("Hello_World_Test.xml" and "PERSONAL.XLSB") when the "WORKBOOK_OPEN" macro runs.
- This will allow the "Macro1" code to execute. However, a double-click on an Excel file icon will usually have only one workbook open. Then the Excel macro "WORKBOOK_OPEN" runs (that workbook is "PERSONAL.XLSB"). So, we will assume for the purposes of this explanation that all calls to Excel come from SAS and the "X" command.
- Figure 12.4.1 above sets up two message boxes to be displayed. The first shows the executing subroutine, and the second shows the code is on the processing loop and the file index counter that is being used. See Figures 12.4.3 and 12.4.4.
- The second time through the workbook count loop, the code for MACRO1 executes and displays word "World". See Figure 12.4.5.
- Figure 12.4.6 is the workbook after the "Workbook_open" processing is complete, and the workbook is ready for you to use.

**Figure 12.4.3: Message Box Displayed Before Entering the Workbook Count Loop.**

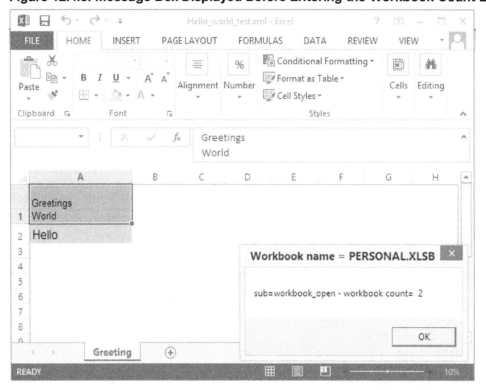

**Figure 12.4.4: Message Box Displayed in the First Workbook Count Loop.**

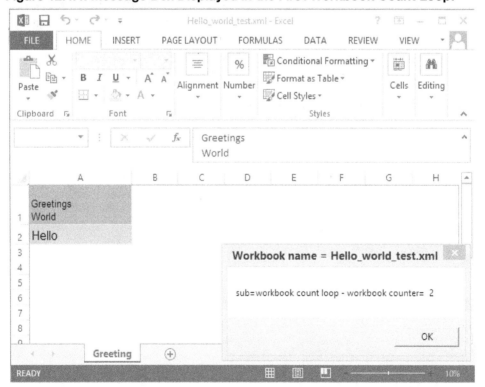

**Figure 12.4.5: Message Box Displayed in the Second Workbook Count Loop.**

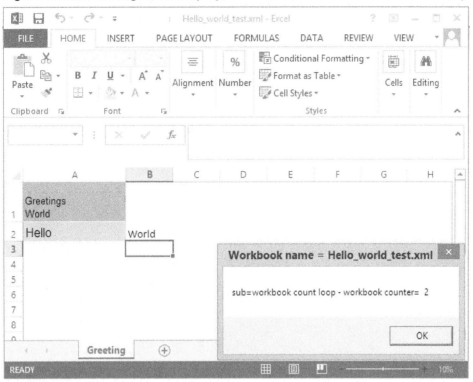

**Figure 12.4.6: After All Workbook_open Processing Is Complete and Ready for You to Use.**

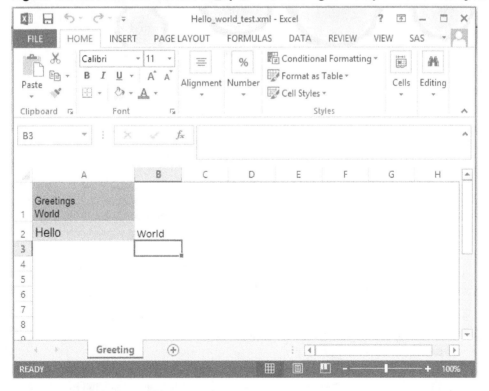

Figure 12.4.6 gives us a fully automated method to run a SAS program and complete a task in Excel. When we take away the VBA message box routines, the full task will run without intervention. Now we can control Excel from our SAS program.

## 12.4.2 Create a SAS Output File with More Data and Control Information

Since we do not want every Excel workbook to have "World" in cell "B2," the next step is to enhance the Excel macro "**WORKBOOK_OPEN**" so that it is more selective when it runs. So, we will take out the message boxes and add in control language to interpret information sent from SAS. By sending commands to Excel from SAS, we can control what actions our Excel macro takes when it gets control of the system.

First, let's upgrade our SAS code.

### Figure 12.4.7a: New SAS Code to Summarize the SASHELP.SHOES Dataset and Send the Data to Excel.

```
** Total sales by region, and put it into a file       **;
Proc summary data= SASHELP.Shoes nway;
weight sales;
var sales;
output out=sales(where=(_stat_ = "SUMWGT"));
by region;
run;
** Create path macro variable to Excel.exe - your path may vary   **;

* 2003 = C:\Program Files\Microsoft Office\OFFICE11\EXCEL.EXE;
* 2007 = C:\Program Files(x86)\Microsoft Office\Office12\EXCEL.EXE;
* 2010 = C:\Program Files (x86)\Microsoft Office\Office14\EXCEL.EXE;
* 2013 = C:\Program Files\Microsoft Office 15\root\office15\EXCEL.EXE;

** Start ODS and name the output xml file             **;
ODS TAGSETS.EXCELXP FILE="c:\my_excel_files\Pie_Chart_test.xml";

** Set up ods options to call the print procedure  **;
ODS TAGSETS.EXCELXP OPTIONS(Sheet_Name='Greeting') Style=minimal;

** Run Proc print using labels and noobs.           **;
PROC PRINT data = sales noobs label;
* add control value to the label value for Excel to use here;
label region = "?graph_1?Region";
var   region sales ;
run;

** Close the ExcelXP tagset to release the output file  **;
ODS TAGSETS.EXCELXP close;

** Choose a version of Excel to run - pick default path **;
** NOTE The file name needs to be quoted if it contains spaces, etc **;

* Excel 2013 64-Bit;
X ' "C:\Program Files\Microsoft Office 15\root\office15\EXCEL.EXE"
    "C:\My_Excel_Files\Pie_Chart_test.xml" ';

RUN;
```

The code in Figure 12.4.7a has been upgraded to do the following:

- Use PROC SUMMARY to create an NWAY summary of the SASHELP.SHOES dataset. The output SAS work file is weighted by the variable SALES.
- Open ODS with the TAGSET EXCELXP to output data to file C:\My_Excel_Files\Pie_Chart_test.xml.
- Use PROC PRINT to write the SAS dataset to the *.xml test file. Set up the label that the phrase "?graph_1?Region" is placed into the Excel cell A1. The "?graph_1? is information we will pass to the Excel macro.
- Use the SAS "X" command to run Excel and open C:\My_Excel_Files\Pie_Chart_test.xml.

The object is to send information to Excel to tell Excel what to do next. It is nice to have a constant location; I always use cell "A1" of the first sheet of the Excel workbook, and bracket the code with characters not normally found in titles or variable names. The same thing can be accomplished by changing the variable name to "_graph_1_Region" for the first column of the first sheet. Your Excel macro will need to remove any control information that you send for the Excel macro.

When we run this SAS code and send the data to Excel, our new output Excel workbook looks something like Figure 12.4.7b.

**Figure 12.4.7b: Excel Results for Output When New SAS Routine Is Executed.**

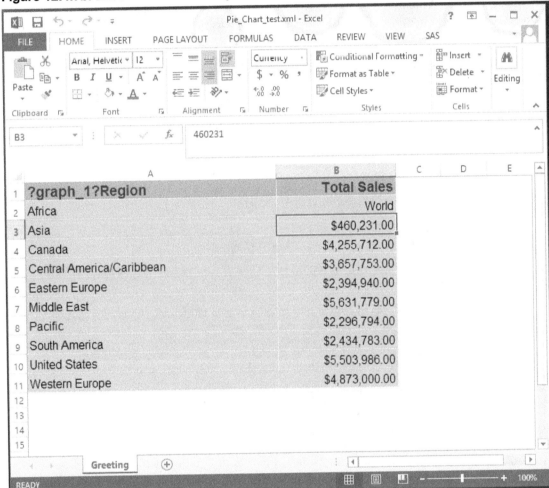

Since we have not re-written the Excel "Workbook_open" macro, we still find the word "World" in cell B2 of the Excel document. Remember that the PERSONAL.XLSB (or *.XLS) workbook is always opened. Since this workbook was opened with the SAS "X" command, the Excel MACRO1 was also executed. You will see the messages boxes described in Figures 12.4.3 to Figure 12.4.6 also were visible. Now that we are confident that our little message boxes are doing their job, we will eliminate them. If you use the code by cutting or copying and then pasting the code, you may not get exactly what you want. This process does not always convert the quotation marks correctly. You may need to retype them.

## 12.4.3 Create an Excel Macro to Process the Output SAS File

Now we have something automated, even though we still have to press Enter when the text boxes appear. We now have a procedure that starts by running SAS code and ends by displaying an active Excel workbook that not only has our data, but has made a change to our data. But, since we do not want to change cell "B2" to the word "World" for every workbook we create, let's change the VBA code for the

"Workbook_Open" VBA subroutine. The change we need to make is to have the Excel Workbook_Open VBA subroutine be selective in choosing which workbooks to modify.

The VBA code in Figure 12.4.8 below shows the code for the new **"Workbook_Open"** VBA subroutine. This macro will test cell A1 of the first worksheet and determine if it contains a special code to trigger processing a VBA macro. The message box calls are also removed from the updated **"Workbook_Open"** VBA subroutine. There is also a new subroutine required for Module1. Here we will just put the code in Figure 12.4.9 as a place holder. This subroutine will display only a message box to indicate that the routine was executed. This example is designed to detect the special code "graph_1" (case-sensitive) after stripping off two question marks from the front and back of the string. Failure to find either question mark will leave the contents of cell A1 undisturbed and return control to Excel. If the question marks are found, the data between the question marks will be removed and the remainder of the cell value restored. This can be changed to suit your needs.

**Figure 12.4.8: Excel Workbook_open Code to Process an Excel Macro Based Upon a Control Field.**

```
Private Sub Workbook_Open()
    Dim my_count As Integer, my_book As String, my_sheet As String
    my_count = Workbooks.Count

    For i = 1 To my_count
        my_book = Workbooks(i).Name
        my_sheet = Workbooks(i).Sheets(1).Name
        If my_book <> "PERSONAL.XLSB" Then

            my_cell_a1 = Workbooks(i).Sheets(1).Range("A1").Value
            my_cell_size = Len(my_cell_a1)

                ' look for a second ? from right side of value  '
                my_flag_size = InStrRev(my_cell_a1, "?") - 1

        ' if my_flag_size is greater than 1 then two ? were found
        ' this is a special processing workbook
        ' so get the info and process the workbook
        ' else ignore and return control to Excel and the user
        If my_flag_size > 1 Then

                ' get left ? character and data between the two ?'s
                my_report = Left(my_cell_a1, my_flag_size)

                ' remove the left ?
                my_data_size = Len(my_report) - 1
                my_report = Right(my_report, my_data_size)

                ' remove control characters from Cell A1 value
                ' and store in cell A1
                Workbooks(i).Sheets(1).Range("A1").Value = _
                Right(my_cell_a1, (my_cell_size - (my_flag_size + 1)))

                ' add new cases to process new reports
                Select Case my_report
                    Case "graph_1"
                        Call Module1.my_graph_1(my_sheet)
                    Case Else
                End Select

            End If
        End If
    Next
End Sub
```

### Figure 12.4.9a: Excel Subroutine in Module1 to Trace Activity.

```
Sub my_graph_1(my_sheet As String)

    Msg = "got to subroutine my_graph_1"
    Title = "Sheet Name = " + my_sheet
    Response = MsgBox(Msg, Style, Title)

End Sub
```

Figure 12.4.9b shows the output of a message box that is used to track the progress of the executing VBA program. This is very much like using PRINT statements or PUT statements in your SAS code to track the progress of the program. The code that produced the message box is in Figure 12.4.9a.

### Figure 12.4.9b: Verification the My_graph_1 Subroutine Was Executed.

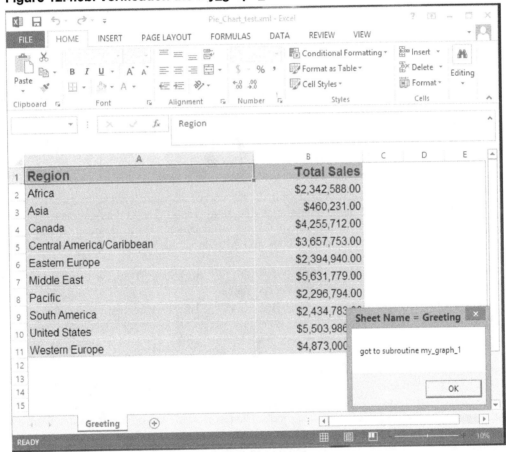

Figure 12.4.9b shows the message box, in the lower right hand corner, that output the trace message when the My_graph_1 subroutine was executed. A description of the new "Workbook_Open" VBA Code Described in Figure 12.4.8. follows to help you understand both the structure and flow of the VBA macro.

- The name of this macro is Workbook_Open. It is a private subroutine stored in the VBAProject area of the workbook called "**PERSONAL.XLS**" (Excel 2003) or "**PERSONAL.XLSB**" (Excel 2007, 2010, 2013). This subroutine has no parameters. Using the keyword "Private" will only allow this subroutine to execute when Excel starts; the macro will not be available at other times to execute. Remove the keyword "Private" to use at other times, like for debugging.

- This subroutine defines several local variables. They are named here.

  - Variable my_count is defined as an integer variable. This variable is set to the number of workbooks that Excel currently has open. The VBA "For" loop is similar to the SAS "DO" loop, and will execute once for each open workbook. Information about workbooks and sheets is stored in arrays,
  - Variable my_book is defined as string variable. This variable is set equal to the name of the currently open workbook Information about workbooks and sheets are stored in arrays.
  - Variable my_sheet is defined as string variable. This variable is set equal to the name of the currently active worksheet. The name of the workbook is tested to make sure it is not "PERSONAL.XLSB" or "PERSONAL.XLS". If it is one of these workbooks, then no action is taken. If the workbook name test is false, then the rest of the "if-then" condition is executed and the contents of Cell "A1" of sheet one of the current user workbook is placed into the Excel variable my_cell_a1. The VBA code shown will reference only the "PERSONAL.XLSB" Excel files.
  - Variable my_cell_1 is a string variable. This variable is a work space for processing cell A1 in the worksheet. The Excel variable my_cell_a1 is tested for a "?" by looking at all of the characters in the variable except the first character and loading the location of the "?" into the variable my_flag_size. If the length of my_flag_size is greater than 1, then there is more than one character between the two question marks on the left side of cell A1 of the worksheet. If there is no question mark on the left side of the cell A1 value, then the value of variable my_report generated later will not match a "CASE" test value.
  - Variable my_cell_size is an integer variable. This variable is set to the number of characters that are in Excel variable my_cell_a1.
  - Variable my_report is derived by getting the left part of the value without the question mark and calculating the size of the temporary value of my_report. To get the final contents of the variable my_report with no question marks, the left question mark is removed. This returns the original value of cell A1 to the workbook cell A1 without the left most control characters.

- This code looks first to find a non "PERSONAL.XLSB" workbook. If you are using Excel 2003, then code "PERSONAL.XLS" as the workbook name.
- This "Select Case" command works much like a SAS "SELECT" statement, "Select Case my_report", which is the same as "Select When(my_sas_variable_name)". The 'Case "graph_1"' value and the next line are the same as the 'When("value") SAS_command;' code. The "Case Else" command is the same as the SAS "Otherwise" command. This code structure is also case sensitive like the SAS SELECT statement.
- This VBA subroutine is sensitive to parameters passed to Excel in cell "A1" of the Excel workbook that is opened. This code is looking for exactly "?graph_1?" in the first nine characters of cell "A1" of the first sheet of the workbook. The test is a case-sensitive test and anything except "?graph_1?" will fail the test. No other processing is done to the workbook when the test fails; the workbook is just displayed for you to examine or update the workbook.
- When SAS writes variable name labels as the first row of the XML file sent to Excel, the label of a variable can contain special characters and spaces that are not valid in SAS variable names. The Excel VBA code can search for these special characters and use their presence or absence to control actions in the Excel VBA subroutine. The control information ("?graph_1?) that we inserted into the label of the "Region" variable ("?graph_1?Region") will be stripped off to become the value of the VBA variable "my_report", and it will be used as the test case in the VBA "Select Case" command. The question marks are removed, too. The rest of the SAS variable names or labels are output in row one and used as column header names.
- The Select command from the end the "Workbook_Open" VBA subroutine and is currently set up to process one report type, namely graph_1. It should be easy enough to add more. The Excel my_graph_1 subroutine called is the subject of the nest section of this chapter.

## 12.4.4 Build an Excel Graph Using an Excel Macro

The object of this task is to create a simple graph. In Section 12.2.3 of this chapter I described how to record an Excel macro. The same process was used here to record an Excel macro to create a graph. The steps are listed here to aid you in creating an Excel macro to build a graph.

- Open Excel.
- Turn on Macro recording.
- Select all of the data in columns A and B.
- Select the pie chart graph from the Excel toolbar.
- Add a title.
- Add a legend.
- Apply the minor name change edits described in Figure 12.4.10.
- Save the macro in your Personal Macro Workbook (PERSONAL.XLSB) Module1 as "my_graph_1".
- This graph is neither pretty nor optimized; it always uses cells A1:B11 and must be the first graph in the workbook. As noted, only the Chart1 name was changed.
- The four lines that begin with Activesheet.shapes may not copy correctly with a "Copy and Paste" and may need to be adjusted with either a continuation character ("_") at the end of the line or by moving the data back one line for each of the four lines affected.

### Figure 12.4.10: My_graph_1 Excel Macro Code.

```
Sub my_graph_1(my_sheet As String)
'
' my_graph_1 Macro
'
' This VBA macro generates a pie chart with minimal number of extras
' The code is reproduced here exactly as it was produced by the
' Excel Macro Record feature - except the name values were adjusted

    Range("A1:B11").Select
    Charts.Add
    ActiveChart.ChartType = xl3DPie
    ActiveChart.SetSourceData Source:=Sheets(my_sheet).Range("A1:B11"), _
PlotBy :=xlColumns
    ActiveChart.Location Where:=xlLocationAsObject, Name:=my_sheet
    With ActiveChart
        .HasTitle = True
        .ChartTitle.Characters.Text = "My_new_Pie_chart"
    End With
    ActiveChart.HasLegend = True
    ActiveChart.Legend.Select
    Selection.Position = xlBottom
    ActiveChart.Legend.Select
    Selection.AutoScaleFont = True
    With Selection.Font
        .Name = "Times New Roman"
        .FontStyle = "Regular"
        .Size = 11
        .Strikethrough = False
        .Superscript = False
        .Subscript = False
        .OutlineFont = False
        .Shadow = False
        .Underline = xlUnderlineStyleNone
        .ColorIndex = xlAutomatic
        .Background = xlAutomatic
    End With
    Selection.Position = xlLeft
```

```
    ActiveChart.ChartArea.Select
    With ActiveChart
        .Elevation = 35
        .Perspective = 30
        .Rotation = 0
        .RightAngleAxes = False
        .HeightPercent = 100
    End With

    ' changed "Chart 1" to the numeric location of the chart - not the chart
name
    ActiveSheet.Shapes(1).ScaleHeight 1.25,
        msoFalse, msoScaleFromTopLeft
    ActiveSheet.Shapes(1).ScaleHeight 1.5,
        msoFalse, msoScaleFromBottomRight

    'ActiveSheet.Shapes("Chart 1").ScaleHeight 1.25,
        msoFalse, msoScaleFromTopLeft
    'ActiveSheet.Shapes("Chart 1").ScaleHeight 1.5,
        msoFalse, msoScaleFromBottomRight

    Application.CommandBars("Task Pane").Visible = False
End Sub
```

The Excel macro code in Figure 12.4.10 is described here.

- The code module name is my_graph_1. It is a public subroutine stored in the "Module1" area of the workbook called "PERSONAL.XLSB" Excel 2007, 2010, 2013 or "PERSONAL.XLS" (Excel 2003). This subroutine has one parameter that is created in the "ThisWorkbook" area of "PERSONAL.XLSB".
- The code from Figure 12.4.8 above "Call Module1.my_graph_1(my_sheet)" executes this macro from within the "Select Case" statement. The value of "my_sheet" is the name of the Excel sheet to process.
- The selected range of cells is always A1:B11 on the first sheet of the workbook.
- This data comes from the output from the SAS program, and the range is "Hard-Coded" to be exactly the data that was output by the PROC PRINT command. Because this is a proof-of-concept program, no effort was put into making the subroutine flexible enough to accept a variable size graph space. This Excel command adds a chart to the workbook.

The ActiveChart commands shown in Figure 12.4.10 define several features of the chart.

**Table 12.3.3: Active Chart Actions from Figure 12.4.10.**

| ActiveChart Excel Object names | Function |
| --- | --- |
| ChartType | Defined the chart as a pie chart. |
| SetSourceData | Define the sheet name, data range, how to plot the data. |
| Location | Define where to place the chart. |
| HasTitle | Toggle a title on. |
| ChartTitle.Characters.text | The text of the title. |
| HasLegend | Toggle a legend on. |
| Position | Put legend on the bottom of the chart. |
| AutoScaleFont | Toggle automatic font scaling on. |
| Font | Select several items about the font. |
| ChartArea.Select | Set several items that relate to the screen location of the charts. |

**Figure 12.4.11: Final Chart and Graph.**

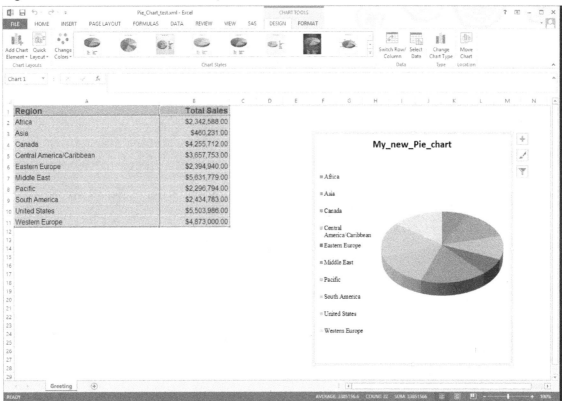

The graph in Figure 12.4.11 is the graph produced by the VBA subroutine my_graph_1. The SAS style option STYLE=minimal produced the Excel worksheet with no border lines. While it is not a pretty graph, it is proof that a SAS code routine can send data to Excel and direct Excel to run an Excel routine, and that is what I wanted to do.

## 12.5 Conclusion

This example is a proof-of-concept example, and has shown that, with relatively simple tools, it is possible to extend the reach of the programmer's ability to automate output. It is my hope that this process can be an insight into the possibilities that are available. Of course, any Excel code can be accessed using this method, and this truly means any code. Excel uses Visual Basic for Applications code and has access to routines that can output any file format that Excel can write. The user-coded macros can also include code to read files, run "Microsoft Word", "Microsoft Access", or even SAS applications. With proper control structures, the formatting of multiple page workbooks can be automated, page titles can be added, column formatting and highlighting can be done, and of course anyone can make a better graph.

# Chapter 13: Building a System of Microsoft Windows Scripts to Control Excel Macros

| | |
|---|---|
| 13.1 Introduction | 211 |
| 13.2 Purpose | 212 |
| 13.3 Guidelines for Building and Using a VBS/VBA Macro Library | 214 |
|     13.3.1 Create Naming Conventions for Storing and Executing VBS/VBA Macros | 214 |
|     13.3.2 Set Up Workstation Options | 215 |
|     13.3.3 Where to Store VBS/VBA Scripts and Macros | 217 |
|     13.3.4 SAS Code to Execute a Visual Basic Script | 219 |
|     13.3.5 Build a Parameter-Driven VBS Script to Control the Execution of Excel | 220 |
|     13.3.6 Build a Control Macro for Each Excel Report | 223 |
| 13.4 Conclusion | 229 |

## 13.1 Introduction

Chapters 12, 13, and 14 each define separate, but increasingly powerful tools that combine SAS and Excel features and that integrate your ability to transfer data between SAS and Excel. At the beginning of each of these chapters I have placed a short list of all of the tools--just in case you look at only one of the chapters today.

Computer systems and software packages like SAS and Excel are really great at crunching numbers. They even consider letters to be a number. An eight-bit computer byte can contain a letter in the form of a number between 0 and 255. Most software packages start out as stand-alone packages. After they mature and become accepted, the software developers work to interface with other software systems. Operating systems, on the other hand, need to communicate with the applications and must have methods of passing data between the different software packages. On Windows, one of those methods of passing and controlling data movement between computer systems, software packages, and the operator is called Visual Basic Scripting (VBS). SAS can communicate with this feature through the "X" command.

The Visual Basic Scripting capability, included in the Windows operating systems, is very powerful. It can open, manipulate, control, and close an Excel program and other Microsoft products. This power permits libraries of Excel macros to be stored as individual code modules such as files with the extensions *.vba, *.vbs, *.bas, *.cls, *.frm, *.frx, and others. When stored in a common directory, the code can be used by any team member on a project. The ability to call these macros at any time allows the creation of a standardized set of macros in a library for report formatting. Community-level routines can be stored in a macro library for departmental use, while report-specific macros can be stored in separate code files, thereby creating a departmental set of standards and reports that can survive personnel changes.

**Table 13.1.1: Tools Described in Chapters 12, 13, and 14.**

| Tool | Chapter | Description |
|---|---|---|
| **Personal Workbook Tool** SAS tool to run personal Excel macros, already included in Excel (under Macros ▶ Record Macro ▶ Store macro in). | 12 | This tool uses the SAS "X" command to execute Excel macros in your Personal Excel "Xstart" directory. It allows Excel workbooks to be delivered without embedded macros. |

| Tool | Chapter | Description |
|---|---|---|
| **Macro Library Tool**<br>SAS tool to run externally stored Excel macros. You can use the "Macro Library Tool" within Excel to take the same macro you would have created in the Personal Workbook Tool and export it to a departmental library. | 13 | This tool uses the SAS "X" command and the Windows operating system scripting language to control processing of Excel macros. It allows Excel workbooks to be delivered without embedded macros. |
| **Excel Workbook Tool**<br>An Excel tool to store parameters for SAS programs and either execute the SAS code or place the code into a directory for execution.<br>(My_Excel_Tool, available in the example code and data folder on the author's SAS Press page) | 14 | This Excel workbook tool will save parameters for a SAS program and either execute or copy the code to a directory. It uses features of the other two tools and allows for storage of SAS code in a production-type area so the original code is not modified when the reports are processed.. |

## 13.2 Purpose

In Chapter 12 the concept of maintaining control of the execution of a task was introduced. The procedures shown in Chapter 12 introduced the idea that a programmer can create a system of software commands to control the capability of multiple software packages. The ability to control and cause multiple tools to work together enables you as a programmer to create far more powerful tools using what you already know. This chapter will build upon the concepts of Chapter 12 and enable us to create a system of tools that will be expandable and flexible enough for departmental rather than simply individual use.

Here is the basic program model, as described in Chapter 12.

- Excel pre-processing setup
    - Open Excel.
    - Write/Store Excel macro to process control information.

- SAS processing
    - Start SAS.
    - Run SAS program.
    - Set up control information for Excel.
    - Create *.xml file.
    - Use "X" command to start Excel and open the generated workbook.

- Excel processing
    - Locate the control information.
    - Execute the Excel macro.
    - Release control to Excel.

SAS features we have seen:

- PROC EXPORT produces Excel files in native Excel format, but the number of columns and rows that can be output are limited by the Microsoft JET and ACE engines. PROC EXPORT has limited file formatting options.
- DDE can use a template to build an output Excel file with great (pre-assigned) formatting, but running multiple reports can cause timing issues and the job stream may fail. The template *.xls file may have Excel macros that need to be removed.
- ODS output files that are generated with TAGSET TEMPLATE processes (CSVALL, HTLM, EXCELXP, and others) whether using PROC PRINT, PROC REPORT, or other processes generate *.xml files (not *.xls files).
- Methods of placing graphs into Excel files are limited, and some formatting options are either not available to SAS programs, or customers need to purchase software packages that may exceed their needs.
- Many Excel formats are not available in SAS. While it is possible to generate code with SAS PROC TEMPLATE, it is not for the faint of heart. Programmers unfamiliar with PROC TEMPLATE may find the code difficult to update or write.
- Some companies do not use all SAS products that are available and therefore do not have access to some of the more powerful features of SAS. The routines described here do not require SAS Enterprise Guide, SAS Business Intelligence, JMP, or SAS Add-In for Microsoft Office.

## New Concepts to Be Introduced

This chapter builds upon Chapter 12, which introduced the program model in which SAS produces an output file with control information. Excel uses that information to determine which Excel macro to execute. When Excel starts, the Excel macro is executed. The tool developed in this chapter is based upon a model that passes control information to a Windows operating system script. That script then controls all of the Excel processing. The script opens Excel, loads the macros, executes the macros, unloads the macros, and closes the Excel workbook. This tool can even reformat the Excel output--say from an xml file--to an xls or xlsx file format. The Excel macros that the tool uses can be stored in a departmental directory rather than on an individual computer.

## The Updated Program Model

- Excel pre-processing setup
  - Open Excel.
  - Write/Store Excel macro in a disk file *.bas.

- SAS processing
  - Start SAS.
  - Run SAS program.
  - Define the file name and path of a Windows *.xbs script to execute.
  - Define the file name and path of generated *.xml file.
  - Define the file name and path of final *.xls output file.
  - Define the file name and path of stored Excel macros (*.bas code).
  - Define the name of the Excel macro to execute.
  - Create *.xml file.
  - Use "X" command to start running a VBS routine to process the generated workbook.

- VBS routine
    - Open Excel.
    - Locate and load VBA code into Excel workbook.
    - Execute the VBA code modules.
    - Remove the VBA code modules.
    - Close and convert the *.xml file to a *.xls (or *.xlsx) file.
    - Terminate Excel.
    - Terminate the VBS script.

This chapter will explain this new method and why building a system of directories, *.BAS files, and *.VBS code is important. Then, I will show how to execute these macros from SAS to format Excel reports. A basic working knowledge of Excel Visual Basic for Applications (VBA) is assumed. The rest of the concept will be sketched out here to allow you to build and expand upon your project needs. Previously, PROC EXPORT, ODS, DDE, the EXCELXP tagset template, and other methods of creating an Excel file have been discussed and examples provided. The concepts shown here will allow you to process Excel and other types of files in many different ways.

## 13.3 Guidelines for Building and Using a VBS/VBA Macro Library

When individuals have a tool that enables them to easily process multifaceted computer programs, they can be more productive to their company. Linking tasks together with software can eliminate human interaction with the data. This can eliminate errors related to manually entering or moving data. But, as we saw in Chapter 12, some software can be restricted to use on one computer. The XLSTART directory used by Excel is available only on the computer where Excel is installed. Hardware failure or upgrades can cause the loss of all of these tools.

The process described in this chapter stores the Excel macros in a directory as separate code modules that are used only while the macro is processing the workbook sheets. These macros can be stored in a Read-only directory with limited Update access that will make them more secure and allow a wider access to the report processing when new data is available. These VBA macros will allow you to take control of Excel or other products as a computer programmer and make it do your bidding. The process is simple and can be set up using the following general guidelines.

- Establish a disk directory where the VBA and VBS code modules can be stored. If multiple users need access, it works best to have it available somewhere like a server location, where everyone can read the directory.
- Publish a standard set of parameters that the VBS routine users can use to access the VBA code modules.
- Unique code modules for individual reports can be placed into the directory and called from the VBS script by using parameters set up by a SAS routine.
- SAS can use the "X" command to start the VBS routine and assign the parameter values to process the report.
- SAS can process the data files to be used as input to the VBS control module. The output from SAS can be any format that the VBA code modules can write.

### 13.3.1 Create Naming Conventions for Storing and Executing VBS/VBA Macros

Let's think about naming conventions for a minute. We can ask the simple questions about them like a newspaper reporter would ask--questions like "Who", "What", "When", "Where", and "Why".

Table 13.3.1: Questions and Answers About Why You Need Naming Conventions.

| Question | Answer |
|---|---|
| Who | All data systems should have naming conventions. |
| What | These are written definitions of where data, code, and output files are stored. |
| When | Naming conventions should be defined before any code or data is prepared. |
| Where | These should apply to all source data providers, programmers, and users of the system. |
| Why | So you can find what you need when you need it. |

Of course, Table 13.3.1 is just a little light-hearted humor about naming conventions. But, those simple phrases speak volumes. The proper naming conventions can make or break a software system. When you walk into a programming job where the system has no naming conventions, it may be hard to find two people who think the data is in the same place. Some guidelines that I use when setting up a system of naming conventions is to look at how and when the data is processed. I look for the following.

- How often is the data processed? Is it updated hourly, daily, weekly, monthly, quarterly, semi-annually, annually, or even every *x* number of years?
- How is the data grouped? Is it by task, file, report, company, department, individual, state, county, zip code, country, or not at all?
- Are project data and code files stored in development directories, source directories, work directories, data directories, production directories, archive directories, or any other directories?
- Has the project team created either a project task flow or a written definition of the project? Sorry, I know I am just dreaming.

When working on a team to define naming conventions, I tend to place parts of the naming convention that are most stable in the first directory position and work from there. Common code source libraries should be separate, but project- or report-level source code should be near the project. It is hard to build a naming convention that is wrong if first you think about how your data inputs, master data files, and project output flow. But, remember that some operating systems have limits on the number of characters that can be in a file path and name.

## 13.3.2 Set Up Workstation Options

Why is there a section in this book about setting up workstation options? The reason is simple; the code is not simple. The Macro Library Tool reaches beyond where most programmers ever venture in their programming experience. Occasionally, you may see a job posting asking for someone who knows something about the "Internal Workings" of system *x* or software *y*. What they are really asking for is someone who can make the system do what they want, not someone who knows how to use a system to get output. I know that sounds almost the same, but let me explain. Getting SAS to write a *.csv file, and then manually opening Excel to read the file and create a graph is getting output from a set of programs. But, writing a SAS program that will pass data to Excel and maintain control while Excel creates a graph that is ready to print is making a system do what you want.

The features described here were accessible to me both on my own system and at locations where tools similar to these were installed for departmental use. That does not mean that your system will allow all of these changes. Digital signatures can be added to programs you write for use on your own computer. This section is designed to prepare you for some of the actions you may need to perform to make these tools work properly.

Other messages that may appear could be something like the ones in Figure 13.3.2. This message talks about trusting the Visual Basic Project code. If you see a similar message, it may be slightly different depending upon the version of Excel and the operating system you are using. Figure 13.3.2 is a message displayed when Excel was called from a *.VBS script. Figures 13.3.3, and 13.3.4 show you what actions you need to take to correct this issue. Once you get this fixed, you should not see it again on the same computer.

### Figure 13.3.2: Windows Message About the Visual Basic Project Not Being Trusted.

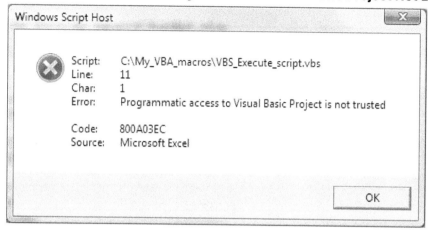

You may be able to eliminate this message by changing the Trust Center options for your computer. If your system administrator has restricted these commands from being modified, then you will need to ask him or her for help. Select the "Trust Center Settings" button shown in Figure 13.3.3 to continue.

### Figure 13.3.3: The Windows Trust Center Window.

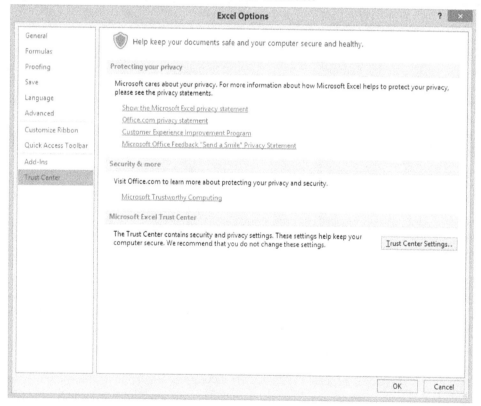

Figure 13.3.4 is shown with the checkbox "Trust access to the VBA project object model" selected. This setting will allow the macros that you create to execute. Remember that Figure 12.3.3a in Chapter 12 describes "Trusted Locations". You may be required to add the location of your VBS/VBA scripts and macros to the list of trusted locations. Use the "Add new location", "Remove", and "Modify" buttons shown in Chapter 12, Figure 12.3.3a, to adjust your list of trusted locations and their subfolders. When you are using these buttons, a pop-up screen will be shown to permit the updates.

**Figure 13.3.4: Windows Trust Center Macro Settings Window.**

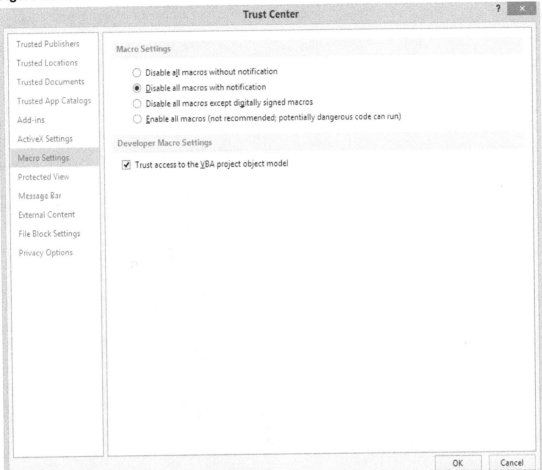

On the Excel screen shown in Figure 13.3.4, selecting the "Trust access to the VBA project object module" checkbox will turn on access to running of VBA code on your computer. It is always wise to verify if you are permitted to modify these settings. Some companies have strict policies about modifying PC settings without permission. Most software that is written or executed from trusted directories on your computer will be allowed to execute. These adjustments allow you to create tools that can control several steps and eliminate manual steps. These tools should be behind strong firewalls. You might even consider Write-protecting the macro code to prevent the code from unauthorized changes.

### 13.3.3 Where to Store VBS/VBA Scripts and Macros

So far, in this chapter we have discussed naming conventions, and workstation setup options to allow macros to run. Before we look at how we create visual basic scripts, Excel data files, or VBA macros, let's look at how to store and reference them. As was mentioned before, naming conventions should be defined before any code is written. No programmer wants to be the person named when something breaks. It can be hard on your job or working relationship within the company or department. Consider including both a "Production" and a "Development" environment into the naming conventions. In addition, to prevent code changes from promoting code from the development environment to the production environment, the naming convention should have something like "PROD" or "DEVL" built into the names.

The type of tool that we are building in this chapter is a little different from the one in Chapter 12. The tool in Chapter 12 pushed control information from SAS directly to an Excel macro. The Excel macro then had to decide whether or not to process a predefined set of Excel macros or do nothing. That tool was built for one user and had to be duplicated if multiple users wanted to use the tool. This type of a tool lends itself to being modified for each copy of the tool. The following table compares the two tools.

**Table 13.3.2: Comparison of Tools Described in Chapter 12 and Chapter 13.**

| Function Program | Chapter 12 Personal Workbook Tool<br>Individual Macro Storage | Chapter 13 Macro Library Tool<br>Excel Macros Stored in Departmental Directory |
|---|---|---|
| SAS | 1. Build *.xml file.<br>2. Set control value in cell A1 label of first sheet.<br>3. Use "X" command to send output file name to Excel for execution by Excel. | 1. Build *.xml file or xls file.<br>2. Define path and file for *.xml file.<br>3. Define path and file for *.xls file.<br>4. Define path and name of *.vbs file.<br>5. Define path and name of *.BAS file.<br>6. Define name of Excel macro to execute.<br>7. Send path, file, and macro names to windows *.vbs. |
| VBS | NO VBS routines are used | 1. Open Excel.<br>2. Load report specific macro.<br>3. Load project common macros.<br>4. Execute report specific macro.<br>5. Delete all Excel macros loaded.<br>6. Save updated file.<br>7. Close Excel.<br>8. End VBS script. |
| Excel workbook built by SAS | First sheet, Cell A1, contains control information | Any xml, xls, or other file that can be read into Excel at Excel start-up time. |
| Excel Personal Workbook | 1. The code module ThisWorkbook stored in the user's "Personal Macro Workbook" contains VBA code to select the requested macro processing. Each computer may have different code here.<br>2. An Excel "module" is used to store recorded Excel macros. | NOT USED |
| VBS code library | NONE | 1. Stores the VBS code.<br>2. Write-protect the file if necessary.<br>3. Store on departmental drive.<br>4. Is accessible to many users. |
| BAS code library | All BAS code is stored in the "Personal Macro Workbook" on the user' computer. Each user's macros are independent of any other user's. | 1. Stores the VBA code.<br>2. Write-protect the file if necessary.<br>3. Store on departmental drive.<br>4. Is accessible to many users. |

As shown in Table 13.3.2, the tools that are defined in chapters 12 and 13 store the Excel macros in different places. Chapter 12 uses the Personal Workbook that is stored in a system-defined location on the computer that is currently being used. However, the macros used by the Macro Library Tool in this chapter can be stored neatly anywhere. This difference also allows better control of the actual code used in the macros. So, it really does not matter where you store the macros; just leave a place in your naming convention for them.

### 13.3.4 SAS Code to Execute a Visual Basic Script

The SAS code shown in Figure 13.3.5 does not need to appear in the order listed in this figure. The only prerequisite for it to operate is that the "X" command has to have everything else defined before it can work.

**Figure 13.3.5: SAS Code to Prepare Data and Parameters and Then Call a VBS Script Module.**

```
proc sort data=sashelp.class out=class;          ❶
by sex Height;
run;

ods tagsets.ExcelXP body='C:\MY_Excel_Files\my_sorted_class_data.xml';   ❷
proc print data=class noobs;
run;

ods tagsets.ExcelXP close;        ❸

* VBA subroutine name to execute                                    ;
%let vbs_code    = C:\My_VBA_macros\VBS_Execute_script.vbs       ;  ❹

* Full File path and Input file name                                ;
%let Input_Excel  = C:\MY_Excel_Files\my_sorted_class_data.xml;     ❺

* Full File path and Output file name                               ;
%let output_excel = C:\MY_Excel_Files\my_sorted_class_data.xlsx;    ❻

* Full File path Location of bas file                               ;
%let bas_code_path= C:\My_VBA_Macros\Chapter_13\;                   ❼

* VBA Module name (without the bas)                                 ;
%let vba_module  = Class_Graph;                                     ❽

* VBA subroutine name to execute                                    ;
%let vba_code    = Class_Graph;                                     ❾

X "'&VBS_code.' ""&Input_Excel"" ""&output_excel""
   ""&bas_code_path"" ""&vba_module"" ""&vba_code""  ";             ❿
```

Here is an explanation of the code in Figure 13.3.5.

❶ Use any SAS code to create a dataset to work with.

❷ This step creates a *.xml file. Any code that can produce a dataset that will open using Excel will work here.

❸ This line of code closes the ODS output routine and releases the output file for use by other programs.

❹ This defines the path and file name of the VBS module that will process the Excel macros and file formatting. The details of the code referenced here are explained in Figure 13.3.7a.

❺ This is the name of the file to be input to Excel and processed by the macros.

❻ This is the name of the output file given to the saved output when the Excel macros have finished processing the input file. If this file path and name are the same as the input file, then the file is saved and replaced the original file.

❼ This is the path to the directory where the Excel macros are stored. Only a path is defined here so that many different macro files (*.bas) can be stored here. This file is a candidate for being placed into a secure location and Write-protected.

❽ This is the name of the VBA code module to load into Excel and execute. The code described in Figure 13.3.7a also reads in a file of common reusable VBA macros. This other file holds reusable code that can be called from the module in item ❼ above to aid programming by having standardized formatting routines.

❾ This is the name of the macro to execute in the module described in item ❽.

❿ This command uses the SAS "X" command to execute the VBS routine listed in item ❹.

## 13.3.5 Build a Parameter-Driven VBS Script to Control the Execution of Excel

Let's dig a little deeper to see how those guidelines can be used to make a working system of VBS/VBA macros. In Chapter 12 we saw the SAS "X" command demonstrate its power to control Excel. In addition to running Excel, the SAS "X" command can execute nearly any operating system command by starting it from within the SAS program code. Directories can be listed, other SAS jobs can be started, Excel workbooks can be opened, and files can be copied or deleted. Now we will discuss how a Windows built-in operating system function can be used to control Microsoft products like Excel and Word.

The VBS scripting language is similar to the VBA code language, but it does have a few minor differences. You can execute the commands stored in a VBS code module (any_file_name.vbs) simply by double-clicking on the filename. For instance, if you double-clicked a file named "broken_command_file.vbs" that contained the words "this does not do anything and produces an error message", the windows operating system would produce something like the following output.

**Figure 13.3.6: VBS Error Message Output Window.**

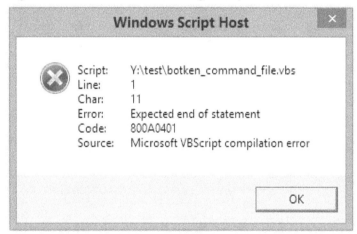

We will look at the details of a VBS script shown in Figure 13.3.7a a little later in the chapter. In Chapter 12, we saw that a VBA macro can be built by recording the macro from within the Excel workbook. These recorded macros generally have a lot of default commands embedded into the macro. Some of these commands are not required for a final macro. A recorded macro may also have a lot of workbook- and worksheet-specific cell references that may be too specific for a general purpose macro.

SAS has the feature of being a programming language, which prepares the code to execute and then starts running at the first executable instruction. What that means to me is that code has to be defined before it can be used. Code is presented to SAS as a text stream, and each macro, DATA step, and procedure call must be executed one group at a time. Code at the bottom of the text file is not executed until the step it is contained within is executed, usually at end of the job. However, object-oriented programming languages like VBA read in the whole set of code routines, compile them, and then pass control of the program to a routine that waits for something to happen. This is similar to a user moving a mouse pointer, clicking on a menu, or pressing a key on the keyboard: each object has its own list of things you can make it do or do to it. These are beyond the scope of this book, but some simple things will be explored to show how to start building a set of your own VBA macros to use to create your reports.

Virtually anything that you can do using Excel you can do using VBA. After all, Excel was written using the VBA language. So, you can modify cells by adding data to them, outlining them, moving them, copying them, or clearing them. You can manipulate rows or columns of data. Files can be read, written, or converted from one format to another. By writing the VBA code yourself, you control the order of the actions that Excel takes. The intent here is to show you how to create formatted Excel output files that are ready for delivery to your user to execute in minutes instead of spending much more time doing it manually yourself. Making them high quality will be left to you. Of course, you need to write the VBA code.

## A Sample VBS Script to Open Excel; Load, Execute, and Delete a VBS Macro

Now, for the fun part, we need a VBS routine to do the work. We need it to open the XML file, load our VBA macros, build our graph, and save our data in the XLS format. The code is in Figure 13.3.7a, with a detailed description of how it works following the code.

**Figure 13.3.7a: The Contents of the VBS Script File VBS_Execute_script.vbs.**

```
Dim Input_Excel, output_excel, bas_code_path, vba_module, vba_code, objxl, objwk,
vbCom, myMod  ❶

Input_Excel   = WScript.Arguments(0)  'Full File path and Input file name
output_excel  = WScript.Arguments(1)  'Full File path and Output file name
bas_code_path = WScript.Arguments(2)  'Full File path Location of .bas file
vba_module    = WScript.Arguments(3)  'VBA module name (without the .bas)
vba_code      = WScript.Arguments(4)  'VBA subroutine name to execute  ❷

set objxl = CreateObject("Excel.Application") 'Start Excel  ❸
set objwk = objxl.Workbooks.Open(Input_Excel) 'Open the input file  ❸

'Activate special software
set vbCom = objxl.ActiveWorkbook.VBProject.VBComponents  ❹

objxl.DisplayAlerts = wdAlertsNone    'Turn off error messages  ❺

if vba_module <> "" then  ❻
  vbCom.Import ("" & bas_code_path & vba_module & ".bas") 'Import VBA Code
  vbCom.Import ("" & bas_code_path & "Common.bas")  'Import Common Routines

  objxl.Run "" & vba_module & "." & vba_code & ""      'Run VBA Code

  'Remove VBA modules
  Set myMod = objxl.ActiveWorkbook.VBProject.VBComponents("" & vba_module & "")
     objwk.VBProject.VBComponents.Remove myMod

  Set myMod = objxl.ActiveWorkbook.VBProject.VBComponents("Common")
     objwk.VBProject.VBComponents.Remove myMod

end if  ❻

'Save as Excel *.xls workbook
if Input_Excel <> output_excel then objxl.ActiveWorkbook.SaveAs output_excel, 51    '
56=xlExcel8, 51=xlsx format  ❼

if Input_Excel = output_excel then objxl.ActiveWorkbook.Save  ❽

objxl.Workbooks.Open(output_excel).Close  ❾
objxl.Quit
set objxl = nothing
set objwk = nothing
```

Here is a step-by-step explanation of the VBS code in Figure 13.3.7a. These descriptions highlight the function of the VBS command without regard to the syntax. The general form of the commands that begin with "SET" are to create a user-named object or instance of whatever is on the right side of the equals sign. Then the name created on the left side of the equals sign can be used as a shorthand definition of the thing on the right side. This is similar to the way a LIBNAME or FILENAME statement works in SAS. Other code (with an equals sign) is an assignment of values. The code in Figure 13.3.7a must be delivered to the Visual Basic Script processor on one line. Some of the code (like lines (1) and (7) wrapped unintendedly and should be on one line.

❶ The code starts by defining the names of input parameters, object names, or variables to be used by the VBS routine. This is done without giving these names any data type or usage definition at the beginning of the routine.

❷ The five lines describing the WScript arguments include a number in parentheses that defines the input parameters number. We will send these parameters to the VBS routine as part of the "X" command when the VBS code is executed from SAS. The parameters are numbered starting from zero and will have the usage as described in the comments on the right of each line.

❸ The VBS command CreatObject actually starts running Excel, and the Workbooks.open command opens the Excel workbook named in the parameter sent from SAS. The workbook is called "Input_Excel". The SAS program provided the full path and filename of the *.xml file created by the ODS tagset template EXCELXP. You still do not see anything; it is all hidden.

❹ The SET command that creates vbCom is providing access to extra Windows software available from the operating system. This is required for some of the other commands to work.

❺ The "objxl.DisplayAlerts = wdAlertsNone" command turns off all messages that VBS may send.

❻ The VBS code from the IF statement to the END IF statement is described here.

    a. If the parameter called vbs_module, which is passed in argument 3 of the SAS X command, is not empty, then the code between the "IF" and "END IF" commands is executed. When the parameter has a value, it is the name of a *.bas file to include into the processing. The parameter called bas_code_path contains the location of the *.bas code modules.

    b. The vbCom.Import VBS commands each open a *.bas module. The code in parentheses concatenates characters to build the full path and file name of the *.bas files.

    c. The first *.bas file read is the Excel macro you want to execute.

    d. The second *.bas module is where some common macro subroutines are stored. This code exists optionally to provide standardized and optimized code to all of the users of your system. This could be considered optional, but it is well worth the effort to create this library of basic Excel macros to perform small tasks. This frees your time and makes writing a program faster.

    e. The objxl.run command executes the Excel macros.

    f. When the VBA macros finish running, the ".VBProject.VBComponents.Remove" command deletes the Excel macros from the workbook.

    g. The "END IF" command finishes the block of VBS code that executes the Excel macro. Steps 7 through 12 are optional. If no *.bas module is passed, this routine can be used to just rename a file or convert the input file into a *.xlsx Excel file format. This routine converts to a *.xlsx output file, but the ActiveWorkbook.SaveAs command described in Step 7 can be modified to do other conversions.

❼ If the Input_Excel and Output_Excel parameter values are the same, then the Output_Excel file is just saved in its updated form. The design of this VBS code routine was to accept a *.xml output from the SAS ODS tagset EXCELXP. After processing the Excel VBA macros, the VBS code module will provide an Excel *.xlsx file in native Excel format.

❽ This step is executed if the Input_Excel parameter equals the Output_Excel parameter. If the parameters are equal than no conversion has been requested. If the file paths and names are the same, the input file is saved in the same format as it was opened..

❾ After you save any changes made, the Excel workbook is closed and the Visual Basic Script is shut down. Control of the computer is returned to the next process. When this *.vbs code module is called from SAS using the "X" command and when the SAS options XWAIT and XSYNC are active, the SAS program will wait for the VBA code to complete and solve some timing issues that could otherwise cause errors.

Depending upon the size of your Excel XML file and how well your VBA macros are written, it can take from a few seconds to a couple of hours to format your file. But, for this process to take a couple of hours it will need a workbook with several hundred pages or very poorly written VBA macros. The good thing is that the next time you process the same file, the formatting will be exactly the same, with no more effort or programming on your part to make it happen that way.

### 13.3.6 Build a Control Macro for Each Excel Report

The result of running the SAS code in Figure 13.3.5 is shown here. We are going to build a macro to produce a bar chart showing all five variables listed in this data.

**Figure 13.3.7b: The SASHELP.CLASS Dataset Sorted by SEX and HEIGHT, Displayed in Excel.**

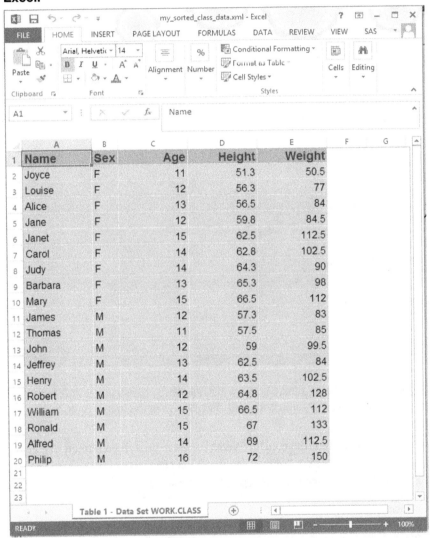

We will start here by recording a macro and saving it into any of the available workbooks. We can save it anywhere because we are going to export the macro to the disk file C:\My_VBA_macros\ with the name Class_Graph.

### Figure 13.3.8: The Record Macro Excel Window.

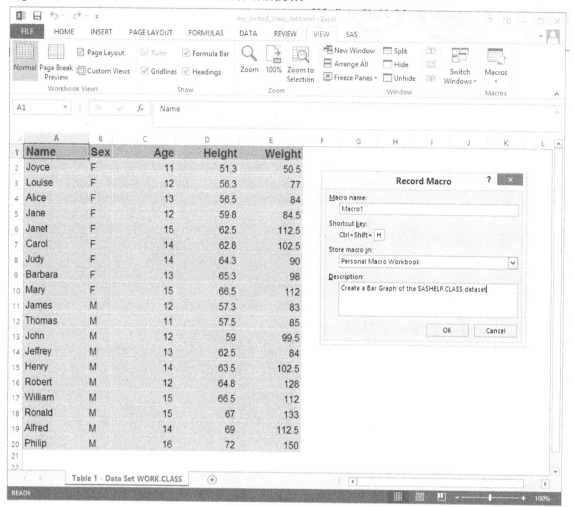

Figure 13.3.8 shows the macro name as being Macro1. We will see how to change that a little later. Once you press the "OK" button, the recording of your actions will start. We want the macro to do the following things.

- Select all of the data cells and adjust the column widths so that each is the minimum size.
- Select all of the data fields and create a 3-dimensional column chart.
- Change the location of the chart to be near the top right of the screen.
- Widen the Excel window.
- Scale the graph so that legends are clear. Sometimes the graph will not scale correctly the first time you execute the code. However, the second time you execute the macro it may appear better because the Excel window was updated when the macro ran the first time.

The graph produced in Figure 13.3.9 was produced by the Excel macro shown in Figure 13.3.10a. The Macro Library Tool executed the code to produce the graph before the SAS program finished running. When the SAS options XWAIT and XSYNC are active, the SAS programs wait for the Windows command window to finish executing before returning control to SAS.

**Figure 13.3.9: Finished Graph of SASHELP.CLASS Data.**

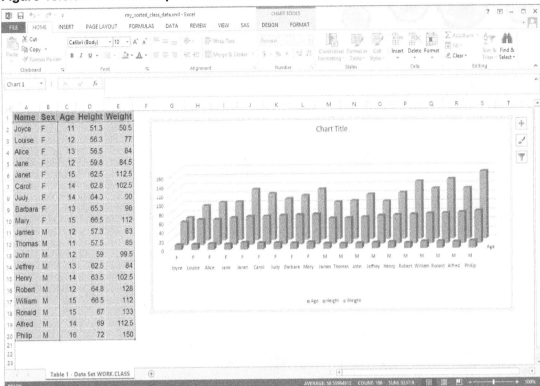

**Figure 13.3.10a: Finished Visual Basic Code Module.**

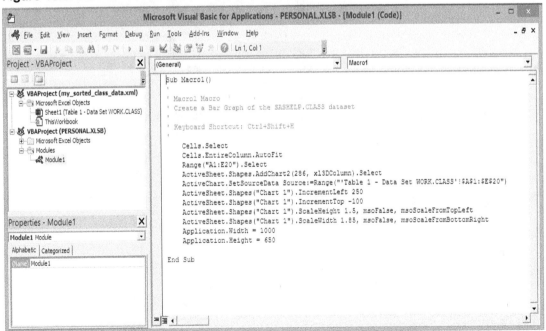

Also notice that the name of the workbook and/or the worksheet 'Table 1 – Data Set WORK.CLASS' may appear embedded in the Excel macro code as part of the range of cells used as the data source for the chart shown in Figure 13.3.10a. As shown above, the workbook name is not present, but the worksheet name is shown. This will also cause the macro to force the debug mode to open and cause the macro to fail to finish building the graph if the page name ever changes. When the workbook name (and the "!") are removed, the reference defaults to the active worksheet in the current workbook. In an effort to keep things simple here, we will not delete the worksheet name--just remember that it could become an issue. Here is the caveat for this paragraph and Figure 13.3.10a. This paragraph will never adequately describe what you need to do to build a graph you would like. So, just turn on the Excel macro recording and make your favorite graph using an Excel graph wizard and see what you have. There will be extra lines of code because the Excel macro recording will generate code to set all of the default values. You can throw away a lot of the code. Just do one line at a time by making the line a comment. Then, check to see if you still have what you want.

**Figure_13.3.10b: Excel Property Window for the Class Graph VBA Code Module.**

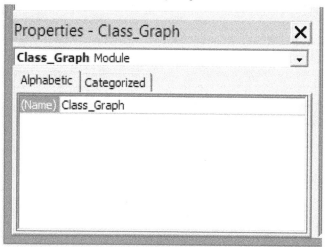

To correct this module name issue, we need to change the name of the macro from Module1 to something easy to remember, such as Class_Graph. We do that by typing a new name into the Properties window in the name attribute field. See Figure 13.3.11 below. Once we have renamed the VBA code module, we can export this Excel VBA code into a macro library for later use. To change the name, we edit the name property (if it is not already named correctly) on the left side to the Visual Basic for Applications window, as shown in Figure 13.3.10b. By using the "Export file" option on the File menu, we can place the new Class_Graph macro into any directory we want to use. The one used here is the Chapter_13 macro directory created for the example.

**Figure 13.3.11: Export Window Showing the Export of File ClassGraph.bas.**

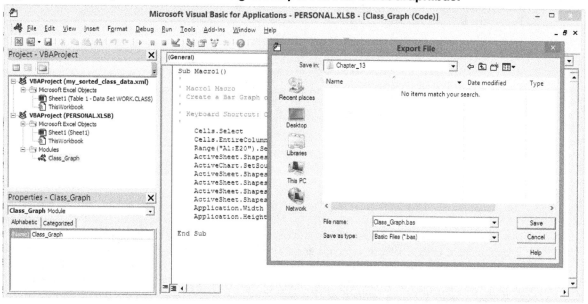

The last step here in Excel will be to record another macro called "**make_thick_red_outside_border**". The module name is changed to "Common". While this VBA code module has only one VBA subroutine, it could contain many subroutines, since all VBA modules can contain many subroutines. This code is to show you how to establish a pattern of how to build these modules and subroutines for generic use. If you can do it for one module, you can do it for many.

**Figure 13.3.12: Define Macro to Make a Thick Red Border Around the Data.**

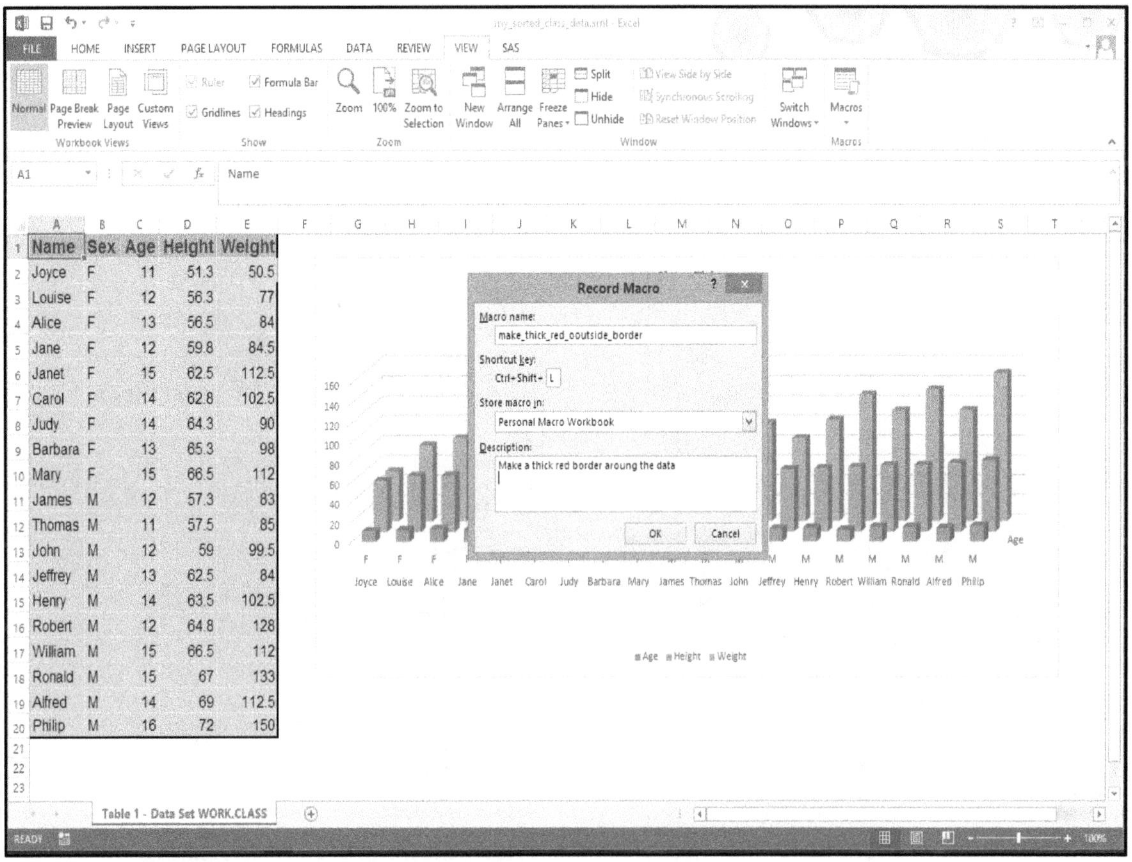

This macro shown in Figure 13.3.13 does the following:
- Selects the data cells (A1:E20).
- Sets the inside diagonal borders to no line.
- Sets the outside top, bottom, left, and right edges to a medium thickness red line.
- Sets the inside vertical and horizontal lines to a thin white line.
- Selects cell A1 when done.

**Figure 13.3.13: The Macro to Make a Thick Red Border Around the Data.**

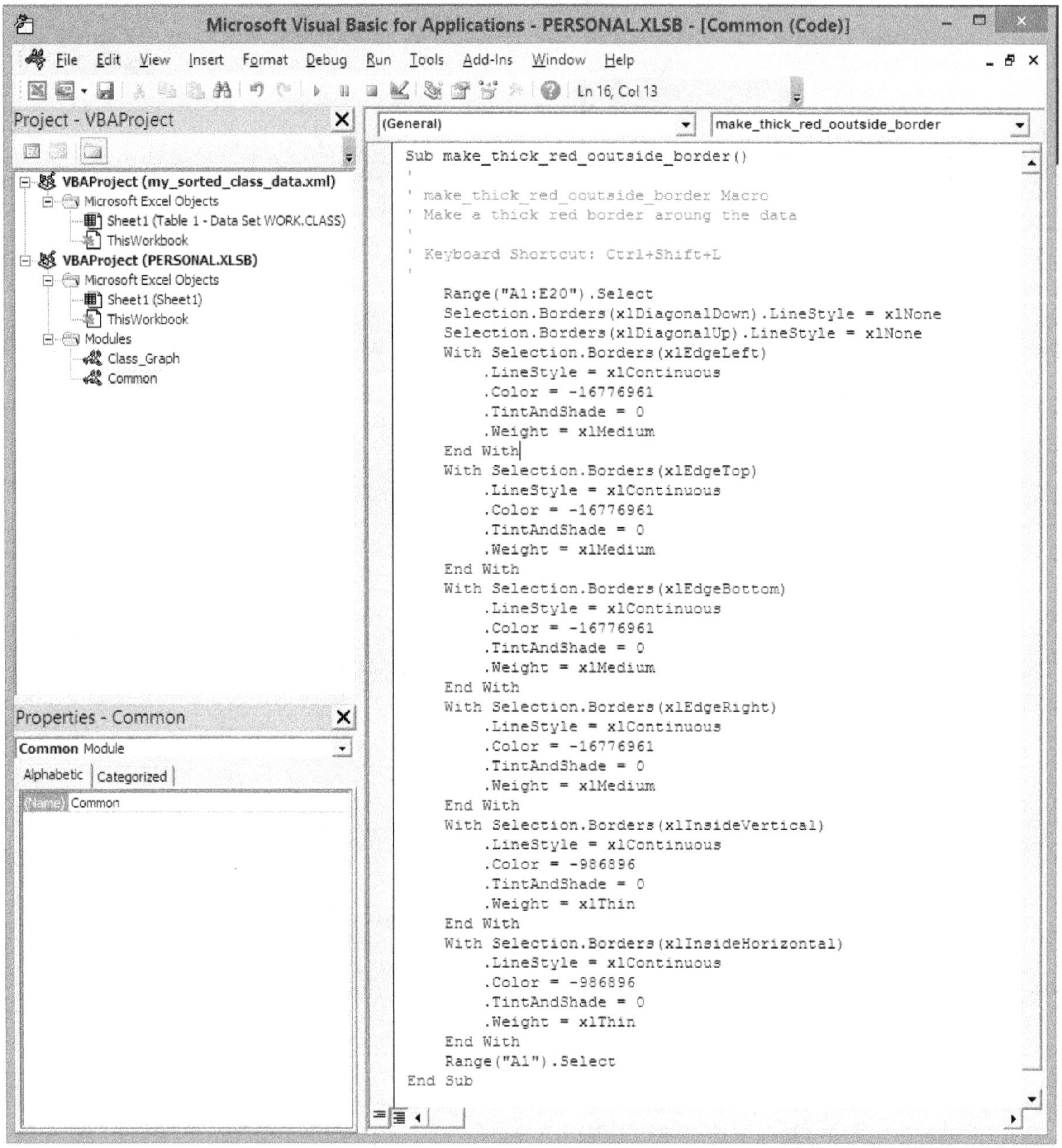
Microsoft Visual Basic for Applications - PERSONAL.XLSB - [Common (Code)]
File Edit View Insert Format Debug Run Tools Add-Ins Window Help
Ln 16, Col 13
Project - VBAProject
VBAProject (my_sorted_class_data.xml)
Microsoft Excel Objects
Sheet1 (Table 1 - Data Set WORK.CLASS)
ThisWorkbook
VBAProject (PERSONAL.XLSB)
Microsoft Excel Objects
Sheet1 (Sheet1)
ThisWorkbook
Modules
Class_Graph
Common
Properties - Common
Common Module
Alphabetic
Categorized
(Name)
Common
(General)
make_thick_red_ooutside_border
Sub make_thick_red_ooutside_border()
'
' make_thick_red_ooutside_border Macro
' Make a thick red border aroung the data
'
' Keyboard Shortcut: Ctrl+Shift+L
'
Range("A1:E20").Select
Selection.Borders(xlDiagonalDown).LineStyle = xlNone
Selection.Borders(xlDiagonalUp).LineStyle = xlNone
With Selection.Borders(xlEdgeLeft)
.LineStyle = xlContinuous
.Color = -16776961
.TintAndShade = 0
.Weight = xlMedium
End With
With Selection.Borders(xlEdgeTop)
.LineStyle = xlContinuous
.Color = -16776961
.TintAndShade = 0
.Weight = xlMedium
End With
With Selection.Borders(xlEdgeBottom)
.LineStyle = xlContinuous
.Color = -16776961
.TintAndShade = 0
.Weight = xlMedium
End With
With Selection.Borders(xlEdgeRight)
.LineStyle = xlContinuous
.Color = -16776961
.TintAndShade = 0
.Weight = xlMedium
End With
With Selection.Borders(xlInsideVertical)
.LineStyle = xlContinuous
.Color = -986896
.TintAndShade = 0
.Weight = xlThin
End With
With Selection.Borders(xlInsideHorizontal)
.LineStyle = xlContinuous
.Color = -986896
.TintAndShade = 0
.Weight = xlThin
End With
Range("A1").Select
End Sub

Now that we have a main routine to build our graph and a common routine that will put a red border around selected cells, we should update our VBA macro to use both pieces of code. The simple way to execute the subroutine to make red borders is to call it by name as in the following command:

***"CALL make_thick_red_outside_border"***

All we need to do is insert this command into our original code, as shown in Figure 13.3.14.

**Figure 13.3.14: Updated Class_Graph Macro to Process the Thick Red Border.**

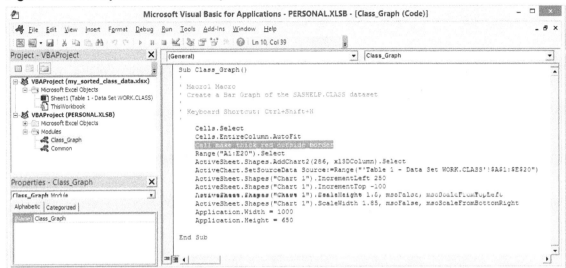

## 13.4 Conclusion

Like Chapter 12, this chapter presents a complete tool that allows you to control all aspects of a task. But this chapter shows you how to expand a single user tool and make one that anyone in the department can use. Some of the features of the Macro Library Tool include the following:

- The Macro Library Tool can easily be included when creating naming conventions.
- The output file has no macros and can be saved as a *.xlsx file.
- The SAS code is limited and flexible. It can be placed into just about any SAS program.
- The VBS code is limited and also has some flexibility with respect to saving files and renaming output files.
- The Excel macros can be easily created by recording macros and exported to your macro library.
- Optimized VBA macros can be stored and available for general use by anyone in your department. These do not need to do big things. For example, some might just create small red boxes.

New concepts that can be made available because of the Macro Library Tool include, but may not be limited to, some of these options. Feel free to dream up some of your own.

- The Macro Library Tool has no manual intervention steps, and the output file is closed when the job is finished. This can allow other data to be added to the file and other macro processing to occur.
- The macros can be Write-protected and saved in a secure location.
- A set of standardized VBA macros can be used to increase programmer productivity.
- Each report or task can have a different set of macros, but if data is formatted, the same one macro can process several reports.
- The VBS routines are not restricted to processing only Excel files.

# Chapter 14: Create an Excel Workbook That Runs SAS Programs

**14.1 Introduction..................................................................................................... 231**

**14.2 Purpose............................................................................................................ 232**

**14.3 Guidelines for Building an Excel User Form Interface ................................. 233**
    14.3.1 Common Excel and Excel User Form Terms ........................................................233
    14.3.2 Introduction to the Integrated Development Environment (IDE) .....................235
    14.3.3 Using the Integrated Development Environment (IDE) Toolbox Menu ............236
    14.3.4 Building a Sample Integrated Development Environment (IDE) Menu ............237
    14.3.5 Linking the Integrated Development Environment (IDE) Menu and the Data ..239
    14.3.6 Storing Control Information in the Excel Workbook Worksheets .....................240
    14.3.7 Set Up Control Variables to Access Data Stored in the Workbook ..................241
    14.3.8 Learn How to Make the Excel UserForm Execute................................................245

**14.4 Excel VBA Routines to Make the Workbook UserForm Active.................... 248**
    14.4.1 Initialize the User Form...........................................................................................248
    14.4.2 Write the User Parameters to a File in a Working Directory ...........................253
    14.4.3 Copy Source Program from a Production Directory to the Working Directory..................................................................................................................253
    14.4.4 Verify the Output Batch File Points to the Correct SAS Run Time Module ......254
    14.4.5 A Routine to Save the Changes and Exit the Program .......................................255
    14.4.6 Directory Structure Associated with the Processing............................................255
    14.4.7 Common Issues That Might Occur. ......................................................................257
    14.4.8 Prepare a VBA Macro to Process Your Output Report........................................258

**14.5 Conclusion ..................................................................................................... 259**

---

## 14.1 Introduction

Chapters 12, 13, and 14 each define separate, but increasingly powerful tools that combine SAS and Excel features and that integrate your ability to transfer data between SAS and Excel. At the beginning of each of these chapters I have placed a short list of all of the tools--just in case you look at only one of the chapters today.

This chapter will describe and show you how to build a simple set of Excel forms and macros. It could provide a starter system to allow you to store parameters in an Excel workbook and run SAS code using only Base SAS and Excel. The tool described here is not intended to replace any SAS software and has some limitations that do not exist in SAS applications. The intended user audience for this Excel workbook application will need to know how to program in both SAS and Microsoft Visual Basic languages. Some of the limitations of this Excel workbook are listed below.

- If the workbook is on a user's computer, only the user has access to the workbook.
- If the workbook is on a server drive, many people can use the workbook one at a time.
- If the workbook is on a server drive, any changes are visible by the next user.
- Knowledge of VBA and VBS is required to install the workbook and add new programs.
- Setting up the workbook for someone new to use requires working knowledge of Excel options.

**232** *Exchanging Data between SAS and Microsoft Excel*

**Table 14.1.1: Tools Described in Chapters 12, 13, and 14.**

| Tool | Chapter | Description |
|---|---|---|
| **Personal Workbook Tool** SAS tool to run personal Excel macros, already included in Excel (under Macros → Record Macro → Store macro in). | 12 | This tool uses the SAS "X" command to execute Excel macros in your Personal Excel "Xstart" directory. Allowing Excel workbooks to be delivered without embedded macros. |
| **Macro Library Tool** SAS tool to run externally stored Excel macros. You can use the "Macro Library Tool" within Excel to take the same macro you would have created in the Personal Workbook Tool and export it to a departmental library. | 13 | This tool uses the SAS "X" command and the Windows operating system scripting language to control processing of Excel macros. Allowing Excel workbooks to be delivered without embedded macros. |
| **Excel Workbook Tool** An Excel tool to store parameters for SAS programs and either execute the SAS code or place the code into a directory for execution. (My_Excel_Tool, available in the example code and data folder on the author's SAS Press page) | 14 | This Excel workbook tool will save parameters for a SAS program and either execute or copy the code to a directory. It uses features of the other two tools and allows for storage of SAS code in a production-type area so the original code is not modified when the reports are processed. . |

## 14.2 Purpose

This approach starts with the following assumptions:

1. You store all of your "Good" [think departmental final report level] code in a source directory. You may want to Write-protect this directory to prevent accidental updates.
2. You have a set of periodic working libraries that you can use to save your current monthly version of code files and output data for each report.
3. The input data is also stored in a set of monthly directories that have an established naming convention that allows systematic access with minimal path name changes. Multiple monthly libraries can be used to write quarterly or annual reports by accessing these data files from within the individual report programs.
4. The code in your source library uses macro variable values to provide the data file path names, dates, variable file names, or other data to the individual reports at run time.
5. Data paths within the report programs are defined with macro variables to prevent the "hard coding" of information that can change. This way, the programs do not change; only the names of the data files for input or output change.
6. You write your programs using standardized macro variables, so that the Excel file or the options on the Excel form can pass macro variables with known names and common data values to each program. This will help standardize the programs and aid in making maintenance easier.
7. The Excel program can store segments of SAS code to set up report-unique macro variables that will be inserted into the output files.
8. The Excel tool is set up to store SAS code in a directory. The code can then be executed by either running SAS and executing the program or by double-clicking on a Windows batch job file (*.bat) to execute the SAS code.
9. Each program has been set up to store the output log and list files into a disk file for audit and job validation processing.

10. The department or programming group has set up a SAS source code library with common routines that can be reused by all of the programmers in the department. The types of SAS code routines that can be stored here include code to generate formats, perform tests, read data files, write data files, and any other code that is used more than once. By placing code here, you ensure that, if the system requirements change, only one code file needs to change and all of the programs are updated. It is also important to verify that changes made are universally acceptable.

This system described here is a simple set of Excel user forms, Excel VBA macros, SAS code files, and Windows *.bat files. The task of building a system to meet your needs is left up to you. This workbook stores and executes the parameters to run SAS programs, but the number of common parameters and extra features installed into the tool is limited.

## 14.3 Guidelines for Building an Excel User Form Interface

### 14.3.1 Common Excel and Excel User Form Terms

So, let's get started. The first step will be to create an Excel Graphical User Interface (GUI). That may sound important and hard, but it really is not hard at all. First there's a description of what we need to do, and then we will look at some screen shots that will show how to perform the tasks. We need to create an Excel file into which we can place a macro. That means that either an *xls, *.xlsb, or *xlsm Excel formatted file will be created. Excel 2007, 2010, and 2013 files (*.xlsx) do not permit macros. This can be done by opening a new Excel workbook since we do not need any special things in the output file. We can build everything we need in a few minutes. Before we get started, we need a few simple term definitions for the Excel structures that will be in the tool. See Figure 14.3.1.

Note: A detailed explanation of this process was presented in Benjamin, William E., Jr. 2013. "Give the Power of SAS to Excel Users Without Making Them Write SAS Code." *Proceedings of the SAS Global Forum 2013 Conference.* Cary, NC: SAS Institute Inc. Available at http://support.sas.com/resources/papers/proceedings13/010-2013.pdf

**Figure 14.1: Glossary of Common Terms Used When Building an Excel UserForm and GUI.**

| Term | Structure | Use in the Tool |
|---|---|---|
| Safe Mode | A method of executing Excel from a command line within a Windows *.bat file. | This will start Excel using command switches that turn off some of the features of Excel. |
| Integrated Development Environment (IDE) | Excel screen that appears when you type the Alt+F11 keys with a spreadsheet open. | This is the Excel tool we will use to build our reporting tool to run our SAS programs. |
| User Form (spelled UserForm) | A program "Object" used by Excel to store tool components. | Used to display the components of the tool. |
| Label | A component of a UserForm that stores text that cannot be changed. | UserForm field describing another Form component; used here to describe screen fields. |
| TextBox | A component of a UserForm that stores text that can be changed. | A box on the UserForm that contains a value that can be changed; this value is used later. We will enter fields or show data we update. |
| ComboBox | A component of a UserForm that stores a list of text values that can be selected. It provides lists of valid option values. | We will use this to store a list of valid programs and prevent the user from picking an invalid value. |

| Term | Structure | Use in the Tool |
| --- | --- | --- |
| CommandButton | A component of a UserForm that, when pressed, activates an action. | The Command buttons will be used to initiate some action requested by the user (You). |
| CheckBox | A component of a UserForm that toggles between the on and off states. | Here we will use this to toggle the on and off states of a requested condition. |
| Properties Window | A window on the Integrated Development Environment (IDE) that shows the state of each element of an item on the UserForm. | We will use this window to associate data elements on an Excel spreadsheet with text fields on the UserForm. |
| PropertySheet | A window on the Integrated Development Environment (IDE) that shows the state of each element of an item on the UserForm. | Property sheets associate Excel spreadsheet data elements with text fields on the UserForm, or apply values to the item. |
| ControlSource | A field on the PropertySheet that points to an Excel cell that is used to store the contents of the item at run time. | We will use this to point to where the data value for this UserForm item is stored in the Excel spreadsheet. |
| Name | A field on the PropertySheet that describes the name of a component. | We will not use this field for most items on the UserForm, making the component items and types clearly identifiable. However, some items on the UserForm will have the name changed, so we can reference specific spreadsheet values, cell, or other items without special coding. |

The object of this workbook tool is to be able to double-click on an icon on your desktop to start Excel, store parameters, and access information about SAS programs. This tool can be selected to create a user copy of a departmental report program. When the SAS code runs, it will produce the requested output based upon periodic execution parameters entered or stored in the Excel workbook. The tool presented here will be a simple implementation with a minimum number of features to demonstrate the concept and show you some basic steps that may be needed to build your own enhanced version of the tool. The general process we will use is the following.

1. Identify the IDE and how to use it to build an Excel UserForm.
2. Identify the Excel objects needed to be placed upon the UserForm.
3. Build storage locations within the Excel tool to store program-unique parameters.
4. Set up objects to access system-global and program-unique data stored in the workbook.
5. Learn how to make the Excel UserForm run.
6. Link the data from worksheet "Common_Parm_Text" to the UserForm1 data fields.
7. Build Excel VBA routines to make the tool work.
8. Build Excel VBA routines to format the output files for printing.
9. Learn how to enter data and execute commands stored in the VBA code supporting the tool.
10. Build data structures that will support the input, output, and working directories.
11. Set up the programs to store output log and list files for audit purposes.
12. Show a step-by-step process to execute the tool and produce a report.
13. Build a command file (*.bat) to execute the tool by clicking on an icon.

The workbook will be built to contain two jobs that can be set up to run periodically. One is called JOB_01 that will write the SASHELP.Shoes file to an Excel workbook and then format it using the VBS/VBA tools described earlier in the book. The second program, called JOB_02, will read an Excel file that contains the output from JOB_01 and place that data into a SAS dataset. Of course, the object of this chapter is not JOB_01 or JOB_02, but rather the tool to run the jobs.

This process can be set up using Base SAS and any version of Excel that is written using Visual Basic for Applications (VBA). Excel should also be able to both store a macro and contain more than one Excel worksheet. By writing this code and creating the UserForm yourself, you can answer the question, "Did you download this from the web?" with a resounding, "NO!" This is another reason why this version of the tool is simple.

OK, let's get started. The version of Excel used here is Excel 2010, and the workbook type is an *xlsm workbook. Next, we will open Excel and save the workbook as "My_Excel_Tool_To_Run_SAS_Jobs.xlsm". We will save the workbook in directory "Q:\My_Excel_Tool".

### 14.3.2 Introduction to the Integrated Development Environment (IDE)

We start by opening the new Excel file (this version is Excel 2010 with the ribbon minimized to conserve space on the page).

**Figure 14.2: A Blank Excel Macro-Enabled Workbook with the Ribbon Minimized.**

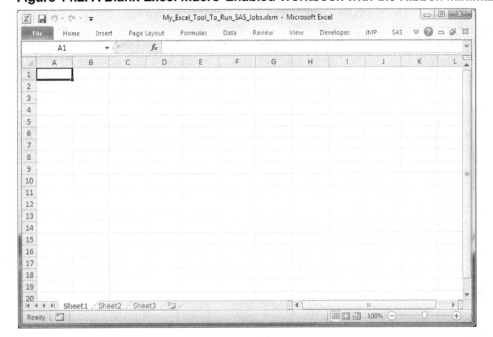

To locate the Integrated Development Environment (IDE), first select the worksheet, and then press the keypad buttons "ALT" and "F11" at the same time. This will open another screen, shown below, which is virtually the same for all versions of Excel that can open the IDE.

**Figure 14.3: A Blank Excel Workbook with the Integrated Development Environment (IDE) Visible.**

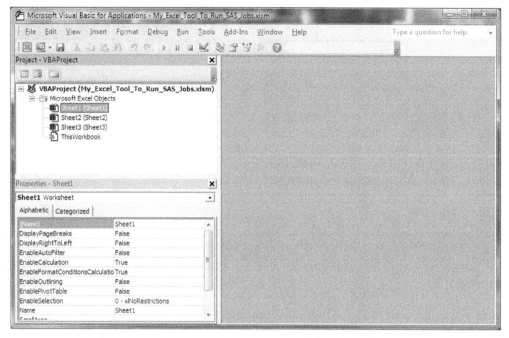

## 14.3.3 Using the Integrated Development Environment (IDE) Toolbox Menu

From here, we need to add a UserForm. That is done by selecting the Insert toolbar option and, on the drop-down menu, selecting the UserForm object to add. The result will look something like the next figure.

**Figure 14.4: A Blank Excel UserForm and Excel Toolbox.**

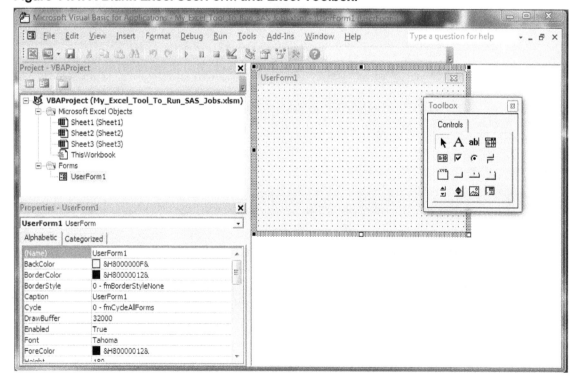

In the Toolbox pop-up window, there are 16 pictures of control items. Here, for simplicity, we can identify them in a 4-by-4 grid that is the same as the layout of the pop-up menu. (See Figure 14.6a below.) To install an object onto the UserForm1 object, click on one of the toolbox items, and then move the cursor to the UserForm. Hold the left mouse button down, and drag the mouse until the object is the size that you want it to appear. The IDE allows you to position and size items on the UserForm with a great deal of ease. If the item is not exactly how you want it, then highlight it and press Delete.

**Figure 14.5: The Excel UserForm Toolbox.**

**Figure 14.6a: An Explanation of the Icons on the Excel Toolbox Shown in Figure 14.5.**

| IDE Controls on the Toolbox Pop-up Menu | | | |
|---|---|---|---|
| Menu option to select a control item | Label | TextBox | ComboBox |
| ListBox | CheckBox | OptionButton | ToggleButton |
| Frame | CommandButton | TabStrip | MultiPage |
| ScrollBar | SpinButton | Image | RefEdit |

## 14.3.4 Building a Sample Integrated Development Environment (IDE) Menu

Our little tool will be using only a few of these controls to make the GUI work. We will use the Label, TextBox, ComboBox, CommandButton, and the CheckBox controls. The next screen has one of each control item on the UserForm1 work area. When you click on the blank UserForm1 work area, it becomes the active component and you can click on the corner and expand the form. Controls are added to the UserForm1 workspace by clicking on the Tool icon in the toolbox, right-clicking on an anchor spot in the UserForm1 space, and enlarging the control before releasing the right mouse button. Oh yes--and the Toolbox menu floats around the screen, so you can put it anywhere.

**Figure 14.6b: The Graphical User Interface UserForm Shown in the IDE.**

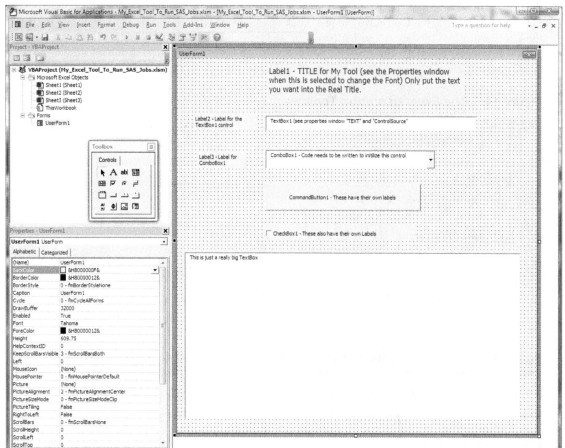

The next screen image has all of the controls for the tool added to the UserForm1 Excel object. This will look similar to the finished GUI form and will not have the control names in the image. The final tool will have two of the TextBox objects hidden. These will be used by the VBA code to make the program easier to code. Note that Figure 14.6b has some notes about the Font, ControlSource, Text values, and the need to initialize the ComboBox. These notes refer to the Properties window contents. This window shows the properties for the UserForm. Most of the items on the Properties window can be changed either by typing over the current value, or right-clicking of the right side of the line item you want to change. We will also use some VBA code to change some of the ControlSource values to make our coding easier. The following items are ones we are likely to change for this tool. All of these items can also be changed using VBA code.

1. Name – This property is the name of the control. If you change the object name, then when referencing the object within the VBA code you should refer to the object (Label1) using the new label name (like – My_New_Label_Name),
2. Font – The font property allows you to right-click on the property item and select the type face, font size, and other font items.
3. Text – Information can be typed here to be displayed on the GUI.
4. ControlSource – This property can be set for the cell in a spreadsheet that contains the current value of the control. No matter how that cell is changed, it will be the current value of the control. We will use this feature to permit direct updating of the program-unique values (parameters, program working directly, and the program SAS name) directly into the spreadsheet where this information is stored.
5. TextAlign – Positions the text on the control (left, right, center).

## 14.3.5 Linking the Integrated Development Environment (IDE) Menu and the Data

Figure 14.7 displays final UserForm layout for your tool. We are now ready to dig into the details about how to link the data to the UserForm boxes and create variable values usable directly by VBA code routines.

### Figure 14.7: The Graphical User Interface UserForm Shown in the IDE.

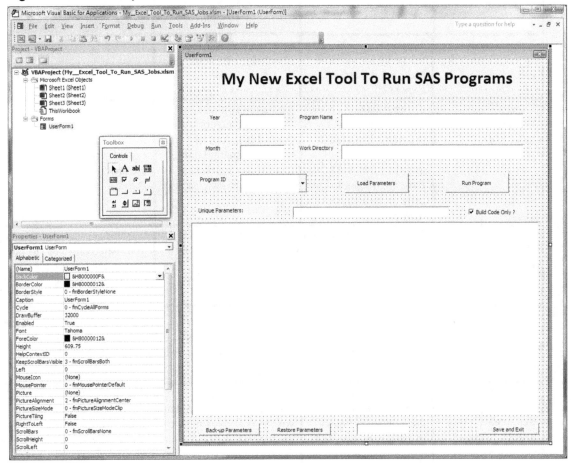

### Identify the Excel objects needed to be placed on the UserForm

Figure 14.7 is our main menu, so let's look at the menu layout (Figure 14.8) to examine the components on the menu and how they will be used. We will identify each element in Figure 14.8, and describe each control by its relative position as shown in the table below (as in TextBox1 or CommandButton2).

**Figure 14.8: A Text Description of the Items on the UserForm, Relative to the Location on the UserForm.**

| Label1 – text -- Contains the title, and is not used for anything else. | | | |
|---|---|---|---|
| Label2 – text "Year", no other use. | TextBox1 – Contains the value for variable Year linked to an Excel cell. | Label4 – "Program Name", no other use. | TextBox3 – Renamed to Program_name. Contains the value for variable Program_Name linked to an Excel cell. |
| Label3 – text "Month", no other use. | TextBox2 – Contains the value for variable Month linked to an Excel cell. | Label5 – "Work Directory", no other use. | TextBox4 – Contains the value for variable Work_Directory linked to an Excel cell. |
| Label6 – text "Program ID", no other use. | ComboBox1 – Contains the list of programs installed in the tool; this is initialized by VBA code and linked to cells in the Excel workbook. | CommandButton1 – "Load Parameters"; this control is linked to a VBA routine that copies data from a worksheet to TextBox5 below. | CommandButton2 – "Run Program"; this control is linked to a VBA routine that executes the VBA code to do the processing. |
| Label7 – text "Unique Parameters", no other use. | TextBox7 – Renamed to Program_directory so that we can take advantage of the properties of a TextBox without writing special code to do the work; we will not make this visible on the GUI screen. | | CheckBox1 – A flag that is true (when checked) to be used in the VBA code to control processing. |
| TextBox5 – Renamed 'Program_Parms' to take advantage of TextBox properties without special code; this Excel control (like TextBox3, TextBox6, and TextBox7) points to cells on the JOB_xx_PARMS worksheets. The contents of the ControlSource field will be pointed to the current JOB_xx_Parms spreadsheet. | | | |
| CommandButton3 – "Back-up Parameters"; this control is linked to a VBA routine that copies data from TextBox5 to a backup worksheet. | CommandButton4 – "Restore Parameters"; this control is linked to a VBA routine that copies data from a backup worksheet to TextBox5 | TextBox6 – Renamed to Program_Number so that we can take advantage of the properties of a TextBox without writing special code to do the work. We will not make this visible on the GUI screen. | CommandButton5 – "Save and Exit" this control is linked to a VBA routine that saves the contents of the workbook and exits Excel. Because VBA runs a macro there are always workbook changes. |

## 14.3.6 Storing Control Information in the Excel Workbook Worksheets

Now that we know what the menu looks like (see Figure 14.7) and have some idea about how to put it to work, let's define some worksheets to use for the data storage of parameters and control information for the tool. For this version of the tool, we will need six worksheets in our workbook. The Excel sheet names (Sheet1, Sheet2, etc.) do not need to be in order because we will be using the user-assigned names of the sheets for all of the references to the sheets. The list below describes their functions:

1. Sheet1 – "Control_Info"       Storage for tool information like variable values.
2. Sheet2 – "Common_Parm_Text"   Storage for text output by the tool to control jobs.
3. Sheet3 – "Job_01_Parms"       Storage for Job 01 parameters.
4. Sheet4 – "Job_01_Backup"      Backup Storage for Job 01 parameters.
5. Sheet5 – "Job_02_Parms"       Storage for Job 02 parameters.
6. Sheet6 – "Job_02_Backup"      Backup Storage for Job 01 parameters.

The next figure shows that the VBA Project window has been updated to show the new worksheets. These worksheets were added by creating a new worksheet in the workbook and renaming them to the user-assigned names.

**Figure 14.9: The Graphical User Interface UserForm Shown in the IDE with Data Linked to the Boxes.**

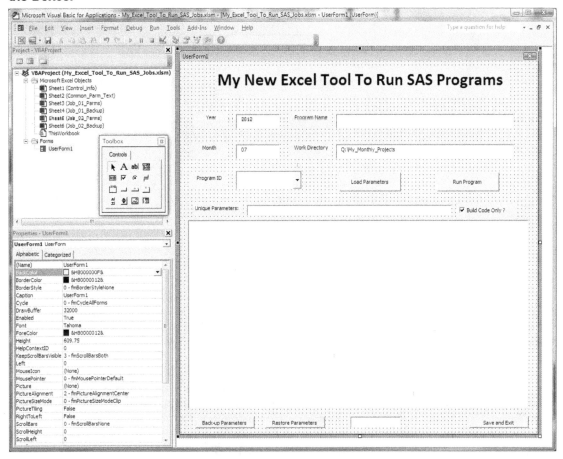

## 14.3.7 Set Up Control Variables to Access Data Stored in the Workbook

### Step 1 – The Global Parameter worksheet – "Control_info"

The next figure is a suggested way to lay out the "**Control_info**" worksheet of our new tool, followed by a description of the layout and cell contents. This purpose of this part of the project is to determine the different aspects of the project you will need. If you have more common variables like the quarter number or specific source file names, you may want to allow more room in the common parameter section. Also, realize that as you add more programs to the tool, you will need more space under the JOB_*xx* list area that is used to fill the ComboBox with job names. So, make sure you have room to allow your tool to grow without redesigning it too soon. Remember, this "**Control_info**" worksheet was specifically designed to fit into one screen shot for this page.

**Figure 14.10: The control_Info Worksheet and the Data Layout of the Control Information.**

| | A | B | C | D |
|---|---|---|---|---|
| 1 | Store Common Parameters in this column and link the control objects to these cells. | Comment about the parameter | List of Installed Programs, and their descriptions | |
| 2 | 2012 | The Current Project Year - Stored as a text value | Live combobox value in cell c/23 | |
| 3 | 07 | The current Project Month - Stored as a text value | JOB_01 | Write SASHELP.SHOES to an Excel file |
| 4 | | | | start Valid Program ids with a blank space |
| 5 | Build code Only Flag source value | | JOB_01 | Write SASHELP.SHOES to an Excel file |
| 6 | TRUE | | JOB_02 | Read My_DATA.SHOES from an Excel file |
| 7 | | | ### | use ### to end the list |
| 11 | Q:\My_Monthly_Projects | The Current Project System working directory where the SAS Code and Unique Parameters will be output. This is loaded from the current Job_xx_Parms worksheet | | |

On Figure 14.10 there are four areas highlighted with bold lines around the cells. These bold boxes are not needed for the tool, but are used here to visually group the data so that it can be easily discussed. Also, most of the information on this worksheet consists of comments about what the data usage is for the cells. The following describes what function the groups of cells perform.

### Control_info Worksheet – Group 1 – Cells A1 to B3.
- The cells A2 and A3 are the only cells that are not comments.
- Cell A2 is linked to TextBox1 and holds the value for the current processing year.
- Cell A3 is linked to TextBox2 and holds the value for the current processing month.

### Control_info Worksheet – Group2 – Cells A5 to A6.
- Cell A5 is a comment.
- Cell A6 is the value for the CheckBox1 control on the screen.

### Control_info Worksheet – Group 3 – Cells A11 to B11.
- Cell A11 is not a comment.
- Cell A11 is linked to TextBox4 and holds the first part of the value for the current User Work Project directory path.
- The value in this textbox is the root directory of the output path where the code files produced by this tool will be sent.
- This directory is used along with the year, month, and JOB number, as well as program name values to preserve the SAS code, logs, list, Excel output, and local job files for the JOB_*xx* program when it runs.

- This value is used when the VBA code processes the outputs when the "Run Program" (CommandButton2) is clicked.
- Cell B11 is a comment.

### Control_info Worksheet – Group 4 – Cells C3 to D7.
- Cells C3 to C7 are not comments.
- Cell C3 is cell that holds the information about the unique parameters that are loaded and for the program that will be executed.
- Cells C4 is part of the ComboBox list of programs and is the initial value of no program loaded.
- Cell C4 to C7 is the full list of active programs that you can choose.
- Cell C7 is an end of list marker but still part of the list area.
- Cells D4 to D7 are mixed use; some cells are used by the tool, and some cells are comments.
- Cell D3 is the cell that holds the description of the program that will be executed.
- Cell D4 is a comment.
- Cells D5-D6 are program descriptions.
- Cell D7 is a comment.

One of the features of this tool that makes it stand out is that the code is written as two *.sas files. One file is written by the tool that contains the following:

- SAS code to preset several system options.
- SAS code to create several macro variables.
- SAS code to display the macro variable values on the SAS log.
- PROC PRINTTO statements to save the log and list outputs.
- A program-unique audit value that is set in the JOB_xx…..sas code file and tested in the "Job_xx_Parm_Code.sas" file to prevent using the wrong "Job_xx_Parm_Code.sas" and JOB_xx…..sas code file together.
- SAS Macro "Init_Global_Macro_Variables".
- Code at the end of the "Job_xx_Parm_Code.sas" includes and executes the JOB_xx…..sas code file.

The other *.sas file is the program that uses the definitions from the "Job_xx_Parm_Code.sas" file to process the current data files. When you use this method, you will find that once a program is written, tested, and debugged if it fails in the future it is usually because the data value ranges, variables, or contents have changed.

Figure 14.11 shows the contents of the Common_Parm_Text worksheet of the tool. This worksheet contains the beginning and the end of a macro definition (the init_Global_Macro_Variables macro). This macro is run at the start of each program to install both the common and program-unique SAS macro variables. By running this code, you will ensure that the changeable code parts of the program are defined by the user before the code is set aside for execution.

### Figure 14.11: Contents of the "Common_Parm_Text" Worksheet of the Reporting Tool.

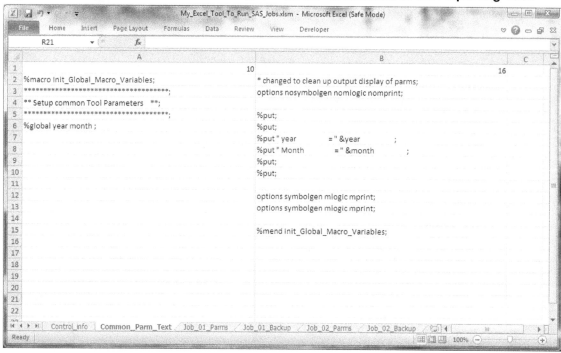

Each of the programs installed into the tool (as it is designed here) has two spreadsheets: one for current data, and one for backup data. The names here are simple Job_xx_Parms and Job_xx_Backup. We will install a routine associated with the CommandButtons 3 and 4 to back up and restore data using the Job_xx_Backup worksheets.

### Common_Parm_Text Worksheet – Part 1

The data on this worksheet is used to build a piece of SAS code that is used as the starting routine of the program. This is a SAS macro that defines the global parameters that are used to define the data this project uses. Since we are writing code to avoid having it change, we will use this macro to define the code and directory changes needed to run the code for the current processing period, without modifying the production code.

### Column A of this worksheet

- Cell A1 has a number that is the number of lines to write to a text output file named Job_xx_Parm_Code.sas.
- Cells A2 - A11 will be written out to the Job_xx_Parm_Code.sas file. Yes, cells A7 to A11 are blank; I like white space.

### Column B of this worksheet

- Cell B1 has a number that is the number of lines to write to a text output file named Job_xx_Parm_Code.sas.
- This column of cells names the macro and the common parameters. You can put as many common parameters here as you like. There are separate %Global commands so that the housekeeping is simple and so that every line can be easily enclosed in a quoted string.
- Cells B2-B16 will be written out to the Job_xx_Parm_Code.sas file. (Yes, Cell B16 is blank; I like white space).
- Cell B3 in this column of cells turns off the macro print options (symbolgen, mlogic, and mprint), and then prints the values of the common macro variables to the log. Turning off the macro print options allows the macro variable values to be spotted easily within the log listing. You can put as many common parameters here as you like. There are separate %put commands so that the housekeeping is simple and so that every line can be easily enclosed in a quoted string.

- Cells B12 and B13 at the end of the code segment turns on the macro print options to aid with debugging and audit requirements for the code. Remember, these jobs can be run in a Windows batch mode without a user interface. So, we need to be able to determine what really happened. The line is duplicated because the first line turns the option back on, and the second line echoes the command on the screen. It would not be visible without the second command.
- The last line (cell B15) ends the macro definition.

### 14.3.8 Learn How to Make the Excel UserForm Execute

**Learn how to make the Excel UserForm run**

Now we have this neat GUI interface that is inside of an Excel workbook, but how do you use it? The following VBA code segment will cause the UserForm1 to execute when the spreadsheet is opened. This code is stored only in this workbook, so it is not visible to any other Excel workbook. This code is placed into the "ThisWorkbook" area of the Excel workbook.

**Figure 14.12: The VBA Macro "Workbook_open" with VBA Code to Load and Show a VBA UserForm.**

```
'''''''''''''''''''''''''''''''''''''''''''''''''''''''
''  My company name and code header here
'''''''''''''''''''''''''''''''''''''''''''''''''''''''
''   Subroutine function:
''
''   Load and show the Excel User Form "UserForm1"
''   on the full screen
'''''''''''''''''''''''''''''''''''''''''''''''''''''''
''   last update:
'''''''''''''''''''''''''''''''''''''''''''''''''''''''
Private Sub Workbook_Open()
    Load UserForm1
    UserForm1.Zoom = 100
    UserForm1.Show
End Sub
```

**Figure 14.13: Workbook_open VBA Subroutine to Open the UserForm1 GUI Shown in the Excel IDE.**

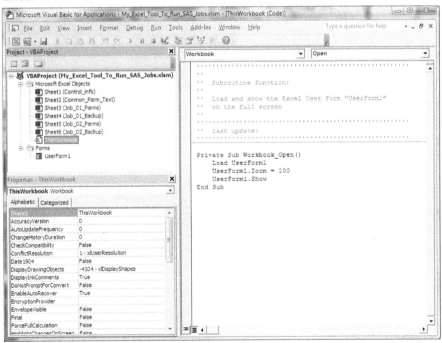

Notice that the item "ThisWorkbook" is highlighted in the IDE Project window. That indicates that it is the active window, and you can enter your VBA code into the workspace on the right. Once that code is entered, you can click on the title of the subroutine and select Run>Run Macro option on the IDE toolbar. You can also save the workbook, exit and re-open the workbook, or press the "F5" function key to make the form execute. If the UserForm1 image does not appear, then check your security settings to ensure that you have Excel macros enabled. Once the workbook opens, the form should appear on your screen and look similar to Figure 14.14.

**Figure 14.14: UserForm1 Is the Graphical User Interface Displayed While the Tool Is Running.**

This form looks great all by itself, but it still does not do anything. So, let's work on getting the data from the worksheet to the form.

### Link the data from worksheet "Common_Parm_Text" to the UserForm1 data fields

Select TextBox1 by clicking on the empty field next to the word "Year", and then enter "Control_info!A2" in the Properties window in the "ControlSource" field. The data in Cell A2 of worksheet "Control_info" will become linked to the UserForm1 control called TextBox1, and will be visible. We'll do the same thing with the other text boxes. See Figure 14.10 to find out what cell to assign to each text box.

**Figure 14.15: UserForm1, Showing How to Set the ControlSource Value.**

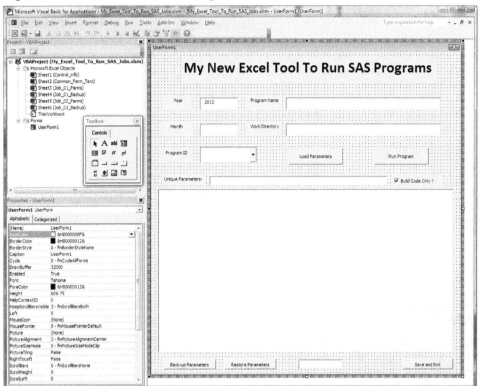

When complete, the form should look something like Figure 14.16.

**Figure 14.16: UserForm1 Running After All Five ControlSource Values Are Set.**

The information in the form in Figure 14.16 will be used to create the Working directories for executing the tool reports. The fields "Work directory", "Year", and "Month" will be used in the naming structure where the Tool will send the working copies of the SAS code file, Tool-Unique Parameter file, and the *.bat job file that can execute the SAS code in a batch mode. Listing output options for SAS 9.3 and later could send the listing output to the "Results" window, which may prevent the listing from going to the List output file. This is due to a change in the default shipping value on the listing output from list to html in SAS 9.3. You can change this value in SAS on the Tool ribbon by selecting "Options/Preferences/Results".

**Figure 14.17: Directory Path for the Working Copies of the Report Code and Temp Files.**

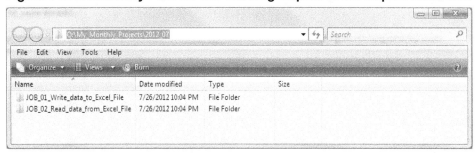

## 14.4 Excel VBA Routines to Make the Workbook UserForm Active

Now comes the fun part--creating VBA routines to make all of these switches and buttons work. The initial task is to figure out what routines we need to write. Since Visual Basic is an object-oriented language," we do not really need to worry about the order of the tasks. We just need to write them so that they are independent of each other. By looking at the GUI, we can get a good idea of the types of code routines we need to have. Let's give it a whirl. Here are some of the things we need to do.

1. Initialize the UserForm.
2. Write the user parameters to a file in a working directory.
3. Copy source programs from a production directory to the working directory.
4. Verify the output batch file points to the correct SAS run-time module.
5. Write a routine to save the changes and exit the program.
6. Set up the directory structure associated with the processing.
7. Prepare for common issues that might occur.
8. Prepare a VBA macro to process your output report.

Each of these groups of VBA routines will have several subroutines and can run in any given order. They need to operate independently because the user (you) will have the ability to select any part of the screen to update or execute in any given order. The descriptions in the rest of this section will give general guidelines about what the code will be designed to accomplish. The actual code will be available for download in the example code and data folder, accessible from the SAS Press author page at http://support.sas.com/publishing/authors/benjamin.html

### 14.4.1 Initialize the User Form

#### 14.4.1.1 Install VBA Code to Start the UserForm When Excel Begins

First we need to gain control of the Excel routines and start our UserForm. The following VBA in Figure 14.18 shows how to gain control of Excel.

**Figure 14.18: Excel VBA Code to Start the UserForm1 Software When Excel Starts.**

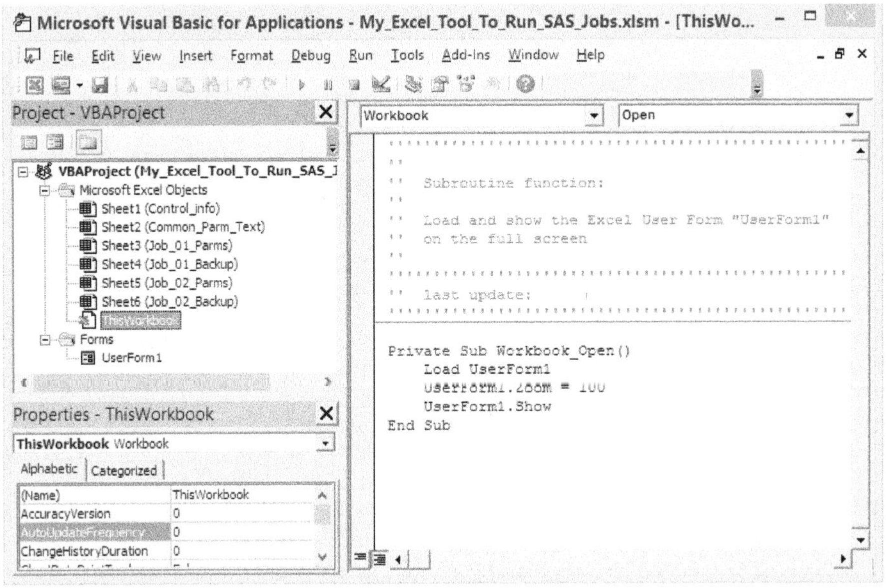

```
Subroutine function:

Load and show the Excel User Form "UserForm1"
on the full screen

last update:

Private Sub Workbook_Open()
    Load UserForm1
    UserForm1.Zoom = 100
    UserForm1.Show
End Sub
```

The process of initializing the UserForm is a multi-step process that requires several types of actions. The following is a list of the type of actions required for setup. Some actions are passive, which means they are set up when the UserForm is built. Other actions must be coded to occur when the workbook opens and the UserForm executes.

## UserForm Initialization Processing

| Initialization Type | Field | Function | Source |
|---|---|---|---|
| Passive | Year | Part of directory path | Control Source cell |
| Passive | Month | Part of directory path | Control Source cell |
| Passive | Work directory | Full work directory path | Control Source cell |
| Passive | Build Code Only | Boolean flag | Control Source cell |
| Active | Program ID | Program ID value | Loaded into ComboBox |
| Active | Program Name | Program name value | Loaded from selected worksheet |
| Active | Unique Parameters | Unique run-time parameters | Loaded from selected worksheet |

### 14.4.1.2 Initializing Passive UserForm Values

Passive parameters are set up when the UserForm is designed. Figure 14.19 shows how the year value is assigned to the worksheet named "Control_Info" in cell A2. If something is typed into the TextBox1 area of the UserForm when the tool is running, the value is placed directly into cell A2 of worksheet Control_info.

250 *Exchanging Data between SAS and Microsoft Excel*

**Figure 14.19: How to Assign a Passive Parameter to a UserForm and Worksheet Cell.**

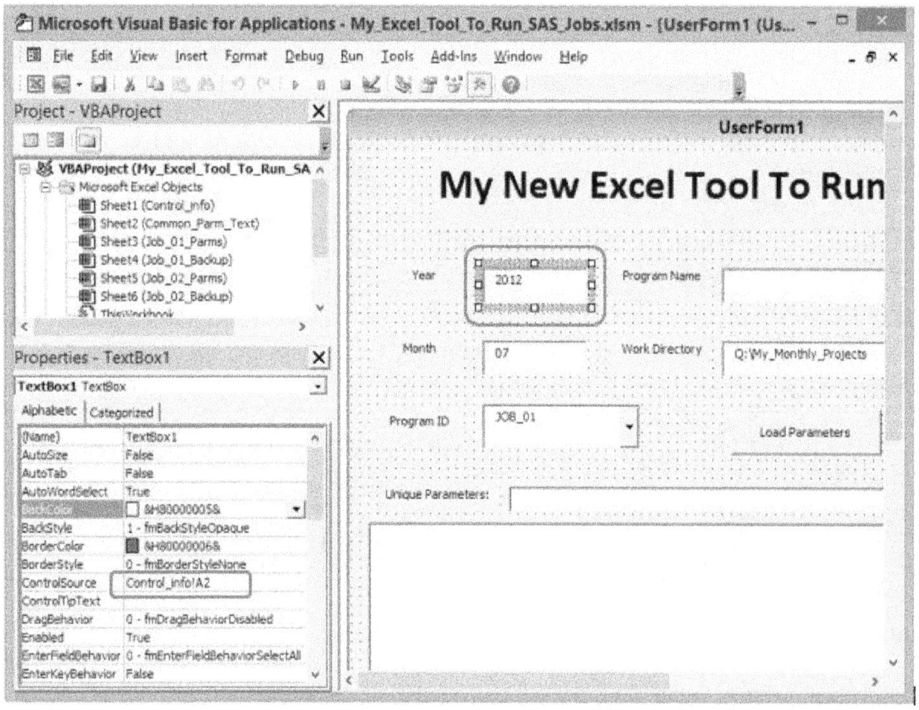

### 14.4.1.3 Initializing Active UserForm Values

Active parameters are set up when the UserForm is designed. Figure 14.20 shows the active value of ComboBox1. The value is "JOB_01". The value is entered into the ComboBox using VBA code when the workbook is activated. The current value is stored in the "Control_Info" worksheet in cell C3, as shown by the property value for the "ControlSource".

**Figure 14.20: How to Assign an Active Parameter to a UserForm and Worksheet Cell.**

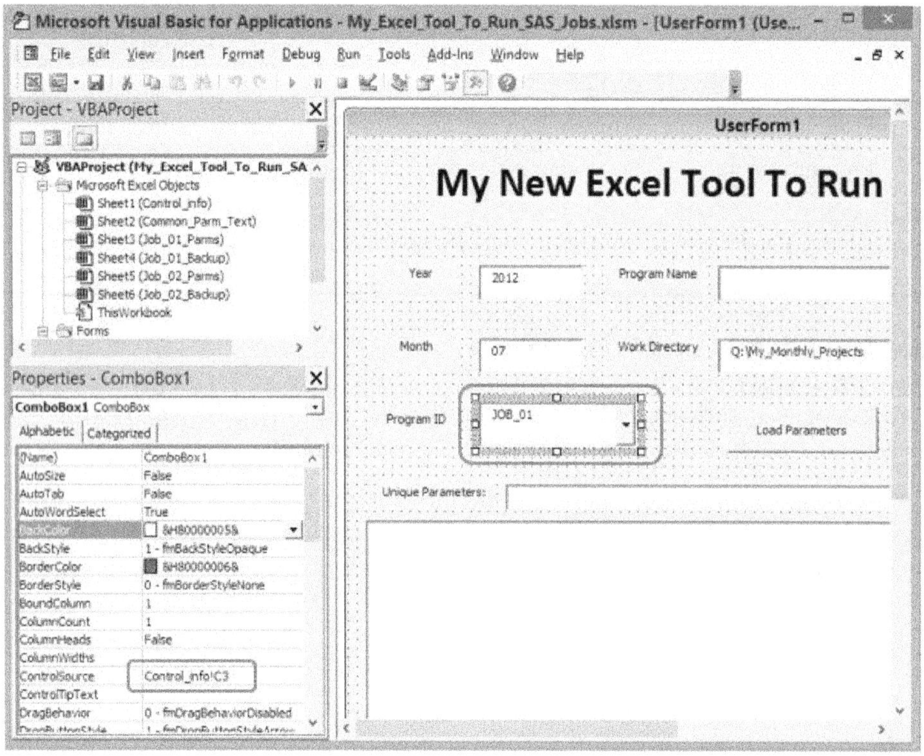

### Figure 14.21: VBA Code Counts Number of Valid ComboBox Entries and Loads the VBA Object.

```
' count the number of job names in the list for ComboBox1
Max = 0
Do Until test = "###"
 idx = 4 + Max
 i = Trim(Str(idx))
 test = Worksheets("Control_info").Range("C" + i).Value 'skip first space
 Max = Max + 1
 If test = "" And idx > 4 Then test = "###"
Loop

' Load the job names into ComboBox1
idx = 4           ' Cell C4 - the first ComboBox1 value
Do Until idx > Max + 2
 i = Trim(Str(idx))
 UserForm1.ComboBox1.AddItem(Worksheets("Control_info").Range("c"+i).Value)
 idx = idx + 1
Loop
```

### 14.4.1.4 Loading Data to the UserForm from a Selected Worksheet

One of the most useful things about the VBA code that supports this workbook is that the VBA "ControlSource" fields of the workbook can be modified at run time. That means that the workbook has the ability to change where the ControlSource value points, and therefore what is displayed on the UserForm. So the ability to store several sets of SAS program variables allows the user (you) to select and store code to be executed into different working directories. The figures below (14.22 to 14.24) show the actions.

### Figure 14.22: UserForm1 Before Any Action and After Initialization.

**Figure 14.23: UserForm1 After Pressing the Load Parameter Button.**

**Figure 14.24: UserForm1 After Selecting JOB_02 and Pressing the Load Parameter Button.**

### Figure 14.25: Worksheet Job_01_Parms, Source of Parameters Shown in Figure 14.23.

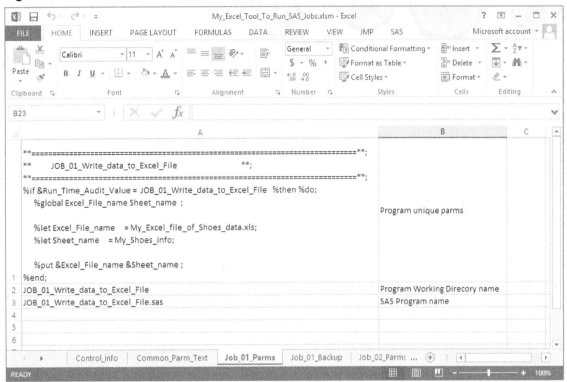

## 14.4.2 Write the User Parameters to a File in a Working Directory

The result of pressing the "RUN PROGRAM" button is that the workbook will write three output files. All three of these files are placed into the same directory. A sample output file name for the tool as designed here is "Q:\My_Monthly_Projects\2012_07\JOB_01_Write_data_to_Excel_File". The "WORK_DIRECTORY", "YEAR", and "MONTH" variables from the input fields on the UserForm become part of the directory name. The last part of the directory name is the SAS program name without the ".SAS" extension.

| Output files Sent to the Working Directory | |
|---|---|
| **File Type** | **Function** |
| User Parameter SAS file (.sas) | Pass user-defined parameters from the Excel workbook UserForm to the User SAS Program. Set up output log and list files and execute the User SAS Program. |
| User Program SAS file (.sas) | This is the production code for the process. No periodic processing parameters should be in this program. All variances in LIBNAME statements and period-unique code should be in the User Parameter SAS file. This SAS program can call Excel to send output back for formatting using VBA code. |
| Executable Windows Batch file (.bat) | This is a Windows Batch command file. It contains commands to start SAS and write log and list files. It also starts the SAS System and submits the User Parameter SAS file for execution. Double-clicking on this file will start the process in a batch mode to run SAS and any steps defined in the User Program SAS file. |

## 14.4.3 Copy Source Program from a Production Directory to the Working Directory

Figure 14.26 shows the "Run Program" button and the "Build Code Only ?" check box circled in red. These two buttons control the output of the workbook. When these buttons are pressed, three files are output. Section 14.4.2 describes these files. Two of the files are written directly from the information in the Excel workbook. The third file is copied from a directory that stores production SAS code into the working directory for the current processing period.

**Figure 14.26: UserForm Showing Button and Switch to Controls Copying the Production SAS Code.**

**Figure 14.27: VBA Code in the UserForm That Copies the SAS Code from the Production Directory.**

```
''''''''''''''''''''''''''''''''''''''''''''''''''''''''''''''''''''''''''''''
'' Copy source SAS Code from the production directory to the work area
''''''''''''''''''''''''''''''''''''''''''''''''''''''''''''''''''''''''''''''
    'object.Copy destination[, overwrite]
    Set sas_O = CreateObject("Scripting.FileSystemObject")

    Prod_path = Worksheets("Control_info").Range("a15").Value

    sas_code_file = Prod_path + "\" + Worksheets(Program_Number.Value + _
                "_Parms").Range("a3").Value

    new_code_file = Full_Path + "\" + Worksheets(Program_Number.Value + _
                "_Parms").Range("a3").Value

    overwrite = True
    sas_O.copyfile sas_code_file, new_code_file, overwrite
    Set sas_O = Nothing
```

## 14.4.4 Verify the Output Batch File Points to the Correct SAS Run Time Module

The VBA code in the UserForm1 code module of the workbook contains VBA code to write data out to a file with the command "Print #2, *some text to output*". This code writes the Windows Batch file (*.bat) that you can double-click on to execute the SAS code. The path to the program SAS.exe is embedded in this code and needs to point to the program SAS.exe. Older versions of the path are similar to "C:\Program Files\SAS\SASFoundation\9.2\SAS.exe", but newer versions of SAS software may include your site number in the path. Therefore, I am not including my path here. Contact your IT Department if you are not able to find the correct path to the SAS.exe program. It needs to be installed into the VBA code for UserForm1 in your version of the workbook.

### 14.4.5 A Routine to Save the Changes and Exit the Program

This workbook updates the ComboBox features of the tool and makes changes to the workbook every time the code executes. When you press the "Save and Exit" button highlighted in red and shown in Figure 14.28, the workbook will be saved and closed. If you choose not to save the changes, then click on the "X" in the upper right corner and exit without saving.

**Figure 14.28: The Position of the "Save and Exit" Button.**

The code box shows the VBA code to save the workbook and exit.

```
Private Sub CommandButton5_Click()
    ActiveWorkbook.Save
    Application.Quit
End Sub
```

### 14.4.6 Directory Structure Associated with the Processing

Here is a suggested list of the types of directories you might want to create. Company or departmental standards might dictate more rigid naming structures, but you get the idea. Here are some thoughts about how to use these directories.

| Directory Structure used for Processing |||
|---|---|---|
| Directory | File type | Function |
| Y | My_BAS_Files | This directory stores the VBA and VBS code for the project. Also, Excel will let you export the UserForm1 code module. This would be a good place to store the download. |
| Y | My_Excel_Tool | This directory stores the workbook tool. |
| Y | My_Production_SAS_Code_Source_Directory | This directory stores the SAS code for the individual SAS programs in the project. The fully tested SAS code files can be stored here. Additional directories could be created for backup copies of SAS code that are being developed. |
| Y | My_Monthly_Projects | This directory stores subdirectories that contain the monthly working files. Place a directory here for each of the periodic report groups. We are using monthly directories here, but they could be daily or quarterly. In this periodic directory, place a folder for each of the reports. |
| **Note:** Other directories can be defined as needed to meet your departmental needs. |||
| | My_Department_Macro_code | Here would be a good place to store common SAS files that can be included in more than one program. |
| | Any_Other_Departmental_directories | This is where your data files reside. It would be a good idea to put any variable into the tool that could be used to describe a part of the directory path. The more flexible this part of the tool can be, the more helpful it will be later in the program execution. |

The folder names are not important, as long as the reporting tool knows what they are and how to access them. Most of these file names come directly from data entered into the screens on the tool. The only notable exceptions to that are the name of the folder that holds the source code and the path names to Excel.exe and SAS.exe.

Start by mapping to a directory space that is common for all users of the tool:

```
Q:\My_Tool
        Any_Other_Departmental_Directories
        My_BAS_Files
        My_Department_Macro_code
        My_Excel_Tool
        My_Monthly_Projects
                2012_07
                        JOB_01_Write_data_to_Excel_File
                        JOB_02_Read_data_from_Excel_File
                2012_08
                        JOB_01_Write_data_to_Excel_File
                        JOB_02_Read_data_from_Excel_File
        My_Production_SAS_Code_Source_Directory
                Backup_Code_Copies
```

### 14.4.7 Common Issues That Might Occur.

Because a few people in the world take things because they can, the rest of us have to put up with many annoying things to prevent us from doing something we would not do anyway. One of those items is shown in Figure 14.29. While some systems will allow you to sign a program that you write for use on your own computer, others will not allow this. Some systems will allow you to use the Excel "OPTIONS" to assign "Trusted" status to programs. This will allow VBA macro code to execute on your computer. Other systems will not allow this either. Older systems are not as good at preventing or trapping attempts to run Excel workbooks with embedded *.bas VBA macros. A message such as that in Figure 14.29 appears typically because the workbook you are trying to execute is not trusted. You may have to ask your IT Department to help you solve these issues.

**Figure 14.29: Security Warning, Workbook VBA Code Probably Not Trusted.**

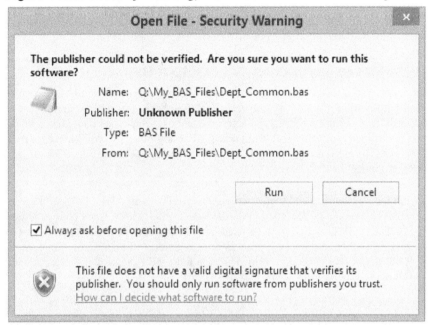

**Figure 14.30: Security Warning, Workbook VBA Code Disabled. Enable Content to Continue.**

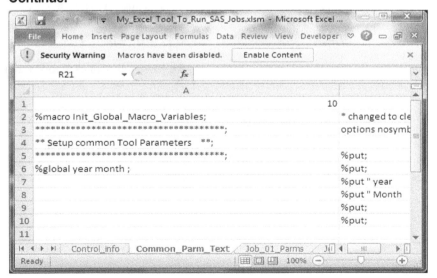

**Figure 14.31: Security Warning: Press "Yes" to Convert the Workbook to a Trusted Document.**

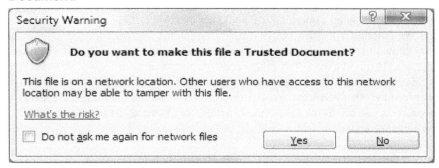

Some versions of Excel will issue this type of a message, however, the newer versions just ask if you want to make the document a trusted document when you open it.

## 14.4.8 Prepare a VBA Macro to Process Your Output Report

This part of the project is really geared toward working with the individual reports. Not all of the reports or processes will need printed output or a formatted spreadsheet. JOB_02 of this group of programs just creates a SAS dataset, while JOB_01 creates a formatted Excel workbook. Other jobs could be created to process data files, print reports directly, or input data from text files or any other job you have ever run on a computer. To work with this file, we need a copy of the output Excel workbook created by JOB_01 so that we can format the report for printing.

This process involves opening the output file and then opening the IDE (with an Alt/F11 keystroke combination). Any Excel VBA macro can be recorded to do the work for us. These VBA code modules can then be exported to our "My_BAS_Files" directory to be used later. Here, we will create several files that will be used to do the processing and make the whole process run from start to finish without intervention. The VBS script "Q:\My_BAS_Files\VBS_Execute_script.vbs" can load and execute VBA macros within an Excel workbook. The following four steps describe that processing.

1. SAS code (to be stored in each SAS program) sets up SAS macro variables and executes an "X" command to link the SAS program to the Excel file using the VBS routine below such as JOB_01_Write_data_to_Excel_File.sas.
2. A VBS program "Q:\My_BAS_Files\VBS_Execute_script.vbs" controls the execution of Excel and the VBA macro code modules to format the report. This process does not leave Excel macros in the final Excel file, so they can be given to anyone as workbooks without embedded macros.
3. The VBA program "Q:\My_BAS_Files\Dept_common.bas" houses commonly used VBA subroutines for departmental use.
4. The VBA program "Q:\My_BAS_Files\JOB_01.bas" stores the VBA subroutines to format the report.

The object of a system of reporting or processing tools like this is to have stable code programs that work right every time. The programs just use different data each time. Big IT departments have been doing this for 50 years, since the early 1960s. Now we can bring the processes to your desktop PC. Of course, stable programs like stable data, so this might not be for everyone. But, front-end processes can make the data ready for the tools. A tool like the one we just described here is simple to operate. After the data is ready, there are just a few steps in the process.

## 14.5 Conclusion

The code in this chapter builds an Excel workbook that brings Programming with SAS and Reporting with Excel together. This chapter presents a unified tool that allows both a programmer and an analyst to use and understand the same processes. It also brings it to a level that gives the company full control of all steps of the process. I hope that the examples and code available for this chapter will be useful to anyone. A summary of the process follows.

1. Run the tool by double-clicking on the desktop icon.
2. Answer any questions about security and enable macros to allow the macros to run.
3. Periodically update the Year and Month fields.
4. Select the program to execute from the ComboBox.
5. Press the "Load Parameters" button.
6. Press the "Run Program" button.
7. Press "OK" in the message box asking if you are really ready to run the program.
8. Press the "Save and Exit" button.
9. Navigate to the directory that the JOB files were sent to by the reporting tool.
10. Double-click on the "run_JOB_*xx*.bat" file to execute the report program.
11. Review the logs and listing output files for accuracy, and correct any errors.
12. Review the output files or reports, and print any reports that are needed.
13. Repeat for the next report.

Most system administrators have implemented a lot of security measures and restrict the use of outside code or embedded macros, especially with tools available to more than one person. If you store this tool with all of its macros on your computer, you can usually get away with leaving the macros enabled. But, that may be a different story on systems where you may need to have more than one person access the tool. If the tool will not run because of security / VBA macro issues, check with your system administrator to get help. If your copy of the tool will run, but gives a message similar to one or more of the figures in Section 14.4.7, Common Issues That Might Occur, it is because your security settings are checking before you run code with macros embedded. Since you wrote the tool and know what is in it and that you can trust it, then follow your company guidelines for how to proceed.

# Index

## A

ABSOLUTE_COLUMN_WIDTH option  130, 134–138
Access Connectivity Engine (ACE)  1, 159
accessing
    data from SAS Explorer window/toolbar  9–18
    data stored in workbooks  241–245
    Excel data from SAS Explorer window  9–18
    Excel files with LIBNAME  48–49
    Excel files with SQL procedure  157–166
    Excel from SAS Enterprise Guide  68
    Excel with OLE DB or ODBC application program interfaces (APIs)  149–156
    Excel with OLE DB/ODBC APIs  149–156
    SAS Export/Import wizards  15–16
activating UserForm  248–258
Add-In for Microsoft Office
    *See* SAS Add-In for Microsoft Office
applying Excel "AUTOFILTER" to selected output columns  136–137
ASCII_DOTS option  131
assigning
    LIBNAME statement  160–161
    Libref  52, 151–152
ATTRIB command  48
AUTOFILTER option
    about  131, 136–137
    applying to selected output columns  136–137
AUTOFILTER_TABLE option  131
AUTOFIT_HEIGHT option  131
automating processing  197–201
AUTO_SUBTOTALS option  131

## B

bar charts
    building using SAS Enterprise Guide  99–100
    exporting as *.srx files  100–102
BLACKANDWHITE option  131
BLANK_SHEET option  132
building
    bar charts using SAS Enterprise Guide  99–100
    Excel macros  189–209
    Excel user form interface  233–248
    VBS scripts  220–223
BY statement  143
BYGROUP setting  143–144
BYLINES option  112

## C

calculating variables within SQL code  165
cell ranges
    defining  171, 172–173
    reading from Excel workbooks  163–165
    reading with Import Data option  80–83
CENTER_HORIZONTAL option  131
CENTER_VERTICAL option  131
changes, saving  255
character fields
    about  2
    saving leading zeros in  123–124
CheckBox  234
client-server environment (Dynamic Data Exchange (DDE))  168–170
Close_Excel macro  171, 174–175
closing Excel workbooks  171, 174–175
COLUMN_REPEAT option  130
columns
    adjusting width with tagset template options  134–135
    hiding  135–136
    selecting  42–43
ComboBox  233
CommandButton  234
commands, issuing to Excel  171, 172
    *See also* specific commands
comparing Excel files  161–162
complex file formats  4
Component Object Model (COM) software package  91
CONNECTION option, LIBNAME statement  50–53
CONNECTION TO statement  158–159
CONTENTS option  132, 139
CONTENTS procedure
    about  57, 165–166
    examining Excel workbook with  60–62
    verifying Excel to OLE DB connection with  154–156
CONTENTS_WORKBOOK option  132
ControlSource  234
CONVERT_PERCENTAGES option  131
"Copy Contents to Clipboard" option (SAS Explorer window)  11–12
copy-and-paste techniques
    about  4
    converting Excel tables to text  9
    converting text data to Excel column data fields  5–7

copying data to SAS Enhanced Editor window 7–8
highlighting cut/copy and then paste 5
saving multiple lines of text 8
copying
    data to SAS Enhanced Editor window 7–8
    source programs 253–254
CREATE TABLE command 166
creating
    Excel files 161–162
    Excel workbooks 231–258
    naming conventions 214–215
    output files 108–148, 202–203
    "ready-to-print" spreadsheet 137–138
    table of contents in Excel workbook 138–139
    worksheets 143–145
CSS_TABLE option 125
CSV option
    about 109, 111–112
    changing delimiters when outputting data with 120–123
    examples 113–123
    file default output differences 113–115
    overview of examples 113
    saving leading zeros in character fields 123–124
    title and footnotes output differences 115–118
    writing currency values as unformatted numbers 118–120
CSVALL option
    about 109, 111–112
    examples 113–123
    file default output differences 113–115
    overview of examples 113
    saving leading zeros in character fields 123–124
    title and footnotes output differences 115–118
    writing currency values as unformatted numbers 118–120
currency values, writing as unformatted numbers 118–120
CURRENCY_AS_NUMBER option 112
CURRENCY_FORMAT option 131
CURRENCY_SYMBOL option 112, 131

## D

data
    access methods for Excel files supported by IMPORT procedure 33–34
    accessing from SAS Explorer window/toolbar 9–18
    copying to Excel files via HTML files with "View in Excel" option 13–14
    copying to SAS Enhanced Editor window 7–8
    exporting to Excel 4/5 format files 23–24
    exporting to Excel files with no column headers 28
    exporting to network Windows computers 28

loading to UserForms 251–253
options for in SAS Add-In for Microsoft Office 96–99
ranges of in Excel 4
reading from Excel to JMP 88–89
sharing with Excel using JMP 85–89
types of 2, 50–51
writing from JMP to Excel 89
writing to external files and Excel workbooks with EXPORT procedure 19–29
Data Base Management System (DBMS) 31
DATAFILE statement 17
dataset options
    processing date and time values with 62–63
    processing variable type conversions with 63–64
datasets, opening 94–96
date values, processing with dataset options 62–63
DBCREATE_TABLE_OPTS option 53
DBDSOPTS= option 39
DBENCODING option 53
DBFORCE option 53
DBGEN_NAME option 53, 163–165
DBLABEL option 53
DBMAX_TEXT option 53
DBMS mode 21–22
DBMS=DLM option
    EXPORT procedure 24–25
    IMPORT procedure 35–37
DBMS=EXCEL option
    EXPORT procedure 25–27
    IMPORT procedure 37–40
DBMS=EXCEL4 option, IMPORT procedure 35
DBMS=EXCEL5 option, IMPORT procedure 35
DBMS=EXCELCS option
    EXPORT procedure 27–28
    IMPORT procedure 40–41
DBMS=XLS option, IMPORT procedure 42–45
DBMS=XLSX option, IMPORT procedure 42–45
DBSASLABEL option 53
DBSASTYPE option 53, 63–64
DBTYPE option 53
DDE
    *See* Dynamic Data Exchange (DDE)
DECIMAL_SEPARATOR option 112, 131
DEFAULT_COLUMN_WIDTH option 130
defining
    cell ranges 171, 172–173
    physical file locations 51–52
DELIMITER option 112
delimiters, changing 120–123
directories, structure of 255–256
DOC option 112, 125, 132
DPI option 131
DRAFTQUALITY option 131

DROP option 54
Dynamic Data Exchange (DDE)
    about 167, 213
    client-server environment 168–170
    examples 177–187
    Hello World project 177–181
    list of examples 177
    macros for 171–177
    purpose of 167–168
    reading and writing to Excel workbooks with 167–187
    syntax of 168–170
    writing "Hello World" to Excel files 182–184
    writing SAS datasets to Excel files 184–187

## E

EMBEDDED_FOOTNOTES option 130
EMBEDDED_TITLES option 130
EMBED_TITLES_ONCE option 130, 146
Enhanced Editor window, copying data to 7–8
Enterprise Guide
    *See* SAS Enterprise Guide
examples
    CSV option 113–123
    CSVALL option 113–123
    Dynamic Data Exchange (DDE) 177–187
    EXCELXP option 133–146
    JMP 87–89
    LIBNAME statement 56–65
    MSOFFICE2K option 126–128
    OLE DB/ODBC APIs 151–156
    SAS Add-In for Microsoft Office 94–105
    SAS Enterprise Guide 69–83
    SQL procedure 160–66
Excel
    *See also* Excel files
    *See also* workbooks (Excel)
    *See also* worksheets (Excel)
    accessing from SAS Enterprise Guide 68
    accessing with OLE DB or ODBC application program interfaces (APIs) 149–156
    building control macros for reports 223–229
    building graphs with macros 207–209
    building macros 189–209
    building user form interface 233–248
    converting tables to text 9
    data ranges 4
    data types 2
    executing UserForm 245–248
    guidelines for building user form interface 233–248
    importing *.srx files into 102–105
    issuing commands to 171, 172
    LIBNAME assignments to access Excel using SQL procedure 160–161
    purpose of 2
    reading data to JMP from 88–89
    selecting ranges 44–45
    sharing data with using JMP 85–89
    sharing methods between JMP and 86–87
    starting 171–172
    verifying to OLE DB connection with CONTENTS procedure 154–156
    workbook formatting groups 3–4
    workbook limitations 2–3
    writing data from JMP to 89
Excel files
    accessing with LIBNAME 48–49
    accessing with SQL procedure 157–166
    comparing 161–162
    copying data to via HTML files with "View in Excel" option 13–14
    creating 161–162
    data access methods for files supported by EXPORT procedure 21–22
    data access methods for files supported by IMPORT procedure 33–34
    exporting data to 23–24, 28
    exporting data to with no column headers 28
    processing with LIBNAME statement 47–65
    processing with pass-through facilities using SQL procedure 162–163
    reading 161–162
    writing "Hello World" to 182–184
    writing SAS datasets to 184–187
EXCEL ODS destination, for writing workbooks 147–148
EXCEL option 109
Excel Workbook Tool 212, 232
Excel-readable files 4
Excel-specific dataset options 53–54
EXCELXP option 109, 189–190
    adjusting column width with tagset template options 134–135
    applying Excel "AUTOFILTER" to selected output columns 136–137
    building worksheets with titles 146
    creating "ready-to-print" spreadsheet 137–138
    creating table of contents in Excel workbook 138–139
    examples 133–146
    generating XML output files with no options 133–134
    hiding columns 135–136
    naming Excel worksheets 140
    overview of examples 132–133
    placing labels in names of Excel worksheets 142–143
    splitting reports onto multiple Excel worksheets 141
    syntax of 128–130

tagset options  130–132
executing
    Excel UserForm  245–248
    VBS/VBA macros  214–215
    Visual Basic Script  219–220
exiting programs  255
Explorer window
    *See* SAS Explorer window
Export method
    outputting graphs or reports with  75–77
    using with SAS Enterprise Guide  69–70
EXPORT procedure
    about  9, 213
    data access methods for Excel files supported by  21–22
    DBMS=DLM option  24–25
    DBMS=EXCEL option  25–27
    DBMS=EXCELCS option  27–28
    examples  23–28
    exporting data to Excel 4/5 format files  23–24
    exporting data to Excel files with no column headers  28
    exporting data to network Windows computers  28
    list of examples  23, 150, 177
    overview of examples  22–23
    purpose of  20
    syntax of  20–21
    writing SAS data to external files and Excel workbooks with  19–29
Export wizard
    accessing  15–16
    selecting from SAS Explorer window "Export" menu  11
    using in 32/64-bit mixed environment  17–18
exporting bar charts as *.srx files  100–102
external files
    reading into SAS with IMPORT procedure  31–45
    writing SAS data to with EXPORT procedure  19–29

## F

file format groups  3
File option (SAS Toolbar)  15–16
file output, opening  192–194
FILENAME statement  50–51, 177–179
FILEREF  173
files
    *See* Excel files
    *See* external files
    *See* HTML files
    *See* output files
FITTOPPAGE option  131
formulas  2
FORMULAS option  131
FROZEN_HEADERS option  130, 137–138

FROZEN_ROWHEADERS option  130

## G

GETNAMES= statement  38, 39, 41, 43–44
graphs, outputting  71–77
GRIDLINES option  131
GUESSINGROWS option  57

## H

hardware configuration  49
HEADER option, LIBNAME statement  49, 50, 56–57
HEADER_DATA_ASSOCIATIONS option  125
HEADER_DOTS option  125
Hello World project  177–181, 189–190
HIDDEN_COLUMNS option  131, 135–136
hiding columns  135–136
highlighting cut/copy and then paste  5
HTML files
    copying data to Excel files via with "View in Excel" option  13–14
    generating with no options  126–127
    generating with Summary_Vars option  127–128
    writing  124–125, 126–128, 133–146
HTML option  109

## I

IDE
    *See* Integrated Development Environment (IDE)
Import Data option (SAS Toolbar)  16–17, 80–83
Import option, reading Excel workbooks with  77–80
IMPORT procedure
    about  9, 31–32
    data access methods for Excel files supported by  33–34
    DBMS=DLM option  35–37
    DBMS=EXCEL option  37–40
    DBMS=EXCEL4 option  35
    DBMS=EXCEL5 option  35
    DBMS=EXCELCS option  40–41
    DBMS=XLS option  42–45
    DBMS=XLSX option  42–45
    examples  35–45
    list of examples  34–35
    overview of examples  34
    purpose of  32
    reading external data files and Excel workbooks into SAS with  31–45
    syntax of  32
Import wizard
    accessing  15–16
    using in 32/64-bit mixed environment  17–18
importing *.srx files into Excel  102–105
IN option  54

INDEX option 132
INFILE statement 7
initializing
    active UserForm values 250–251
    passive UserForm values 249–250
    UserForm 248–249
Insert group (SAS Add-In for Microsoft Office) 93
INSERT_SQL option 53
Integrated Development Environment (IDE)
    about 233, 235–236
    accessing data stored in workbooks 241–245
    building sample menus 237–238
    executing Excel UserForm 245–248
    linking menu and data 238–239
    setting up control variables 241–245
    storing control information in Excel workbook worksheets 240–241
    Toolbox menu 236–237
issuing commands to Excel 171, 172

## J

JMP
    about 85
    examples 87–89
    list of examples 87
    purpose of 85–86
    reading data from Excel to 88–89
    setting preferences 87–88
    sharing data with Excel using 85–89
    sharing methods between Excel and 86–87
    writing data to Excel from 89
Joint Engine Technology (JET) database engine 1, 159

## K

KEEP option 54

## L

Label 233
labels, placing in Excel worksheet names 142–143
leading zeros, saving in character fields 123–124
LIBNAME statement
    about 16, 47–48
    assigning to access Excel using SQL procedure 160–161
    building OLE-DB connection with prompt mode 152–153
    CONNECTION option 50–53
    examining Excel workbooks with CONTENTS procedure 60–62
    examples 56–65
    Excel-specific dataset options 53–54
    Excel-specific features of 48–49
    HEADER option 49, 50, 56–57
    LINUX option 54
    MIXED option 49, 50–51, 57
    overview of examples 55
    PATH option 49, 51–52, 58, 162–163
    processing date and time values with dataset options 62–63
    processing Excel files with 47–65
    processing on 64-bit operating system 64–65
    processing variable type conversions with dataset options 63–64
    PROMPT option 50, 52
    purpose of 48
    64-bit Windows connection option 54
    syntax of 49–50
    UNIX option 54
    uses for 49
    using named literals with 59–60
    VERSION option 50, 52, 58–59
Libref
    assigning 52, 151–152
    assigning to Excel worksheets with OLE-DB dialog box 151–152
LINUX option 54
loading data to UserForms 251–253
LRECL= option 180–181

## M

Macro Library Tool 212, 232
macros
    building Excel graphs with 207–209
    building for Excel 189–209
    controlling with Microsoft Windows scripts 211–229
    for Dynamic Data Exchange (DDE) 171–177
MERGE_TITLES_FOOTNOTES option 130
Microsoft Excel
    *See* Excel
Microsoft Windows scripts, controlling with macros 211–229
MINIMIZE_STYLE option 132
MISSING_ALIGN option 130
MIXED option, LIBNAME statement 49, 50–51, 57
MSOFFICE2K option
    about 109
    examples 126–128
    generating HTML files with no options 126–127
    generating HTML files with Summary_Vars option 127–128
    overview of examples 126
    syntax of output processes 124–125
    tagset template options 125

## N

Name 234
named literals, using with LIBNAME statement 59–60
NAMEROW= statement 43–44
naming conventions, creating 214–215

naming Excel worksheets 140
network Windows computers, exporting data to 28
NEWFILE= option 27, 28
NONE option, SHEET_INTERVAL option 143
NOTAB option 180–181
NOTES option 112
numeric values 2
NUMERIC_TEST_FORMAT option 131

## O

ODS
    *See* Output Delivery System (ODS)
OLE DB LIBNAME, syntax of 150
OLE DB/ODBC APIs
    accessing Excel with 149–156
    assigning Libref to Excel worksheets with OLE-DB dialog box 151–152
    building OLE-DB connection with LIBNAME prompt mode 152–153
    concept of processes 149
    examples 151–156
    list of examples 150
    opening Excel workbooks with OLE-DB init_string 154
    setting up connections 150
    verifying Excel to OLE DB connection with CONTENTS procedure 154–156
OLE-DB dialog box, assigning Libref to Excel worksheets with 151–152
OPEN option, reading Excel workbooks with 77–80
Open_cmd macro 171, 172
opening
    datasets 94–96
    Excel workbooks with OLE-DB init_string 154
    file output 192–194
    report datasets (*.srx) using SAS Add-In for Microsoft Office 99–105
options
    See specific options
ORIENTATION option 130, 137–138
output
    graphs 71–77
    processing 191–192
    processing reports 258
    reports 71–75, 75–77
    sorting 165
    verifying batch files 254–255
Output Delivery System (ODS)
    about 108
    creating output files with 108–148
    ODS tagset compared with ODS destination 111
    purpose of 108–109
    SAS Tagset templates 109–110
    syntax of CSV and CSVALL output processes 111
output files
    creating 108–148, 202–203
    creating with ODS (Output Delivery System) 108–148
    processing 203–206
Out_range macro 171, 172–173

## P

PAGE option, SHEET_INTERVAL option 143
PAGE_BREAK option 125, 131
PAGE_ORDER_ACROSS option 131
PAGES_FITHEIGHT option 131
PAGES_FITWIDTH option 131
"PASS-THROUGH" processing 160
PATH option, LIBNAME statement 49, 51–52, 58, 162–163
PCFILES special query 159, 165–166
PERCENTAGE_AS_NUMBER option 112
PERCENTAGE_FONT_SIZE option 125
Personal Workbook Tool 211, 232
physical file locations, defining 51–52
placing labels in Excel worksheet names 142–143
preferences (JMP) 87–88
PREPEND_EQUALS option 112
PRINT procedure 5, 115, 143, 191, 196, 202, 208
PRINT_FOOTER option 130
PRINT_FOOTER_MARGIN option 130
PRINT_HEADER option 130
PRINT_HEADER_MARGIN option 130
PROC option, SHEET_INTERVAL option 143, 144–145
processing
    automating 197–201
    date and time values with dataset options 62–63
    Excel files with LIBNAME statement 47–65
    Excel files with pass-through facilities using SQL procedure 162–163
    output 191–192
    output files 203–206
    output reports 258
    on 64-bit operating system 64–65
    variable type conversions with dataset options 63–64
PROC_TITLES option 112
production directories, copying source programs to working directories from 253–254
programs, exiting 255
PROMPT option, LIBNAME statement 50, 52
proof-of-concept program 190
Properties Window 234
PropertySheet 234
PUTNAMES=NO option 28, 36

## Q

QUOTE_BY_TYPE option  112
QUOTED_COLUMNS option  112

## R

RANGE= statement  37, 41, 43–44
READBUFF option  53
reading
    cell ranges from Excel workbooks  163–165
    cell ranges with Import Data option  80–83
    data from Excel to JMP  88–89
    Excel files  161–162
    to Excel workbooks with Dynamic Data Exchange (DDE)  167–187
    Excel workbooks with Open or Import options  77–80
    external data files and Excel workbooks into SAS with IMPORT procedure  31–45
    pre-defined cell ranges from Excel workbooks  163–165
    subsets of records from Excel workbooks with SQL procedure  162
    variable names  50
"ready-to-Print" spreadsheet, creating  137–138
reformatting Excel workbooks  194–197
RENAME option  39, 54
report datasets (*.srx), opening using SAS Add-In for Microsoft Office  99–105
reports
    outputting  71–75, 75–77
    splitting onto multiple Excel worksheets  141
ROWCOLHEADINGS option  131
ROW_HEIGHT_FUDGE option  131
ROW_HEIGHTS option  131
ROW_REPEAT option  131
rows, selecting  43–44

## S

Safe Mode  233
SAS Add-In for Microsoft Office
    about  91
    data options for  96–99
    examples  94–105
    list of examples  94
    methods of sharing data using  92–93
    opening datasets using  94–96
    opening report datasets (*.srx) using  99–105
    purpose of  91
SAS Enhanced Editor window, copying data to  7–8
SAS Enterprise Guide
    about  67–68
    accessing Excel from  68
    building bar charts using  99–100
    examples  69–83
    exporting bar charts as *.srx files from using  100–102
    list of examples  68–69
    methods  67–83
    overview of examples  68
    reading cell ranges with Import Data option  80–83
    reading Excel workbooks with Open or Import options  77–80
    "Send To" method  71–75
    using Export method with  69–77
SAS Explorer window
    about  10–11
    accessing Excel data from  9–18
    "Copy Contents to Clipboard" option  11–12
    "Save as Html" option  12–13
    selecting Export wizard from "Export" menu  11
SAS Toolbar
    File option  15–16
    "Import Data" option  16–17
SAS_2_Excel macro  171, 175–176, 184–187
SASDATEFMT option  53
"Save as Html" option (SAS Explorer window)  12–13
Save_Excel macro  171, 174
saving
    changes  255
    Excel workbook contents  171, 174
    leading zeros in character fields  123–124
    lines of text  8
SCALE option  131
SELECT statement, ordering of clauses in  158–159
selecting
    columns  42–43
    data types  50–51
    Excel ranges  44–45
Selection group (SAS Add-In for Microsoft Office)  93
"Send To" method  71–75
SET statement  59
setup, workstation options  215–217
sharing
    data using SAS Add-In for Microsoft Office  92–93
    data with Excel using JMP  85–89
SHEET_INTERVAL option
    about  131
    creating worksheets with BYGROUP setting  143–144
    creating worksheets with PROC setting  144–145
SHEET_LABEL option  132, 142–143
SHEET_NAME option  132, 140, 141
simple file formats  4
64-bit operating system
    processing on  64–65
    using Export/Import wizards in  17–18
    Windows connection option  54
SKIP_SPACE option  132
software configuration  49
sorting output  165
source programs, copying  253–254

splitting reports onto multiple Excel worksheets  141
SQL procedure
    about  57, 157
    accessing Excel files with  157–166
    calculating variables within SQL code  165
    comparing Excel files  161–162
    creating Excel files  161–162
    examples  160–166
    LIBNAME assignments to access Excel using  160–161
    list of examples  160
    "PASS-THROUGH" processing  160
    "PCFILES::" special query  165–166
    processing Excel files with pass-through facilities  162–163
    purpose of  158
    reading Excel files  161–162
    reading pre-defined cell ranges from Excel workbooks  163–165
    reading subsets of records from Excel workbooks with  162
    sorting output  165
    syntax of  158–159
*.srx files
    exporting bar charts as  100–102
    importing into Excel  102–105
Start_Excel macro  171–172
starting Excel  171–172
statements
    See specific statements
storing
    VBS/VBA macros  214–215, 217–219
    VBS/VBA scripts  217–219
structure, of directories  255–256
SUMMARY option  125, 130, 202
SUMMARY_AS_CAPTION option  125, 130
SUMMARY_BYVALS option  125, 130
SUMMARY_BYVARS option  125, 130
SUMMARY_PREFIX option  125, 130
Summary_Vars option, generating HTML files with  127–128
SUPPRESS_BYLINES option  132
syntax
    of CSV and CSVALL output processes  111
    of Dynamic Data Exchange (DDE)  168–170
    of EXCELXP option  128–130
    of EXPORT procedure  20–21
    of IMPORT procedure  32
    of LIBNAME statement  49–50
    of MSOFFICE2K output processes  124–125
    of OLE DB LIBNAME  150
    of SQL procedure  158–159

## T

table of contents, creating in Excel workbook  138–139
TABLE option, SHEET_INTERVAL option  143
TABLE_HEADERS option  112
tables, converting to text in Excel  9
tagset templates
    about  109–110
    EXCELXP option  130–132
    MSOFFICE2K option  125
TEMPLATE procedure  109–110, 113
text
    converting Excel tables to  9
    converting to Excel column data fields  5–7
    saving multiple lines of  8
TextBox  233
32-bit operating system, using Export/Import wizards in  17–18
THOUSANDS_SEPARATOR option  112, 131
time values, processing with dataset options  62–63
TITLE_FOOTNOTE_WIDTH option  130
titles, building worksheets with  146
TITLES option  112
toolbar processing method  10–11
Tools group (SAS Add-In for Microsoft Office)  93

## U

unformatted numbers, writing currency values as  118–120
UNIX option  54
user parameters, writing to files in working directories  253
UserForm
    about  233
    activating  248–258
    initializing  248–249
    loading data to  251–253

## V

variables
    calculating within SQL code  165
    processing type conversions of with dataset options  63–64
    reading names for  50
    writing to Excel workbooks  171, 175–176
VBA routines  248–258
VBS/VBA macros
    executing  214–215
    guidelines for building and using  213–229
    preparing  258
    storing  214–215, 217–219
VBS/VBA scripts
    building  220–223
    storing  217–219
verifying output batch files  254–255
VERSION option, LIBNAME statement  50, 52, 58–59
"View in Excel" option, copying data to Excel files via HTML with  13–14
Visual Basic Scripting (VBS)  211, 219–220

## W

WHERE statement  5, 54, 115
WIDTH_FUDGE option  131
WIDTH_POINTS option  131
Workbook_Open macro  198–201, 202–203
workbooks (Excel)
    closing  171, 174–175
    creating table of contents in  138–139
    creating that runs SAS programs  231–258
    examining contents and structure of  165–166
    examining with CONTENTS procedure  60–62
    formatting groups  3–4
    limitations of  2–3
    opening with OLE-DB init_string  154
    reading into SAS with IMPORT procedure  31–45
    reading pre-defined cell ranges from  163–165
    reading subsets of records from with SQL procedure  162
    reading to with Dynamic Data Exchange (DDE)  167–187
    reading with Open or Import options  77–80
    reformatting  194–197
    saving contents of  171, 174
    writing  147–148
    writing SAS data to with EXPORT procedure  19–29
    writing to with Dynamic Data Exchange (DDE)  167–187
    writing variables to  171, 175–176
working directories
    copying source programs to production directories from  253–254
    writing user parameters to files in  253
worksheets (Excel)
    assigning Libref with OLE-DB dialog box  151–152
    building with titles  146
    creating with SHEET_INTERVAL=BYGROUP option  143–144
    creating with SHEET_INTERVAL=PROC option  144–145
    naming  140
    placing labels in names of  142–143
    splitting reports onto multiple  141
workstation options, setting up  215–217
WRAPTEXT option  130, 137–138
writing
    currency values as unformatted numbers  118–120
    data from JMP to Excel  89
    to Excel workbooks with Dynamic Data Exchange (DDE)  167–187
    "Hello World" to Excel files  182–184
    HTML files  124–125, 126–128, 133–146
    SAS datasets to Excel files  184–187
    user parameters to files in working directories  253
    variables to Excel workbooks  171, 175–176

    workbooks  147–148

## X

X command  179–180, 202, 219–220
XLSTART directory  193–194
XML output files, generating with no options  133–134

## Z

ZOOM option  130, 137–138

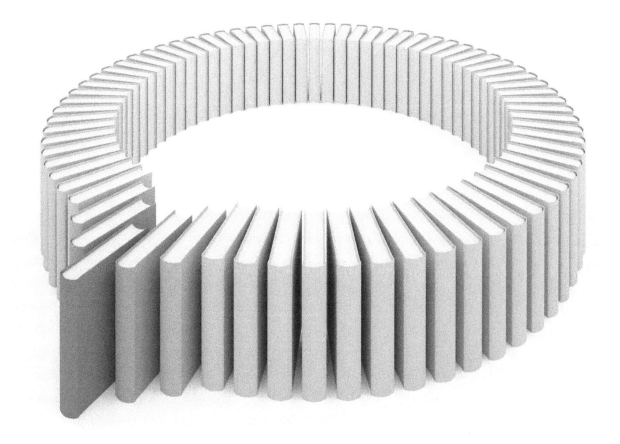

# Gain Greater Insight into Your SAS® Software with SAS Books.

Discover all that you need on your journey to knowledge and empowerment.

SAS and all other SAS Institute Inc. product or service names are registered trademarks or trademarks of SAS Institute Inc. in the USA and other countries. ® indicates USA registration. Other brand and product names are trademarks of their respective companies. © 2013 SAS Institute Inc. All rights reserved. S107969US.0613

CPSIA information can be obtained
at www.ICGtesting.com
Printed in the USA
BVHW09s1958151018
530263BV00008B/211/P